AF361787

# Cell-ECM Interactions for Tissue Engineering and Tissue Regeneration

# Cell-ECM Interactions for Tissue Engineering and Tissue Regeneration

Editors

**Ngan F. Huang**
**Brandon J. Tefft**

Basel • Beijing • Wuhan • Barcelona • Belgrade • Novi Sad • Cluj • Manchester

*Editors*

Ngan F. Huang
Stanford University
Stanford, CA, USA

Brandon J. Tefft
Medical College of Wisconsin
Milwaukee, WI, USA

*Editorial Office*
MDPI
St. Alban-Anlage 66
4052 Basel, Switzerland

This is a reprint of articles from the Special Issue published online in the open access journal *Bioengineering* (ISSN 2306-5354) (available at: https://www.mdpi.com/journal/bioengineering/special_issues/Cell_ECM_Interactions).

For citation purposes, cite each article independently as indicated on the article page online and as indicated below:

Lastname, A.A.; Lastname, B.B. Article Title. *Journal Name* **Year**, *Volume Number*, Page Range.

**ISBN 978-3-0365-9989-2 (Hbk)**
**ISBN 978-3-0365-9990-8 (PDF)**
**doi.org/10.3390/books978-3-0365-9990-8**

# Contents

# About the Editors

**Ngan F. Huang**

Ngan F. Huang, Ph.D. is an Associate Professor in the Department of Cardiothoracic Surgery at Stanford University and a Research Career Scientist at the Veterans Affairs Palo Alto Health Care System. Dr. Huang completed her BS in Chemical Engineering from the Massachusetts Institute of Technology, followed by a PhD in Bioengineering from the University of California Berkeley and the University of California San Francisco Joint Program in Bioengineering. Prior to joining the faculty, she was a postdoctoral scholar in the Division of Cardiovascular Medicine at Stanford University. Her laboratory investigates the interactions between stem cells and the extracellular matrix microenvironment to engineer cardiovascular tissues to treat cardiovascular and musculoskeletal diseases. Her recent research focuses on the role of microgravity during the drug screening of engineered muscle tissue. Dr. Huang has authored over 90 publications and patents, including those in Nat Med, PNAS, and Circ Res. She is the recipient of a Research Career Scientist Award from the Department of Veteran Affairs. She recently received the Society for Biomaterials Mid-Career award and the Alan Hirsch Mid-Career Award in Vascular Medicine from the American Heart Association. She has active or completed projects funded by the NIH, NSF, AHA, Department of Defense, California Institute of Regenerative Medicine, and Department of Veteran Affairs.

**Brandon J. Tefft**

Brandon J. Tefft, PhD is an Assistant Professor in the Joint Department of Biomedical Engineering at the Medical College of Wisconsin and Marquette University. Dr. Tefft completed his BS in General Engineering from the University of Illinois Urbana-Champaign, followed by a MS and PhD in Biomedical Engineering from Northwestern University. He then completed postdoctoral research fellowship training at Mayo Clinic in the Department of Cardiovascular Medicine. His research program focuses on regenerative engineering strategies to treat cardiovascular diseases including coronary heart disease, valvular heart disease, vascular disease, and congenital heart disease. His lab develops devices and tissues to replace or reconstruct blood vessels, cardiac valves, and cardiac tissue. Dr. Tefft has received numerous honors including a NSF Graduate Research Fellowship Program Award, NIH K99/R00 Pathway to Independence Award, and American Heart Association Career Development Award. He has active or completed projects funded by NIH, NSF, AHA, and Advancing a Healthier Wisconsin.

# Preface

The extracellular matrix (ECM) is a dynamic scaffold structure that surrounds all cells in the body and provides structural integrity as well as dynamic signaling cues. These ECM cues influence many aspects of cellular function and phenotype, including cell death, proliferation, cell fate specification, and tissue morphogenesis. Therefore, it is critically important to understand the impact of ECM interactions on disease pathology, tissue engineering, and tissue regeneration. These original research and review articles illustrate the importance of ECM interactions on fundamental cell behavior. Using bioengineering and biomaterial strategies, the effects of the ECM on cell behavior and function can be systematically studied. Cell–ECM interactions are further explored in a variety of tissue and disease models, including cardiovascular, musculoskeletal, lymphatic, and skin applications. These articles were published from 2022 to 2023 in Bioengineering.

**Ngan F. Huang and Brandon J. Tefft**
*Editors*

*bioengineering*

*Review*

# Engineering Smooth Muscle to Understand Extracellular Matrix Remodeling and Vascular Disease

Danielle Yarbrough [1,2] and Sharon Gerecht [1,2,*]

1    Department of Biomedical Engineering, Duke University, Durham, NC 27708, USA
2    Department of Chemical and Biomolecular Engineering, Johns Hopkins University, Baltimore, MD 21218, USA
*    Correspondence: sharon.gerecht@duke.edu

**Abstract:** The vascular smooth muscle is vital for regulating blood pressure and maintaining cardiovascular health, and the resident smooth muscle cells (SMCs) in blood vessel walls rely on specific mechanical and biochemical signals to carry out these functions. Any slight change in their surrounding environment causes swift changes in their phenotype and secretory profile, leading to changes in the structure and functionality of vessel walls that cause pathological conditions. To adequately treat vascular diseases, it is essential to understand how SMCs crosstalk with their surrounding extracellular matrix (ECM). Here, we summarize in vivo and traditional in vitro studies of pathological vessel wall remodeling due to the SMC phenotype and, conversely, the SMC behavior in response to key ECM properties. We then analyze how three-dimensional tissue engineering approaches provide opportunities to model SMCs' response to specific stimuli in the human body. Additionally, we review how applying biomechanical forces and biochemical stimulation, such as pulsatile fluid flow and secreted factors from other cell types, allows us to study disease mechanisms. Overall, we propose that in vitro tissue engineering of human vascular smooth muscle can facilitate a better understanding of relevant cardiovascular diseases using high throughput experiments, thus potentially leading to therapeutics or treatments to be tested in the future.

**Keywords:** tissue engineering; extracellular matrix; vascular smooth muscle cells; cardiovascular disease

**Citation:** Yarbrough, D.; Gerecht, S. Engineering Smooth Muscle to Understand Extracellular Matrix Remodeling and Vascular Disease. *Bioengineering* **2022**, *9*, 449. https://doi.org/10.3390/bioengineering9090449

Academic Editors: Ngan F. Huang and Brandon J. Tefft

Received: 8 August 2022
Accepted: 2 September 2022
Published: 7 September 2022

## 1. Introduction

Cardiovascular disease (CVD) is the leading cause of death worldwide and involves a myriad of conditions evolved from different disease states affecting the behavior of the major cell types in the vascular system [1]. Vascular smooth muscle cells (SMCs) play an integral role in vasoconstriction and blood pressure regulation, and the pathological dysregulation of these vascular properties due to SMC dysfunction can lead to life-threatening conditions. Along with fibroblasts, SMCs secrete and degrade the vessel wall's extracellular matrix (ECM). The components, structure, and mechanical properties of the ECM dictate SMC phenotype via biomechanical and biochemical signals, and the SMCs, in turn, remodel the ECM via secretory factors and direct interaction via the focal adhesions [2]. These interactions are essential to understand how the ECM's crosstalk with the vasculature contributes to life-threatening disease states, many of which are brought on by stiffening arterial walls.

Animal models help us to understand these disease states on a systemic level [3]. The genome of rodents can be altered so that a specific phenotype corresponding to a disease state of interest can be studied. However, there are reported differences in the physiology and overall functionality of human vasculature compared to animal models [4], and induced genetic mutations in rodents may not necessarily result in a relevant phenotype in humans suffering from a similar genetic defect. Large animal models, such as swine and sheep, allow for the pre-clinical testing of potential surgical or therapeutic treatments for severe cardiovascular events [5]. Nonetheless, extensive animal studies are expensive and not amenable to high throughput experimentation.

Tissue engineering with human cells offers a unique opportunity and ability to model biological cues accurately in vitro [6]. Thus, engineered tissues can better recapitulate how cells are likely to behave in the human body and enable a better understanding of diseases and the underlying mechanisms governing them. In this review, we discuss recent studies using traditional in vitro culture systems and animal models that have improved our understanding of how specific properties of the ECM in the vasculature modulate the SMC phenotype, and how the SMCs, in turn, alter the makeup of that ECM. We also discuss how other external cues, such as flow, mechanical strain, and biochemical signals, are used to create three-dimensional (3D) tissue engineered in vitro models of vasculature. These models can help further elucidate the SMC behavior and contribute to a more profound knowledge of cardiovascular diseases resulting from smooth muscle tissue dysfunction.

## 2. ECM Properties Affecting SMC Phenotype

Because SMCs are so vital to healthy vascular function, there is an extensive base of knowledge around the various CVD phenotypes related to SMC behavior. Many of these disease phenotypes result from SMCs exhibiting phenotypic plasticity, meaning they can shift between a more mature or immature phenotype, depending on signals from the ECM. When SMCs are in a more immature state seen during development and vascular disease or injury, known as the synthetic phenotype, they proliferate quickly and secrete more significant amounts of ECM proteins. On the other hand, when the vessels are healthy, SMCs are typically in a more mature state, known as the contractile phenotype [7]. In vivo studies, particularly in rodent models, have pinpointed the specific ECM molecules' role in SMC phenotypic shifting and the overall vascular function resulting from this behavior. The ability to genetically modify these organisms has given investigators insight into how the entire system is affected by the increased expression of a signaling molecule or the absence of a certain protein. The isolation and in vitro culture of SMCs from these in vivo models or human tissues has additionally prompted further insight into some of the mechanisms behind these changes.

When SMCs are cultured on 2D substrates, detailed phenotypic characteristics can be evaluated through a variety of methods. Contractile marker expression can be quantified as protein or RNA transcript [8,9]. The degree of functional contractility can be determined via planar cell surface area (PCSA) measurement before and after the treatment of cells with vasodilating or vasoconstricting agents [10–12]. The calcium and potassium ion channel activity, which directly induces the SMC contractile function, can be evaluated via electrical current measurement across the cell membrane using patch clamp electrophysiology [13–15]. Recent studies using methods such as these and the major findings are discussed below and summarized in Figure 1 and Table 1.

**Table 1.** Summary of recent studies covering ECM factors involved in SMC behavior regulation.

| Extracellular Matrix (ECM) Property | Factor or Pathway of Interest | Experimental Model | Findings |
| --- | --- | --- | --- |
| Stiffness | Lysyl oxidase-like 2 (LOXL2) | Human smooth muscle cells (SMCs) in 2D culture; LOXL2 knockout mice | • LOXL2 promotes vascular stiffening by increasing overall cell and matrix stiffness and SMC contractility [16] |
| | Mineralocorticoid receptor (MR) | MR deleted male mice | • Progression of cardiac fibrosis is mitigated by MR deletion<br>• Therapeutic antagonism of MR produced antifibrotic biomarkers [17] |
| | Circulating molecules | Parabiosis with young and old mice | • In young mice, introduction of blood from old mice upregulated genes related to pathologic wall remodeling [18] |

**Table 1.** *Cont.*

| Extracellular Matrix (ECM) Property | Factor or Pathway of Interest | Experimental Model | Findings |
|---|---|---|---|
| | Focal adhesion kinases (FAKs) and N-cadherin | FAK knockout mice | • Inhibition of FAK activity blocks SMC proliferation and neointimal hyperplasia after injury [9] |
| | | FAK and N-cadherin knockout mice | • N-cadherin, in response to FAK activation, mediates cell–cell adhesion and SMC proliferation rate [19] |
| | Transforming factor beta (TGF-β) signaling pathway | Human SMCs on collagen I (COL1)-coated polyacrylamide (PA) gels | • SMC contractile phenotype is induced as substrate stiffness increased; Inhibition of TGF-beta receptor reversed the stiffness effects [8] |
| | | Human SMCs on polymethylsiloxane (PDMS) | • SMCs expressed highest levels of osteogenic markers on intermediate stiffness gels (0.9 MPa) as opposed to high or low stiffness [20] |
| | | Human SMCs on silk fibroin gels | • Softer gels induced maturation of mesenchymal stem cells into SMCs [21] |
| | DNA methyltransferase I (DNMT1) | Human SMCs on fibronectin (FN)-coated PA gels; acute aortic injury and chronic kidney failure mouse models | • Substrate stiffening induced synthetic phenotype in SMCs<br>• DNMT1 is repressed in stiffening arteries of both mouse models<br>• DNMT1 inhibition facilitates increased arterial stiffening in mice, and cellular stiffening and calcification in vitro [22] |
| | Ascending thoracic aortic aneurysm | Healthy and aneurysmal human SMCs on compliant hydrogels | • Cytoskeletal stiffness was increased as substrate stiffness increased; aneurysmal cells exhibited increased traction forces compared to healthy cells [23] |
| Fibrillar Protein Composition | Fibronectin | Porcine SMCs suspended in COL1-FN gels | • Fibronectin promoted elastin deposition and expression of assembly proteins; gel contraction and elastic modulus were increased in fibronectin-laden gels [24] |
| | Elastin | Human smooth muscle cells on porous collagen-elastin scaffold sheets | • Elastin promoted mechanical and viscoelastic properties similar to native vessels, and contractile SMC phenotype [25] |
| | Collagen 1 and fibronectin | Human smooth muscle cells on ECM-coated polyacrylamide gels | • SMCs on COL1-coated gels showed decreased migration and increased stress fiber orientation, and more organized cytoskeleton on stiffer gels, while the reverse was true for FN-coated gels [26] |
| Non-fibrillar Protein Abundance and Structure | Advanced Glycation End products (AGEs) | Mice models | • AGEs increase vascular stiffness by prompting collagen crosslinking and inflammatory activation [27] |
| | Hyaluronic acid (HA) | Human SMCs cultured on micropatterned and HA/ECM-coated titanium | • HA/ECM surface inhibits excessive SMC proliferation [28] |
| | Rho-related BTB domain–containing protein 1 (RhoBTB1) | Angiotensin-II treated (hypertensive) mice | • RhoBTB1 alleviates arterial stiffness via actin depolymerization, but does not reverse hypertension [29] |
| | Elastin assembly proteins | Knockout mice models | • Genetic deletions of specific elastin polymerization proteins, such as fibulin-4, fibrillin-1, lysyl oxidase, etc. degrade the integrity of elastin-contractile units, resulting in a range of disease phenotypes [30] |

**Table 1.** *Cont.*

| Extracellular Matrix (ECM) Property | Factor or Pathway of Interest | Experimental Model | Findings |
| --- | --- | --- | --- |
| | Small leucine-rich repeat proteoglycans | Human coronary artery bypass patients | • High pulse wave velocity was associated with significant downregulation of these proteoglycans, implicating involvement of collagen fibrillogenesis [31] |
| | Lysyl hydroxylase I (PLOD1), lysyl oxidase (LOX) | Human and mouse SMCs cultured in osteogenic medium | • LOX overexpressing mouse SMCs exhibited increased calcification and increased collagen crosslinking [32] |
| | Post-translationally modified (glycosylated) fibronectin (gFN) | Rat SMCs on fibronectin | • SMC adhesion to glycosylated fibronectin (gFN) was increased compared to native fibronectin, and was integrin independent; RAGE inhibition blocked adhesion to gFN [33] |
| | Protein and lipid phosphatase (PTEN) | Mice with SMC-specific-PTEN knockout; Isolated human atherosclerotic arteries | • PTEN expressed in the SMC nuclei regulates the Serum Response Factor, maintaining the contractile phenotype; and PTEN expression is decreased in human atherosclerotic lesions [34] |
| | Matrix metalloproteinase-12 (MMP12) | MMP12 knockout mice | • Deletion of MMP12 abrogates arterial stiffening by reducing elastin degradation [35] |
| | Matrixmetalloproteinase-9 (MMP9) | Macrophage depleted mice | • Resident macrophages regulate collagen production in SMCs by MMP9 production, mediated by interaction of macrophages with hyaluronan [36] |

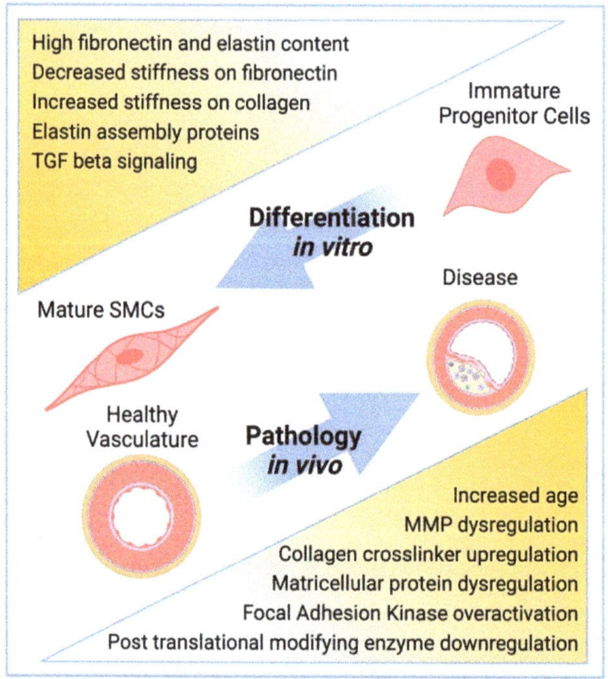


**Figure 1.** Schematic illustration of factors that prompt SMC maturation in culture systems (top) or disease progression in animal models (bottom). Created with BioRender.com.

### 2.1. Stiffness

Stiffness is an important mechanical property of the ECM in vessel walls and tightly regulates the ability of SMCs to control blood pressure. In turn, increased ECM stiffness, associated with many prominent CVDs, can be exacerbated from changes in the expression of specific genes and secreted factors by the cells. Genetically modified mice have been used extensively to identify the mediators of vascular ECM stiffness in SMCs. In a study of age-related vascular stiffening, the depletion of lysyl oxidase-like 2 (LOXL2) results in increased stiffening with age in mice [16], and mineralocorticoid receptor depletion in aged mice and pharmacological inhibition in elderly humans successfully mitigates vascular fibrosis [17]. In addition, young mice introduced to blood flow from old mice via parabiosis show an upregulation of the genes related to the pathologic vascular wall remodeling [18]. Focal adhesion kinases (FAK) and related focal adhesion proteins have also been an area of interest in studying ECM stiffness effects, as SMC focal adhesions are highly dependent on the stiffness of their matrix and regulate the contractile capability of the cells [37]. Transgenic mouse models have revealed that the aberrant activation of FAKs (chemically induced) contributes to increased neointimal hyperplasia via cyclin D1 signaling [9]. Additionally, SMC proliferation and neointima formation were reduced in response to injury when FAK and/or N-cadherin genes are knocked out. Complementary 2D in vitro studies indicated that N-cadherin mediates stronger cell–cell adhesions and increased the proliferation rate in response to increased ECM stiffness and FAK activation [19].

For many in vitro studies, the cells are cultured in two-dimensional (2D) Petri dishes or gels engineered to have tunable chemistry and tightly controlled stiffness properties. This allows for a detailed evaluation of how single variable changes in substrate stiffness or other matrix mechanical properties affect cells. SMCs cultured on collagen I-coated polyacrylamide (PA) gels (stiffness ranging from 1 to 100 kPa) exhibit increased expression of contractile markers on stiffer substrates, mediated through transforming growth factor beta (TGFβ) signaling [8], while SMCs on fibronectin-coated PA gels shifted to a synthetic phenotype through the downregulation of DNA methyltransferase 1 as substrate stiffness increased [22]. The SMCs deposit high levels of calcium and undergo osteogenesis on poly(dimethylsiloxane) (PDMS) substrates of intermediate stiffness (0.91 MPa) when compared to substrates of extreme high (2.33 MPa) or low (0.36 MPa) stiffness [20]. Meanwhile, the actual stiffness of the cytoskeleton in the SMCs can be affected by the properties of the surrounding matrix. Cytoskeletal stiffness can be analyzed using atomic force microscopy (AFM), and this method has been used to show that increased substrate stiffness results in significantly larger traction forces applied by aortic SMCs [23]. Alternatively, an increase in cytoskeletal stiffness brought on by genetic mutation, such as a gain-of function mutation in the hypoxia-inducible factor 2α gene, can lead to a positive feedback loop and signaling cascade, ultimately leading to pathological vessel wall stiffening and hypertension [38].

### 2.2. Fibrillar Protein Composition

The composition of the ECM surrounding the SMCs in the vessel walls is tightly regulated throughout the development and maturation of the vasculature. The most abundant components in the ECM (most notably collagen, elastin, fibronectin, and laminin) make up the fibrillar network and work together to provide support and flexibility to blood vessels that need to expand and contract steadily throughout the human lifespan. When the vasculature is diseased, the ratio of these vital components shifts due to changes in ECM protein secretion and expression of ECM modifiers (crosslinkers and proteases) [39]. In a diseased state, SMCs in their synthetic phenotype exhibit increased collagen expression, decreased expression of mature contractile markers, and increased cell migration and proliferation. On the other hand, healthy SMCs in their mature contractile phenotype produce higher levels of elastin, and express robust contractile markers, such as smooth muscle myosin heavy chain (SM-MHC) [7,40].

To understand their relationship with the ECM, these components are used in vitro to promote either contractile or synthetic phenotype in SMCs. For example, the addition of soluble fibronectin in SMC-laden collagen I gel leads to faster gel contraction, enhanced mechanical tension and compression properties, and upregulation of elastin assembly proteins [24]. Additionally, by incorporating increased proportions of insoluble elastin into fibrillar collagen in SMC-laden gels, the contractile marker expression increases [25].

The synergistic effects of mechanical and biochemical cues of ECM components are also considered. For example, in SMCs cultured on stiffness-tunable gels with ECM coatings, collagen I promotes a less proliferative and more migratory phenotype with increased stiffness, while gels coated in fibronectin induce the opposite effects [26]. Meanwhile, the myogenic differentiation of mesenchymal stem cells cultured on silk fibroin hydrogels with TGF-β1 supplementation is significantly increased on soft gels (6 kPa) compared with stiff (33 kPa) gels [21].

### 2.3. Non-Fibrillar Proteins and Matrix Modifiers

The importance of matricellular proteins has become increasingly apparent in recent studies as the degree of crosslinking, presence of precursor and chaperone molecules, and the post-translational modifications play a vital yet complex role in determining ECM structural properties [2,41]. Numerous matricellular proteins regulate SMC behavior and overall vascular dysfunction, and studies using isolated human cells or genetically modified rodents have helped to pinpoint many of these proteins. Advanced Glycation End products (AGEs) increase vascular stiffness via the crosslinking of collagen and gene expression modulation via inflammatory cascade [27]. Matrix metalloproteinase (MMP) -12 production is induced in SMCs after vascular injury in mice and is accompanied by increased vascular stiffness, while deletion of MMP-12 in mice abrogates arterial stiffening via a reduction in elastin degradation [35]. The resident macrophages in mouse and human aortic tissue interact with the hyaluronic acid ECM produced by SMCs and can modulate SMC collagen expression via MMP-9 production, ultimately preventing arterial stiffness [36]. The Serum Response Factor (SRF) regulates the SMC phenotype through the expression of a protein phosphatase, PTEN: with decreased PTEN expression in SMCs isolated from human atherosclerotic lesions [34]. RhoBTB1 attenuates vascular stiffness via actin depolymerization in mice with angiotensin-II-induced hypertension, though it does not reverse the hypertension [29]. Meanwhile, disruptions in the polymerization of elastin due to various genetic mutations in mice degrade the integrity of elastin-contractile units with SMCs, which lethally affect the material properties of the arterial wall [30].

Again, in vitro studies are used to elucidate certain effects of non-fibrillar proteins that may be difficult to decipher using mice. Tunable gels have been used to show that increased collagen crosslinking induces ECM calcification and osteogenic trans-differentiation in SMCs [32]. Additionally, a hyaluronic-acid micropatterned surface used for culturing vascular cells promotes a contractile phenotype and decreased proliferation rate in SMCs [28]. In an example of the effects of post-translational modifications, SMC adhesion to glycosylated fibronectin is significantly increased when compared to adhesion affinity with native fibronectin [33].

In the interest of clinical applications for these findings, non-invasive methods for studying human vessels, such as pulse wave velocity (PWV) measurements, have been combined with proteome analysis of patient samples. In one study using this method, the matricellular proteins involved in collagen fibril assembly and turnover were significantly downregulated in samples from the patients with high PWV, implicating their role in arterial stiffness and SMC dysfunction [31].

## 3. Engineering Complex In Vitro Models of Smooth Muscle

Traditional cell culture methods and animal models have yielded a vast improvement in the knowledge of how specific aspects of the SMC phenotype are affected by individual changes in the properties and composition of the surrounding matrix. However, it is still a challenge to prompt SMCs to behave in culture as they behave in the body [40], so there is a gap in our understanding of how exactly a specific pathology seen in vivo may be triggered on a cellular and molecular level [42,43]. This gap can be filled by using more complex culture systems incorporating more of the biological cues that SMCs experience in the body. These systems can range from simple mechanical stimulation to co-culture with other cell types and organ-on-a-chip devices to complex tissue engineered vascular grafts, as illustrated in Figure 2.

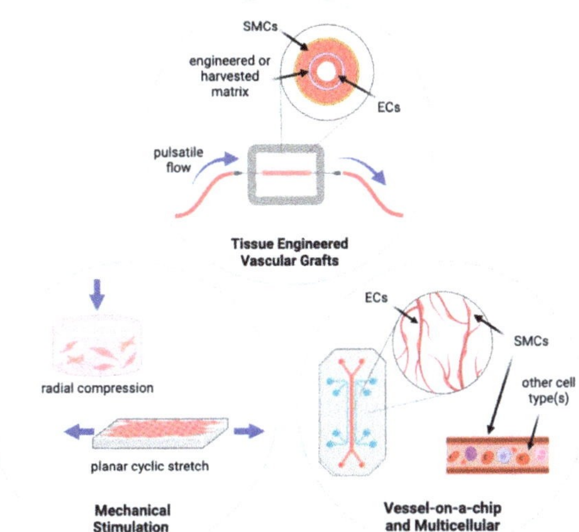


**Figure 2.** Schematic illustration of smooth muscle tissue engineering approaches. (**top**) Tissue Engineered Vascular Grafts, (**left**) Mechanical Stimulation Systems, and (**right**) Vessel-on-a-chip and multicellular culture systems. Created with BioRender.com.

### 3.1. Mechanical Stimulation in Culture Systems

Because SMCs are sensitive mechotransducers, many in vitro models attempted to mimic physiological mechanical forces through various methods, such as axial stretch, vacuum-driven strain, radial compression, or interstitial flow, as illustrated in Figure 2. Using these models, investigators have observed that SMCs behave differently under mechanical strain than in traditional static culture. This indicates that adding mechanical cues to culture systems may better encompass the conditions needed to correlate in vitro results with SMC behavior in vivo [43].

Cyclic strain is a standard method of mechanical stimulation applied to cell culture constructs to induce the desired SMC behavior [44–46]. In one study, incrementally increasing the frequency of cyclic stretch resulted in a higher degree of SMC alignment and denser collagen structure but decreased the expression of many ECM proteins and proteases compared to unstretched controls [47]. The impact of cyclic strain is also directly compared between 2D and 3D SMC cultures. In one example of this, SMCs in 3D constructs align parallel to the direction of strain and shift to a more contractile phenotype, while the cells in 2D align perpendicularly to the direction of strain and showed no changes in contractility [48]. Shear effects have also been compared between 2D and 3D—an increased contractile phenotype was induced in SMCs cultured with interstitial flow in 3D gels compared to SMCs cultured under laminar flow in 2D [49]. Alternatively to cyclic stretch, the centrifugal compression of SMC-laden collagen gels has been used to simulate wall radial strain, and compressed cells exhibited an increased expression of alpha-smooth muscle actin ($\alpha$SMA) and a decreased expression of MMPs and their inhibitors controls [50].

### 3.2. Organ-on-a-Chip and Multicellular Systems

While mechanical stimulation does provide the strain that SMCs would experience natively, the cells may still not respond as they would physiologically if they are not cultured with other tissue cell types that communicate with SMCs and affect their interactions with the surrounding matrix [42]. Thus, more complex models utilizing the 3D environment in combination with co-culture techniques have been used to prompt SMC maturation and more in vivo-like characteristics.

One promising direction for the in vitro modeling of SMC behavior in 3D matrices is organ-on-a-chip technology [51]. Cellular crosstalk was successfully demonstrated between ECs and SMCs when both of the cell types were cultured in a fibrin gel in a microfluidic device modeling vasculogenesis [52]. A versatile device containing four microfluidic parallel channels was used to co-culture SMCs and ECs in an anatomically accurate geometry with tunable channel structures for different flow conditions [53]. Meanwhile, a microfluidic platform was fabricated for the culture and detailed assessment of explanted small arteries [54].

Apart from microfluidics and organ-on-a-chip, other co-culture and 3D studies have shown significant changes in cell behavior due to signaling between SMCs and other cell types. The SMCs cultured with ECs in electrospun fibrin microfibers yielded a robust microvascular structure with significant deposition of collagen I and elastin by SMCs [55]. When culturing monocytes with SMCs at different ratios, a 2:1 and 4:1 ratio resulted in balanced protease activity and ECM deposition rate by SMCs, representing healthy tissue [56]. The SMCs cultured on fibrin discs or strands were treated with adipose stromal cell-secreted factors—this induced increased expression of tropoelastin and elastin assembly proteins and increased high stretch modulus, indicating the deposition of mature collagen [57]. The maturation of mesenchymal cells into SMC-like cells has been achieved through co-culture with fibroblasts in decellularized ECM; this maturation was confirmed using the PCSA analysis [11]. Alternatively, endothelial cell adhesion properties were altered when the cells were exposed to media from SMCs exposed to cyclic stretch, indicating the importance of cell signaling from SMCs to other vascular cells as well [58].

### 3.3. Tissue Engineered Vascular Grafts and Pulsatile flow

As the most complex type of in vitro model, tissue engineered vascular grafts provide the opportunity to recapitulate physiological geometry, biomechanical, and biochemical. However, when constructing these systems, little focus is given to the accuracy of the ECM components, structure, and bioactive nature. Instead, investigators prioritize the matrix mechanical properties of strength and elasticity, with the clinical application of these grafts typically an ultimate goal [6,59].

The common materials used for vascular grafts include synthetic polymer materials such as poly(glycolic acid) (PGA), polycaprolactone (PCL), and polytetrafluoroethylene (PTFE) [60–63]. These polymeric materials are specifically chosen for their proven strength, biocompatibility, and elastic properties, as well as a well-characterized slow rate of degradation in the body [64]. These properties have made them an attractive option for clinical applications in recent years. For example, when PGA scaffolds were seeded with human-induced pluripotent stem cell (iPSC)-derived SMCs and subjected to a radial strain of 1% via pulsatile flow for 7 weeks, SMCs produced robust ECM and acquired a contractile phenotype. These matured tissue constructs remained patent for 30 days after implantation into nude rats [60]. Poly(ethylene glycol)l dimethacrylate/poly(L-lactide) scaffolds were seeded with iPSC-derived SMCs and subjected to pulsatile flow through the scaffold lumen for 6 days, and the cells exhibited increased contractility, increased mature elastin production, and better mechanical properties than the static controls [12].

It is evident through the evaluation of some of these models that the bioactivity of the scaffold material, not just the material properties, significantly affects SMC behavior. The phenotype of SMCs seeded on purely synthetic matrices differs from that on a synthetic/natural mixture or just natural ECM components, and the benefits are generally seen with the addition of natural matrices. For example, TGFβ2-eluting scaffolds composed of PCL and gelatin significantly altered SMC numbers and cell areas on seeded scaffolds by changing the ratio of PCL to gelatin in combination with TGFβ2 dose [65]. Meanwhile, vascular grafts, fabricated via co-electrospinning collagen and hyaluronic acid and implanted in rats, enhanced the host SMC regeneration and proliferation into the graft [66]. In another study, a PCL/collagen I scaffold seeded with endothelial progenitor cells (EPCs) and SMCs, and implanted in sheep, exhibited extensive engraftment to the host vasculature and produced vast amounts of collagen, elastin, and GAGs in a structure resembling native ECM [67].

The most common natural ECM component to use as a scaffold for tissue-engineered small diameter vascular grafts is collagen, but fibrin, elastin, chitosan, and silk are also commonly used, in addition to decellularized blood vessels from animals [61]. In the collagen scaffolds, the fibrillar collagen density has been shown to significantly modulate the secretory factors and cytokines released by iPSC-derived SMCs, with higher densities resulting in the creation of a hypoxic pro-angiogenic and anti-inflammatory microenvironment [68]. In tubular nanofiber collagen scaffolds laden with iPSC-derived ECs internally and iPSC-derived SMCs externally, the collagen fiber alignment prompted cell alignment and consequently reduced the inflammatory response [69]. Meanwhile, in tubular collagen gels with SMCs only, the addition of cyclic strain induced an increased contractile marker expression and mechanical strength [70].

To better facilitate cell incorporation and growth in scaffolds, bioprinting has also been used as an in vitro blood vessel fabrication method, utilizing various combinations of natural and synthetic components and chemical modifications of these components. This has resulted in successful long-term cell viability (120 days) [71], mature cellular behavior [72], and recapitulation of some of the material properties (such as the storage modulus) similar to that of in vivo measurements [73]. In one example, tubular coaxial printing of patient-derived ECM encapsulating ECs or SMCs induced the successful and extensive endogenous deposition of collagen I and some deposition of elastin after 2-week culture with pulsatile flow through the lumen [74]. Separate from bioprinting but in another whole-cloth construction methodology, tissue-engineered rings of iPSC-derived SMCs have also been formed directly from a single cell suspension using clever ring-shaped culture dishes [75], and Raman spectroscopic analysis enabled highly specific characterization of iPSC-derived SMC phenotypic shift in response to growth media formulation [76].

## 4. Engineering of Smooth Muscle to Elucidate Mechanisms of Vascular Disease

Various tissue engineered constructs with human cells have already been applied to model certain vascular diseases to understand the underlying mechanisms of disease progression better. These models can also be used to identify therapeutic targets and to test potential mitigation or treatment strategies in the human mimicry models, thus potentially increasing successful clinical translation.

### 4.1. Hypertension, Pulmonary Arterial Hypertension, and Atherosclerosis

Tissue-engineered models have led to a better understanding of some of the mechanisms involved in vessel wall remodeling, and they have been used to model hypertension and pulmonary arterial hypertension (PAH). Both of these life-threatening pathologies are generally defined by characteristic vessel-wall stiffening and medial thickening in arterial vasculature induced by dysfunctional SMCs and fibroblasts [77,78]. While hypertension is a very common systemic condition usually resulting from external factors or aging and has a myriad of treatment options, PAH is an often incurable chronic condition in which the lung vasculature becomes mechanically obstructed, and the specific cause or disease mechanism is unknown for many patients [79]. Thus, it is important to be able to model both of these conditions in different ways in vitro. In one model of hypertension, explanted rabbit aortas were cultured under perfusion and showed activation of ERK 1/2 signaling cascade via FAK activation under high intraluminal steady stretch, while pulsatile stretch initiated the ERK1/2 signaling without FAK activation [80]. Meanwhile, PAH has been modeled in an artery-on-a-chip device using patient-derived cells. The characteristic PAH phenotype, including intimal thickening and arterial remodeling seen in vivo, was successfully demonstrated in the in vitro model [81].

Atherosclerosis, another common life-threatening cardiovascular condition, is usually brought on by external factors and aging, similarly to hypertension. Yet, the pathology is defined by the formation of intimal plaques that obstruct blood flow. These form due to inflammatory activation of ECs and lipid and immune cell accumulation, and dysfunctional SMCs contribute to plaque formation and progression in several ways [82]. The mechanisms of SMC involvement in atherosclerosis have been elucidated using 3D tissue-engineered models. In one example, a tissue-engineered "flap" laden with SMCs was subjected to complex pulsatile flow to model pathological ECM production as seen in the atherosclerotic plaques, and investigators found that doxycycline treatment abrogated the effects of the complex flow [83]. In another study, the atherosclerotic plaques were modeled with SMCs cultured in collagen gels in calcifying media, revealing that the extracellular vesicles secreted by SMCs in these conditions were calcified, thus further contributing to the plaque formation [84].

*4.2. Insult and Acute Injury*

Invasive procedures, such as plaque removal or vascular graft implantation, can induce acute vascular injury. To model this type of injury and work toward mitigating potential outcomes of restenosis or neointimal hyperplasia in patients, a bioreactor system was developed to culture ex vivo rat and human vessels and subjected them to a controlled injury that vigorously removed the luminal ECs from the vessels. The researchers found that SMC proliferation and neointima formation was reduced by seeding the injured vessels with human umbilical vein ECs prior to subjecting the vessels to simulated arterial flow in a bioreactor. Importantly, this result correlates with the clinical outcomes after vascular graft implantation in animals and humans [85]. Other insults investigated using tissue engineering techniques include radiation and oxidative stress. In one study, reactive oxygen species production induced by the treatment of isolated arterioles with lysophosphatidic acid resulted in significant changes in SMC contractile behavior in response to intraluminal flow when the arterioles were cultured in a bioreactor system [86]. In addition to these examples, a variety of acute insults including specific environmental factors and pharmaceuticals could be readily modeled with similar tissue engineered systems.

*4.3. Genetic Mutations*

Genetic mutations can seriously inhibit the ability of SMCs to secrete ECM and maintain their contractile function properly, which can result in lethal cardiovascular dysfunction in patients [87]. Tissue engineered models can provide deeper insight into the mechanisms through which these genetic mutations manifest vascular dysfunction.

For example, human SMCs from either healthy patients or those affected by genetic mutations resulting in abdominal aortic aneurysms (AAA) were isolated and cultured on poly-lactide-co-glycolide scaffolds for up to 5 weeks, and the differences in cytoskeletal alignment and ECM production were elucidated [88]. Progeria syndrome, a lethal vascular disease resulting from a mutation in the Lamin A gene, was modeled using iPSCs reprogrammed from a patient with progeria. The iPSC-SMCs were then seeded in tubular collagen scaffolds and cultured with ECs under peristaltic flow. The constructs exhibited increased medial thickness, calcification, apoptosis, and diminished vaso-activity, recapitulating the progeria pathology [89]. Another genetic disease associated with SMCs and ECM dysregulation, Marfan syndrome, was modeled using SMCs derived from iPSCs from a Marfan patient. These SMCs were subjected to cyclic stretch and exhibited decreased contractility, upregulated TGFβ signaling, and increased ECM accumulation: all characteristics of the syndrome [90].

## 5. Conclusions and Future Directions

Tissue engineering approaches have shown that SMCs respond differently to various stressors when in a 3D multicellular environment with mechanical stimulation, rather than the traditional 2D static culture systems. Additionally, further mechanistic insight into specific disease phenotypes has been achieved with human cells in tissue engineered in vitro models. A downfall of many of the successful tissue engineered models is that they utilize primary, explanted patient blood vessels or cells, which can inhibit the ability to achieve high throughput and clinically relevant experiments. A promising solution to this is to combine the wide variety of 3D designs and mechanical/chemical stimulation with the use of iPSC-derived SMCs from patients to minimize sample variability and provide a readily available cell source [6,51,75,89–91]. One aspect that could be explored more with these 3D models, particularly in the case of small diameter vascular grafts, is the use of natural matrix materials that better encompass the composition and structure of native ECM. The utilization of these materials has already been shown to induce more native-like behavior in cultured SMCs, as described previously. Further exploration of these materials in vascular grafts could facilitate more clinically relevant models and more insight into CVD mechanisms than is currently possible [11,72]. In addition, computational modeling before designing 3D vascular systems allows researchers to implement better design principles and forego much of the costly empirical experimentation usually needed to optimize these vascular systems [92–94]. Ultimately, combining all of these techniques can move the field forward in analyzing cardiovascular disease mechanisms in more depth and further testing potential future therapeutics or mitigation strategies that could help patients. is a review article

**Funding:** Danielle Yarbrough is a 2022 recipient of the NSF Graduate Research Fellowship Program. Studies of smooth muscle engineering in our lab are partially funded by RAD0102 from the Translational Research Institute through NASA Cooperative Agreement NNX16AO69A.

**Conflicts of Interest:** The authors declare no conflict of interest.

## References

1. Andersson, C.; Vasan, R.S. Epidemiology of cardiovascular disease in young individuals. *Nat. Rev. Cardiol.* **2017**, *15*, 230–240. [CrossRef]
2. Ma, Z.; Mao, C.; Jia, Y.; Fu, Y.; Kong, W. Extracellular matrix dynamics in vascular remodeling. *Cell Physiol.* **2020**, *319*, C481–C499. [CrossRef]
3. Savoji, H.; Mohammadi, M.H.; Rafatian, N.; Toroghi, M.K.; Wang, E.Y.; Zhao, Y.; Korolj, A.; Ahadian, S.; Radisic, M. Cardiovascular disease models: A game changing paradigm in drug discovery and screening. *Biomaterials* **2020**, *198*, 3–26. [CrossRef] [PubMed]
4. Milani-Nejad, N.; Janssen, P.M.L. Small and large animal models in cardiac contraction research: Advantages and disadvantages. *Pharmacol. Ther.* **2014**, *141*, 235–249. [CrossRef]
5. Tsang, H.G.; Rashdan, N.A.; Whitelaw, C.B.A.; Corcoran, B.M.; Summers, K.M.; MacRae, V.E. Large animal models of cardiovascular disease. *Cell Biochem. Funct.* **2016**, *34*, 113–132. [CrossRef] [PubMed]
6. Song, H.-H.G.; Rumma, R.T.; Ozaki, C.K.; Edelman, E.R.; Chen, C.S. Vascular Tissue Engineering: Progress, Challenges, and Clinical Promise. *Cell Stem Cell* **2018**, *22*, 340–354. [CrossRef] [PubMed]
7. Owens, G.K.; Kumar, M.S.; Wamhoff, B.R. Molecular Regulation of Vascular Smooth Muscle Cell Differentiation in Development and Disease. *Physiol. Rev.* **2004**, *84*, 767–801. [CrossRef] [PubMed]
8. Tian, B.; Ding, X.; Song, Y.; Chen, W.; Liang, J.; Yang, L.; Fan, Y.; Li, S.; Zhou, Y. Matrix stiffness regulates SMC functions via TGF-β signaling pathway. *Biomaterials* **2019**, *221*, 119407. [CrossRef]
9. Jeong, K.; Kim, J.H.; Murphy, J.M.; Park, H.; Kim, S.J.; Rodriguez, Y.A.; Kong, H.; Choi, C.; Guan, J.-L.; Taylor, J.M.; et al. Nuclear Focal Adhesion Kinase Controls Vascular Smooth Muscle Cell Proliferation and Neointimal Hyperplasia Through GATA4-Mediated Cyclin D1 Transcription. *Circ. Res.* **2019**, *125*, 152–168. [CrossRef] [PubMed]
10. Ding, Y.; Johnson, R.; Sharma, S.; Ding, X.; Bryant, S.J.; Tan, W. Tethering transforming growth factor β1 to soft hydrogels guides vascular smooth muscle commitment from human mesenchymal stem cells. *Acta Biomater.* **2020**, *105*, 68–77. [CrossRef] [PubMed]
11. Li, N.; Sanyour, H.; Remund, T.; Kelly, P.; Hong, Z. Vascular extracellular matrix and fibroblasts-coculture directed differentiation of human mesenchymal stem cells toward smooth muscle-like cells for vascular tissue engineering. *Mater. Sci. Eng. C* **2018**, *93*, 61–69. [CrossRef]
12. Eoh, J.H.; Shen, N.; Burke, J.A.; Hinderer, S.; Xia, Z.; Schenke-Layland, K.; Gerecht, S. Enhanced elastin synthesis and maturation in human vascular smooth muscle tissue derived from induced-pluripotent stem cells. *Acta Biomater.* **2017**, *52*, 49–59. [CrossRef]

13. Daley, M.C.; Bonzanni, M.; MacKenzie, A.M.; Kaplan, D.L.; Black, L.D., III. The effects of membrane potential and extracellular matrix composition on vascular differentiation of cardiac progenitor cells. *Biochem. Biophys. Res. Commun.* **2020**, *530*, 240–245. [CrossRef]
14. Hamill, O.P.; Marty, A.; Neher, E.; Sakmann, B.; Sigworth, F.J. Improved Patch-Clamp Techniques for High-Resolution Current Recording from Cells and Cell-Free Membrane Patches. *Pflug. Arch.* **1981**, *391*, 85–100. [CrossRef] [PubMed]
15. Bobi, J.; Garabito, M.; Solanes, N.; Cidad, P.; Ramos-Pérez, V.; Ponce, A.; Rigol, M.; Freixa, X.; Pérez-Martínez, C.; de Prado, A.P.; et al. Kv1.3 blockade inhibits proliferation of vascular smooth muscle cells in vitro and intimal hyperplasia in vivo. *Transl. Res.* **2020**, *224*, 40–54. [CrossRef] [PubMed]
16. Steppan, J.; Wang, H.; Bergman, Y.; Rauer, M.J.; Tan, S.; Jandu, S.; Nandakumar, K.; Barreto-Ortiz, S.; Cole, R.N.; Boronina, T.N.; et al. Lysyl oxidase-like 2 depletion is protective in age-associated vascular stiffening. *Am. J. Physiol. Heart Circ. Physiol.* **2019**, *317*, H49–H59. [CrossRef]
17. Kim, S.K.; McCurley, A.T.; DuPont, J.J.; Aronovitz, M.; Moss, M.E.; Stillman, I.E.; Karumanchi, S.A.; Christou, D.D.; Jaffe, I.Z. Smooth Muscle Cell–Mineralocorticoid Receptor as a Mediator of Cardiovascular Stiffness With Aging. *Hypertension* **2018**, *71*, 609–621. [CrossRef] [PubMed]
18. Kiss, T.; Nyúl-Tóth, Á.; Gulej, R.; Tarantini, S.; Csipo, T.; Mukli, P.; Ungvari, A.; Balasubramanian, P.; Yabluchanskiy, A.; Benyo, Z.; et al. Old blood from heterochronic parabionts accelerates vascular aging in young mice: Transcriptomic signature of pathologic smooth muscle remodeling. *GeroScience* **2022**, *44*, 953–981. [CrossRef]
19. Mui, K.L.; Bae, Y.H.; Gao, L.; Liu, S.-L.; Xu, T.; Radice, G.L.; Chen, C.S.; Assoian, R.K. N-Cadherin Induction by ECM Stiffness and FAK Overrides the Spreading Requirement for Proliferation of Vascular Smooth Muscle Cells. *Cell Rep.* **2015**, *10*, 1477–1486. [CrossRef]
20. Chen, J.-Y.; Wang, Y.-X.; Ren, K.-F.; Wang, Y.-B.; Fu, G.-S.; Jia, J. The influence of substrate stiffness on osteogenesis of vascular smooth muscle cells. *Colloids Surf. B Biointerfaces* **2021**, *197*, 111388. [CrossRef]
21. Floren, M.; Bonani, W.; Dharmarajan, A.; Mott, A.; Migliaresi, C.; Tan, W. Human mesenchymal stem cells cultured on silk hydrogels with variable stiffness and growth factor differentiate into mature smooth muscle cell phenotype. *Acta Biomater.* **2016**, *31*, 156–166. [CrossRef]
22. Xie, S.-A.; Zhang, T.; Wang, J.; Zhao, F.; Zhang, Y.-P.; Yao, W.-J.; Hur, S.S.; Yeh, Y.-T.; Pang, W.; Zheng, L.-S.; et al. Matrix stiffness determines the phenotype of vascular smooth muscle cell in vitro and in vivo: Role of DNA methyltransferase 1. *Biomaterials* **2018**, *155*, 203–216. [CrossRef]
23. Petit, C.; Yousefi, A.-A.K.; Moussa, O.B.; Michel, J.-B.; Guignandon, A.; Avril, S. Regulation of SMC traction forces in human aortic thoracic aneurysms. *Biomech. Modeling Mechanobiol.* **2021**, *20*, 717–731. [CrossRef]
24. Pezzoli, D.; Paolo, J.D.; Kumra, H.; Fois, G.; Candiani, G.; Reinhardt, D.P.; Mantovani, D. Fibronectin promotes elastin deposition, elasticity and mechanical strength in cellularised collagen-based scaffolds. *Biomaterials* **2018**, *180*, 130–142. [CrossRef]
25. Ryan, A.J.; O'Brien, F.J. Insoluble elastin reduces collagen scaffold stiffness, improves viscoelastic properties, and induces a contractile phenotype in smooth muscle cells. *Biomaterials* **2015**, *73*, 296–307. [CrossRef]
26. Rickel, A.P.; Sanyour, H.J.; Leyda, N.A.; Hong, Z. Extracellular Matrix Proteins and Substrate Stiffness Synergistically Regulate Vascular Smooth Muscle Cell Migration and Cortical Cytoskeleton Organization. *ACS Appl. Biomater.* **2020**, *3*, 2360–2369. [CrossRef] [PubMed]
27. Senatus, L.M.; Schmidt, A.M. The AGE-RAGE Axis: Implications for Age-Associated Arterial Diseases. *Front. Genet.* **2017**, *8*, 187. [CrossRef]
28. Li, J.; Zhang, K.; Wu, J.; Zhang, L.; Yang, P.; Tu, Q.; Huang, N. Tailoring of the titanium surface by preparing cardiovascular endothelial extracellular matrix layer on the hyaluronic acid micro-pattern for improving biocompatibility. *Colloids Surf. B Biointerfaces* **2015**, *128*, 201–210. [CrossRef] [PubMed]
29. Fang, S.; Wu, J.; Reho, J.J.; Lu, K.-T.; Brozoski, D.T.; Kumar, G.; Werthman, A.M.; Sebastiao Donato Silva, J.; Veitia, P.C.M.; Wackman, K.K.; et al. RhoBTB1 reverses established arterial stiffness in angiotensin II–induced hypertension by promoting actin depolymerization. *JCI Insight* **2022**, *7*, e158043. [CrossRef]
30. Yanagisawa, H.; Wagenseil, J. Elastic fibers and biomechanics of the aorta: Insights from mouse studies. *Matrix Biol.* **2020**, *85–86*, 160–172. [CrossRef] [PubMed]
31. Hansen, M.L.; Beck, H.C.; Irmukhamedov, A.; Jensen, P.S.; Olsen, M.H.; Rasmussen, L.M. Proteome Analysis of Human Arterial Tissue Discloses Associations Between the Vascular Content of Small Leucine-Rich Repeat Proteoglycans and Pulse Wave Velocity. *Arterioscler. Thromb. Vasc. Biol.* **2015**, *35*, 1896–1903. [CrossRef] [PubMed]
32. Jover, E.; Silvente, A.; Marin, F.; Martinez-Gonzalez, J.; Orriols, M.; Martinez, C.M.; Puche, C.M.; Valdés, M.; Rodriguez, C.; Hernández-Romero, D. Inhibition of enzymes involved in collagen cross-linking reduces vascular smooth muscle cell calcification. *FASEB J.* **2018**, *32*, 4459–4469. [CrossRef]
33. Dhar, S.; Sun, Z.; Meininger, G.A.; Hill, M.A. Nonenzymatic glycation interferes with fibronectin-integrin interactions in vascular smooth muscle cells. *Microcirculation* **2017**, *24*, e12347. [CrossRef] [PubMed]
34. Horita, H.; Wysoczynski, C.L.; Walker, L.A.; Moulton, K.S.; Li, M.; Ostriker, A.; Tucker, R.; McKinsey, T.A.; Churchill, M.E.A.; Nemenoff, R.A.; et al. Nuclear PTEN functions as an essential regulator of SRF-dependent transcription to control smooth muscle differentiation. *Nat. Commun.* **2016**, *7*, 10830. [CrossRef]

35. Liu, S.-L.; Bae, Y.H.; Yu, C.; Monslow, J.; Hawthorne, E.A.; Castagnino, P.; Branchetti, E.; Ferrari, G.; Damrauer, S.M.; Puré, E.; et al. Matrix metalloproteinase-12 is an essential mediator of acute and chronic arterial stiffening. *Sci. Rep.* **2015**, *5*, 17189. [CrossRef]
36. Lim, H.Y.; Lim, S.Y.; Tan, C.K.; Thiam, C.H.; Goh, C.C.; Carbajo, D.; Chew, S.H.S.; See, P.; Chakarov, S.; Wang, X.N.; et al. Hyaluronan Receptor LYVE-1-Expressing Macrophages Maintain Arterial Tone through Hyaluronan-Mediated Regulation of Smooth Muscle Cell Collagen. *Immunity* **2018**, *49*, 326–341. [CrossRef] [PubMed]
37. Schiller, H.B.; Friedel, C.C.; Boulegue, C.; Fässler, R. Quantitative proteomics of the integrin adhesome show a myosin II-dependent recruitment of LIM domain proteins. *EMBO Rep.* **2011**, *12*, 259–266. [CrossRef] [PubMed]
38. Chan, X.Y.; Volkova, E.; Eoh, J.; Black, R.; Fang, L.; Gorashi, R.; Song, J.; Wang, J.; Elliott, M.B.; Barreto-Ortiz, S.F.; et al. HIF2A gain-of-function mutation modulates the stiffness of smooth muscle cells and compromises vascular mechanics. *IScience* **2021**, *24*, 102246. [CrossRef] [PubMed]
39. Espinosa, M.G.; Gardner, W.S.; Bennett, L.; Sather, B.A.; Yanagisawa, H.; Wagenseil, J.E. The Effects of Elastic Fiber Protein Insufficiency and Treatment on the Modulus of Arterial Smooth Muscle Cells. *J. Biomech. Eng.* **2014**, *136*, 021030. [CrossRef]
40. Beamish, J.A.; He, P.; Kottke-Marchant, K.; Marchant, R.E. Molecular Regulation of Contractile Smooth Muscle Cell Phenotype: Implications for Vascular Tissue Engineering. *Tissue Eng. Part B Rev.* **2010**, *16*, 467–491. [CrossRef]
41. Ramaswamy, A.K.; Vorp, D.A.; Weinbaum, J.S. Functional Vascular Tissue Engineering Inspired by Matricellular Proteins. *Front. Cardiovasc. Med.* **2019**, *6*, 74. [CrossRef]
42. Mantella, L.-E.; Quan, A.; Verma, S. Variability in vascular smooth muscle cell stretch-induced responses in 2D culture. *Vasc. Cell* **2015**, *7*, 7. [CrossRef]
43. Jensen, L.F.; Bentzon, J.F.; Albarrán-Juárez, J. The Phenotypic Responses of Vascular Smooth Muscle Cells Exposed to Mechanical Cues. *Cells* **2021**, *10*, 2209. [CrossRef]
44. Walters, B.; Turner, P.A.; Rolauffs, B.; Hart, M.L.; Stegemann, J.P. Controlled Growth Factor Delivery and Cyclic Stretch Induces a Smooth Muscle Cell-like Phenotype in Adipose-Derived Stem Cells. *Cells* **2021**, *10*, 3123. [CrossRef]
45. Sato, K.; Nitta, M.; Ogawa, A. A Microfluidic Cell Stretch Device to Investigate the Effects of Stretching Stress on Artery Smooth Muscle Cell Proliferation in Pulmonary Arterial Hypertension. *Inventions* **2019**, *4*, 1. [CrossRef]
46. Yamashiro, Y.; Thang, B.Q.; Ramirez, K.; Shin, S.J.; Kohata, T.; Ohata, S.; Nguyen, T.A.V.; Ohtsuki, S.; Nagayama, K.; Yanagisawa, H. Matrix mechanotransduction mediated by thrombospondin-1/integrin/YAP in the vascular remodeling. *Proc. Natl. Acad. Sci. USA* **2020**, *117*, 9896–9905. [CrossRef]
47. Lévesque, L.; Loy, C.; Lainé, A.; Drouin, B.; Chevallier, P.; Mantovani, D. Incrementing the Frequency of Dynamic Strain on SMC-Cellularised Collagen-Based Scaffolds Affects Extracellular Matrix Remodeling and Mechanical Properties. *ACS Biomater. Sci. Eng.* **2018**, *4*, 3759–3767. [CrossRef] [PubMed]
48. Bono, N.; Pezzoli, D.; Levesque, L.; Loy, C.; Candiani, G.; Fiore, G.B.; Mantovani, D. Unraveling the role of mechanical stimulation on smooth muscle cells: A comparative study between 2D and 3D models. *Biotechnol. Bioeng.* **2016**, *113*, 2254–2263. [CrossRef]
49. Shi, Z.-D.; Tarbell, J.M. Fluid Flow Mechanotransduction in Vascular Smooth Muscle Cells and Fibroblasts. *Ann. Biomed. Eng.* **2011**, *39*, 1608–1619. [CrossRef] [PubMed]
50. Hiroshima, Y.; Oyama, Y.; Sawasaki, K.; Nakamura, M.; Kimura, N.; Kawahito, K.; Fujie, H.; Sakamoto, N. A Compressed Collagen Construct for Studying Endothelial–Smooth Muscle Cell Interaction Under High Shear Stress. *Ann. Biomed. Eng.* **2022**, *50*, 951–963. [CrossRef] [PubMed]
51. Cochrane, A.; Albers, H.J.; Passiera, R.; Mummery, C.L.; Berg, A.v.d.; Orlova, V.V.; Meer, A.D.v.d. Advanced in vitro models of vascular biology: Human induced pluripotent stem cells and organ-on-chip technology. *Adv. Drug Deliv. Rev.* **2019**, *140*, 68–77. [CrossRef]
52. Cuenca, M.V.; Cochrane, A.; Hil, F.E.v.d.; Vries, A.A.F.d.; Oberstein, S.A.J.L.; Mummery, C.L.; Orlova, V.V. Engineered 3D vessel-on-chip using hiPSC-derived endothelial- and vascular smooth muscle cells. *Stem Cell Rep.* **2021**, *16*, 2159–2168. [CrossRef]
53. Cho, M.; Park, J.-K. Fabrication of a Perfusable 3D In Vitro Artery-Mimicking Multichannel System for Artery Disease Models. *ACS Biomater. Sci. Eng.* **2020**, *6*, 5326–5336. [CrossRef]
54. Yasotharan, S.; Pinto, S.; Sled, J.G.; Bolzef, S.-S.; Günther, A. Artery-on-a-chip platform for automated, multimodal assessment of cerebral blood vessel structure and function. *Lab A Chip* **2015**, *15*, 2660–2669. [CrossRef]
55. Barreto-Ortiz, S.F.; Fradkin, J.; Eoh, J.; Trivero, J.; Davenport, M.; Ginn, B.; Mao, H.-Q.; Gerecht, S. Fabrication of 3-dimensional multicellular microvascular structures. *FASEB J.* **2015**, *29*, 3302–3314. [CrossRef]
56. Zhang, X.; Battiston, K.G.; Labow, R.S.; Simmons, C.A.; Santerre, J.P. Generating favorable growth factor and protease release profiles to enable extracellular matrix accumulation within an in vitro tissue engineering environment. *Acta Biomater.* **2017**, *54*, 81–94. [CrossRef] [PubMed]
57. Ramaswamy, A.K.; Sides, R.E.; Cunnane, E.M.; Lorentz, K.L.; Reines, L.M.; Vorp, D.A.; Weinbaum, J.S. Adipose-derived stromal cell secreted factors induce the elastogenesis cascade within 3D aortic smooth muscle cell constructs. *Matrix Biol. Plus* **2019**, *4*, 100014. [CrossRef] [PubMed]
58. Liu, P.; Shi, Y.; Fan, Z.; Zhou, Y.; Song, Y.; Liu, Y.; Yu, G.; An, Q.; Zhu, W. Inflammatory Smooth Muscle Cells Induce Endothelial Cell Alterations to Influence Cerebral Aneurysm Progression via Regulation of Integrin and VEGF Expression. *Cell Transplant.* **2018**, *28*, 713–722. [CrossRef]
59. Naegeli, K.M.; Kural, M.H.; Li, Y.; Wang, J.; Hugentobler, E.A.; Niklason, L.E. Bioengineering Human Tissues and the Future of Vascular Replacement. *Circ. Res.* **2022**, *131*, 109–126. [CrossRef]

60. Luo, J.; Qin, L.; Zhao, L.; Gui, L.; Ellis, M.; Huang, Y.; Kural, M.; Clark, J.A.; Ono, S.; Wang, J.; et al. Tissue-Engineered Vascular Grafts with Advanced Mechanical Strength from Human iPSCs. *Cell Stem Cell* **2020**, *26*, 251–261. [CrossRef]
61. Yalcin, I.; Horakova, J.; Mikes, P.; Sadikoglu, T.G.; Domin, R.; Lukas, D. Design of Polycaprolactone Vascular Grafts. *J. Ind. Text.* **2014**, *45*, 813–833. [CrossRef]
62. Leal, B.B.J.; Wakabayashi, N.; Oyama, K.; Kamiya, H.; Braghirolli, D.I.; Pranke, P. Vascular Tissue Engineering: Polymers and Methodologies for Small Caliber Vascular Grafts. *Front. Cardiovasc. Med.* **2021**, *7*, 592361. [CrossRef] [PubMed]
63. Li, M.-X.; Li, L.; Zhou, S.-Y.; Cao, J.-H.; Liang, W.-H.; Tian, Y.; Shi, X.-T.; Yang, X.-B.; Wu, D.-Y. A biomimetic orthogonal-bilayer tubular scaffold for the co-culture of endothelial cells and smooth muscle cells. *RSC Adv.* **2021**, *11*, 31783–31790. [CrossRef] [PubMed]
64. Serrano, M.C.; Pagani, R.; Vallet-Regí, M.; Peña, J.; Rámila, A.; Izquierdo, I.; Portolés, M.T. In vitro biocompatibility assessment of poly($\varepsilon$-caprolactone) films using L929 mouse fibroblasts. *Biomaterials* **2004**, *25*, 5603–5611. [CrossRef] [PubMed]
65. Ardila, D.C.; Tamimi, E.; Doetschman, T.; Wagner, W.R.; Geest, J.P.V. Modulating smooth muscle cell response by the release of TGFβ2 from tubular scaffolds for vascular tissue engineering. *J. Control. Release* **2019**, *299*, 44–52. [CrossRef]
66. Qin, K.; Wang, F.; Simpson, R.M.L.; Zheng, X.; Wang, H.; Hu, Y.; Gao, Z.; Xu, Q.; Zhao, Q. Hyaluronan promotes the regeneration of vascular smooth muscle with potent contractile function in rapidly biodegradable vascular grafts. *Biomaterials* **2020**, *257*, 120226. [CrossRef] [PubMed]
67. Ju, Y.M.; Ahn, H.; Arenas-Herrera, J.; Kim, C.; Abolbashari, M.; Atala, A.; Yoo, J.J.; Lee, S.J. Electrospun vascular scaffold for cellularized small diameter blood vessels: A preclinical large animal study. *Acta Biomater.* **2017**, *59*, 58–67. [CrossRef]
68. Dash, B.C.; Setia, O.; Gorecka, J.; Peyvandi, H.; Duan, K.; Lopes, L.; Nie, J.; Berthiaume, F.; Dardik, A.; Hsia, H.C. A Dense Fibrillar Collagen Scaffold Differentially Modulates Secretory Function of iPSC-Derived Vascular Smooth Muscle Cells to Promote Wound Healing. *Cells* **2020**, *9*, 966. [CrossRef]
69. Nakayama, K.H.; Joshi, P.A.; Lai, E.S.; Gujar, P.; Joubert, L.-M.; Chen, B.; Huang, N.F. Bilayered vascular graft derived from human induced pluripotent stem cells with biomimetic structure and function. *Future Med.* **2015**, *10*, 745–755. [CrossRef]
70. Bono, N.; Meghezi, S.; Soncini, M.; Piola, M.; Mantovani, D.; Fiore, G.B. A Dual-Mode Bioreactor System for Tissue Engineered Vascular Models. *Ann. Biomed. Eng.* **2017**, *45*, 1496–1510. [CrossRef]
71. Chimene, D.; Peak, C.W.; Gentry, J.L.; Carrow, J.K.; Cross, L.M.; Mondragon, E.; Cardoso, G.B.; Kaunas, R.; Gaharwar, A.K. Nanoengineered Ionic–Covalent Entanglement (NICE) Bioinks for 3D Bioprinting. *ACS Appl. Mater. Interfaces* **2018**, *10*, 9957–9968. [CrossRef] [PubMed]
72. Schöneberg, J.; De Lorenzi, F.; Theek, B.; Blaeser, A.; Rommel, D.; Kuehne, A.J.; Kießling, F.; Fischer, H. Engineering biofunctional in vitro vessel models using a multilayer bioprinting technique. *Sci. Rep.* **2018**, *8*, 10430. [CrossRef]
73. Cao, X.; Maharjan, S.; Ashfaq, R.; Yu, J.S.; Zhang, S. Bioprinting of Small-Diameter Blood Vessels. *Engineering* **2021**, *7*, 832–844. [CrossRef]
74. Gao, G.; Kim, H.; Kim, B.S.; Kong, J.S.; Lee, J.Y.; Park, B.W.; Chae, S.; Kim, J.; Ban, K.; Jang, J.; et al. Tissue-engineering of vascular grafts containing endothelium and smooth-muscle using triple-coaxial cell printing featured. *Appl. Phys. Rev.* **2019**, *6*, 041402. [CrossRef]
75. Dash, B.C.; Levi, K.; Schwan, J.; Luo, J.; Bartulos, O.; Wu, H.; Qiu, C.; Yi, T.; Ren, Y.; Campbell, S.; et al. Tissue-Engineered Vascular Rings from Human iPSC-Derived Smooth Muscle Cells. *Stem Cell Rep.* **2016**, *7*, 19–28. [CrossRef]
76. Marzia, J.; Brauchle, E.M.; Schenke-Layland, K.; Rolle, M.W. Non-invasive functional molecular phenotyping of human smooth muscle cells utilized in cardiovascular tissue engineering. *Acta Biomater.* **2019**, *89*, 193–205. [CrossRef]
77. Stenmark, K.R.; Frid, M.G.; Graham, B.B.; Tuder, R.M. Dynamic and diverse changes in the functional properties of vascular smooth muscle cells in pulmonary hypertension. *Cardiovasc. Res.* **2018**, *114*, 551–564. [CrossRef]
78. Harvey, A.; Montezano, A.C.; Lopes, R.A.; Rios, F.; Touyz, R.M. Vascular Fibrosis in Aging and Hypertension: Molecular Mechanisms and Clinical Implications. *Can. J. Cardiol.* **2016**, *32*, 659–668. [CrossRef] [PubMed]
79. Guignabert, C.; Tu, L.; Girerd, B.; Ricard, N.; Huertas, A.; Montani, D.; Humbert, M. New Molecular Targets of Pulmonary Vascular Remodeling in Pulmonary Arterial Hypertension: Importance of Endothelial Communication. *Chest* **2015**, *147*, 529–537. [CrossRef] [PubMed]
80. Lehoux, S.; Esposito, B.; Merval, R.; Tedgui, A. Differential Regulation of Vascular Focal Adhesion Kinase by Steady Stretch and Pulsatility. *Circulation* **2005**, *111*, 643–649. [CrossRef]
81. Al-Hilal, T.A.; Keshavarz, A.; Kadry, H.; Lahooti, B.; Al-Obaida, A.; Ding, Z.; Li, W.; Kamm, R.; McMurtry, I.F.; Lahm, T.; et al. Pulmonary-arterial-hypertension (PAH)-on-a-chip: Fabrication, validation and application. *Lab A Chip* **2020**, *20*, 3334–3345. [CrossRef]
82. Seidman, M.A.; Mitchell, R.N.; Stone, J.R. Pathophysiology of Atherosclerosis. In *Cellular and Molecular Pathobiology of Cardiovascular Disease*; Monte, S., Willis, J.W.H., Stone, J.R., Eds.; Academic Press: Cambridge, MA, USA; Elsevier: Amsterdam, The Netherlands, 2014; pp. 221–237. [CrossRef]
83. Hosseini, V.; Mallone, A.; Mirkhani, N.; Noir, J.; Salek, M.; Pasqualini, F.S.; Schuerle, S.; Khademhosseini, A.; Hoerstrup, S.P.; Vogel, V. A Pulsatile Flow System to Engineer Aneurysm and Atherosclerosis Mimetic Extracellular Matrix. *Adv. Sci.* **2020**, *7*, 2000173. [CrossRef] [PubMed]

84. Hutcheson, J.D.; Goettsch, C.; Bertazzo, S.; Maldonado, N.; Ruiz, J.L.; Goh, W.; Yabusaki, K.; Faits, T.; Bouten, C.; Franck, G.; et al. Genesis and growth of extracellular-vesicle-derived microcalcification in atherosclerotic plaques. *Nat. Mater.* **2016**, *15*, 335–343. [CrossRef]
85. Kural, M.H.; Dai, G.; Niklason, L.E.; Gui, L. An Ex Vivo Vessel Injury Model to Study Remodeling. *Cell Transplant.* **2018**, *27*, 1375–1389. [CrossRef] [PubMed]
86. Staiculescu, M.C.; Ramirez-Perez, F.I.; Castorena-Gonzalez, J.A.; Hong, Z.; Sun, Z.; Meininger, G.A.; Martinez-Lemus, L.A. Lysophosphatidic acid induces integrin activation in vascular smooth muscle and alters arteriolar myogenic vasoconstriction. *Front. Physiol.* **2014**, *5*, 413. [CrossRef]
87. Lacolley, P.; Regnault, V.; Segers, P.; Laurent, S. Vascular Smooth Muscle Cells and Arterial Stiffening: Relevance in Development, Aging, and Disease. *Physiol. Rev.* **2017**, *97*, 1555–1617. [CrossRef]
88. Bogunovic, N.; Meekel, J.P.; Majolée, J.; Hekhuis, M.; Pyszkowski, J.; Jockenhövel, S.; Kruse, M.; Riesebos, E.; Micha, D.; Blankensteijn, J.D.; et al. Patient-Specific 3-Dimensional Model of Smooth Muscle Cell and Extracellular Matrix Dysfunction for the Study of Aortic Aneurysms. *J. Endovasc. Ther.* **2021**, *28*, 604–613. [CrossRef] [PubMed]
89. Atchison, L.; Zhang, H.; Cao, K.; Truskey, G.A. A Tissue Engineered Blood Vessel Model of Hutchinson-Gilford Progeria Syndrome Using Human iPSC-derived Smooth Muscle Cells. *Sci. Rep.* **2017**, *7*, 8168. [CrossRef]
90. Granata, A.; Serrano, F.; Bernard, W.G.; McNamara, M.; Low, L.; Sastry, P.; Sinha, S. An iPSC-derived vascular model of Marfan syndrome identifies key mediators of smooth muscle cell death. *Nat. Genet.* **2016**, *49*, 97–109. [CrossRef]
91. Luo, J.; Qin, L.; Park, J.; Kural, M.H.; Huang, Y.; Shi, X.; Riaz, M.; Wang, J.; Ellis, M.W.; Anderson, C.W.; et al. Readily Available Tissue-Engineered Vascular Grafts Derived From Human Induced Pluripotent Stem Cells. *Circ. Res.* **2022**, *130*, 925–927. [CrossRef]
92. Lashkarinia, S.S.; Coban, G.; Kose, B.; Salihoglu, E.; Pekkan, K. Computational modeling of vascular growth in patient-specific pulmonary arterial patch reconstructions. *J. Biomech.* **2021**, *117*, 110274. [CrossRef] [PubMed]
93. Mousavi, S.J.; Jayendiran, R.; Farzaneh, S.; Campisi, S.; Viallon, M.; Croisille, P.; Avril, S. Coupling hemodynamics with mechanobiology in patient-specific computational models of ascending thoracic aortic aneurysms. *Comput. Methods Programs Biomed.* **2021**, *205*, 106107. [CrossRef] [PubMed]
94. Blum, K.M.; Zbinden, J.C.; Ramachandra, A.B.; Lindsey, S.E.; Szafron, J.M.; Reinhardt, J.W.; Heitkemper, M.; Best, C.A.; Mirhaidari, G.J.M.; Chang, Y.-C.; et al. Tissue engineered vascular grafts transform into autologous neovessels capable of native function and growth. *Commun. Med.* **2022**, *2*, 3. [CrossRef] [PubMed]

 *bioengineering*

*Review*

# Anti-Inflammatory and Anti-Thrombogenic Properties of Arterial Elastic Laminae

**Jeremy Goldman [1],*, Shu Q. Liu [2],* and Brandon J. Tefft [3],***

[1]  Department of Biomedical Engineering, Michigan Technological University, Houghton, MI 49931, USA
[2]  Biomedical Engineering Department, Northwestern University, Evanston, IL 60208, USA
[3]  Department of Biomedical Engineering, Medical College of Wisconsin & Marquette University, Milwaukee, WI 53226, USA
*  Correspondence: jgoldman@mtu.edu (J.G.); sliu@northwestern.edu (S.Q.L.); btefft@mcw.edu (B.J.T.)

**Abstract:** Elastic laminae, an elastin-based, layered extracellular matrix structure in the media of arteries, can inhibit leukocyte adhesion and vascular smooth muscle cell proliferation and migration, exhibiting anti-inflammatory and anti-thrombogenic properties. These properties prevent inflammatory and thrombogenic activities in the arterial media, constituting a mechanism for the maintenance of the structural integrity of the arterial wall in vascular disorders. The biological basis for these properties is the elastin-induced activation of inhibitory signaling pathways, involving the inhibitory cell receptor signal regulatory protein α (SIRPα) and Src homology 2 domain-containing protein tyrosine phosphatase 1 (SHP1). The activation of these molecules causes deactivation of cell adhesion- and proliferation-regulatory signaling mechanisms. Given such anti-inflammatory and anti-thrombogenic properties, elastic laminae and elastin-based materials have potential for use in vascular reconstruction.

**Keywords:** elastin; inflammation; thrombosis; intimal hyperplasia; arterial reconstruction

Citation: Goldman, J.; Liu, S.Q.; Tefft, B.J. Anti-Inflammatory and Anti-Thrombogenic Properties of Arterial Elastic Laminae. *Bioengineering* **2023**, *10*, 424. https://doi.org/10.3390/bioengineering10040424

Academic Editors: Kurt Pfannkuche and George A. Truskey

Received: 17 December 2022
Revised: 7 March 2023
Accepted: 16 March 2023
Published: 28 March 2023



## 1. Introduction

The wall of arteries consists of extracellular matrix components, including collagen matrix and elastic laminae. The essential functions of the extracellular matrix are to organize vascular endothelial cells, smooth muscle cells, and fibroblasts into the intima, media, and adventitia of the arterial wall, respectively; provide mechanical strength and elasticity to the arterial wall; and participate in cell signal transduction involved in vascular development and pathogenic processes such as inflammation, thrombosis, and atherosclerosis. Elastic laminae work with the collagen matrix in an antagonistic manner to control vascular cell and leukocyte adhesion, proliferation, and migration, which are cell activities directly influencing inflammatory, thrombogenic, and atherogenic processes [1–4]. Whereas the collagen matrix stimulates these cell activities, enhancing inflammatory, thrombogenic, and atherogenic processes [5–7], the elastic laminae exert an opposite effect, suppressing these pathogenic processes [8–21]. The antagonistic action of the elastic laminae helps prevent excessive inflammatory responses and vascular disorders [8,9,11–21]. In arterial reconstruction, these elastic lamina properties can prevent intimal hyperplasia, a process leading to restenosis and failure of arterial grafts. This paper reviews the role of the arterial elastic laminae in controlling inflammatory responses, thrombosis, and neointima formation in reconstructed arteries.

Inflammation in reconstructed arteries is a series of processes activated in response to surgery, mechanical injury, and exposure to biomaterials [1,22–26]. In the host artery near the junction with a reconstructed artery, several inflammatory processes can occur, including elevation in the endothelial permeability, interstitial edema, cytokine expression and secretion, leukocyte adhesion to injured endothelial cells, smooth muscle cell and fibroblast proliferation, extracellular matrix overproduction, and fibrosis. At the junction of the host artery and the reconstructed artery, blood coagulation may be induced in response to injury and hemorrhage. These processes contribute to thrombosis and intimal hyperplasia, resulting in the formation of neointima. In reconstructed arteries, inflammation and thrombosis occur during the early phase with the level dependent on the material at the blood-contacting surface [27–29]. For instance, in autologous vein grafts, the causes and pathological processes described above for the host artery occur. In synthetic material-based arterial grafts, blood coagulation can be induced rapidly, resulting in thrombosis (note that synthetic grafts can only be used under high flow conditions, which reduce the rate of thrombogenesis, and it is necessary to use anti-coagulants to minimize the risk of thrombus development). Following this phase, smooth muscle cells can proliferate and migrate from the host artery into the thrombus of the reconstructed artery, contributing to the development of neointima, which can cause restenosis of the reconstructed artery [30–38]. In autologous vein-based arterial constructs, smooth muscle cells in the venous wall can also proliferate and migrate into the thrombus to form neointima. Thus, a critical concern in arterial reconstruction is how to prevent inflammatory responses, thrombosis, and neointima formation. As the arterial elastic laminae exert an inhibitory effect on inflammatory responses, thrombosis, and smooth muscle cell proliferation and migration [8,9], this extracellular matrix and elastin-based materials can potentially be used as a surface material for arterial reconstruction.

## 2. Molecular Structure of Elastic Laminae

### 2.1. Elastin Gene

Elastin is a polymer, and its precursor, tropoelastin, is a protein encoded by the *ELN* gene in humans [39]. The *ELN* gene is a 45 kb segment within chromosome 7q11.1. It is comprised of 34 exons and nearly 700 introns [40]. Elastin consists of alternating hydrophobic and hydrophilic domains. The hydrophobic domains are rich in hydrophobic amino acids such as glycine and proline. These domains are important for the self-assembly of supramolecular structure. The hydrophilic domains are rich in lysine residues. These domains are important for crosslinking to form a highly stable, insoluble structure.

Elastin production occurs primarily during fetal development and postnatal growth and is negligible by early adulthood [41–43]. *ELN* gene transcription is steady throughout the lifespan and elastin production is primarily regulated by posttranscriptional destabilization of tropoelastin mRNA in mature tissue [44,45]. The low synthesis and turnover of elastin in adults has important implications in aging and disease.

There are at least 11 isoforms of elastin as a result of alternative splicing of the *ELN* pre-mRNA [46,47]. These isoforms result in tissue-specific variants of elastin with unique properties [48]. The structure and function of these isoforms are subjects of ongoing research.

### 2.2. Tropoelastin

Tropoelastin is the soluble protein precursor to elastin and has a molecular weight of 60–70 kDa [49–51]. Once exported from the cell, tropoelastin molecules reversibly self-assemble into globular aggregates of elastin. Self-assembly is caused by interactions between the hydrophobic domains in a process known as coacervation [52–54]. This is followed by irreversible crosslinking of lysine residues within the hydrophilic domains. The crosslinking process involves the formation of desmosine and isodesmosine covalent crosslinks by the enzyme lysyl oxidase (LOX) [55]. Both inter- and intra-chain crosslinks are formed. Important for its mechanical characteristics, tropoelastin is sufficiently structured to self-assemble, yet sufficiently flexible to maintain elasticity. Elastin's remarkable extensibility arises from the coil region near the N-terminus. This region acts like a spring, allowing tropoelastin to stretch up to eight times its resting length when free of crosslinks [56].

### 2.3. Elastic Fibers and Elastic Laminae

Elastic fibers are composed of amorphous elastin and fibrous microfibrils (mainly fibrillin-1 and/or -2) [57]. The microfibrils are attached to the cell surface via integrins and the elastin aggregates are integrated along the microfibril scaffold in a process known as elastogenesis [58]. Once crosslinked, mature elastic fibers are insoluble and highly durable, exhibiting a half-life of 74 years [59].

Elastic fibers in the medial layer of arteries are primarily produced by vascular smooth muscle cells [60]. These elastic fibers orientate circumferentially and organize into fenestrated sheets called elastic laminae. The internal elastic lamina defines the boundary between the intima and the media, and the external elastic lamina defines the boundary between the media and the adventitia. Larger arteries have multiple concentric layers of elastic laminae between the internal and external laminae.

## 3. Mechanical Properties of Elastic Laminae

Elastin is highly elastic and imparts unique mechanical properties to elastic laminae within the arterial wall. Elastic laminae will stretch circumferentially when a load is applied and then return to their original configuration when the load is removed. Energy loss is minimal during the loading and unloading cycle, estimated to be 15–20% [61]. The elastic laminae allow for pressure wave propagation in arteries to help the flow of blood. The strain energy stored during systole allows blood to continue flowing downstream during diastole as the elastic arteries recoil. This is especially important for the coronary circulation, which is perfused during diastole.

Atomic force microscopy (AFM) measurements have determined that single elastic fibers have Young's moduli in the range of 0.3–1.5 MPa [62,63]. Elastin from aortic tissue has a Young's modulus in the range of 0.1–0.8 MPa and ultimate strain in the range of 100–120% [64,65]. Elastin is several orders of magnitude more compliant than collagen, which has a Young's modulus in the range of 300–1200 MPa [66,67].

Elastic fibers in the aorta of rabbits are oriented circumferentially (i.e., perpendicular to blood flow) with the exception of the internal elastic lamina, where elastic fibers are oriented longitudinally (i.e., parallel to blood flow) [68]. The circumferentially oriented elastic fibers are able to support the circumferential mechanical stress that arteries experience during systole. The longitudinally oriented elastic fibers are finer and more fenestrated to act as a semi-permeable membrane.

The circumferential mechanical properties of elastin in the descending thoracic aorta of pigs are position-dependent [69]. Elastin is 30% stiffer and 54% stronger near the diaphragm compared to that near the aortic arch. This was explained by a progressive increase in circumferential alignment of elastic fibers along the length of the aorta. The study has also found that the circumferential strain of the aortic wall is relatively constant along the aorta for a given pressure, indicating location-dependent variations in cellular and matrix compositions, the arterial wall and lumen dimensions, and the distribution of arterial wall stress.

The micromechanics of elastic laminae in arteries are determined by reversible structural changes: the folding and unfolding of elastic fibers/elastic laminae at the micro-level and stretching and recoiling of elastin at the nano-level [70]. At an arterial blood pressure level, the elastic fibers and laminae are unfolded, whereas at zero blood pressure (when an arterial specimen is removed), these structures are folded. Likewise, the elastin molecules are elongated at an arterial blood pressure level, whereas these molecules recoil at zero blood pressure. Interestingly, the elastic laminae near the inner portion of the arterial wall are wavier than those in the outer portion of the arterial wall and can therefore unfold to a larger degree when blood pressure increases because of the presence of a more negative stress (a higher compressive stress) in the inner portion. This is a mechanism to accommodate the larger circumferential stretch experienced by the inner portion of the wall in response to an increase in arterial blood pressure. This results in approximately even stretch of the elastic laminae and even stress distribution throughout the arterial wall, avoiding inner-wall stress concentration, which is a condition that potentially causes inner-wall cell injury. These findings were confirmed by a subsequent study using synchrotron-based phase-contrast imaging [71].

A study of biaxial mechanical properties of human arteries demonstrated that most common carotid arteries, subclavian arteries, thoracic aortas, abdominal aortas, and common iliac arteries are stiffer in the longitudinal direction, while most renal arteries are stiffer in the circumferential direction [72]. Elastic fibers were primarily circumferentially oriented and localized to the medial layer in the common carotid artery, subclavian artery, thoracic aorta, and abdominal aorta, reflecting the higher elasticity in the circumferential direction. Elastic fibers were primarily longitudinally oriented and localized to the external elastic lamina in the renal artery, reflecting the higher elasticity in the longitudinal direction. Interestingly, the elastic fibers were also primarily longitudinally oriented and localized to the external elastic lamina in the common iliac artery, opposing the higher elasticity in the circumferential direction. The authors speculated that the longitudinally aligned elastic fibers in the common iliac artery may be necessary to accommodate bending and compression as the hip moves.

A study of human left anterior descending (LAD) coronary arteries found that the intimal and adventitial layers are stiffer in the longitudinal direction compared to the circumferential direction, whereas the reverse is true for the medial layer [73]. These findings are consistent with the predominantly longitudinal orientation of collagen and elastin fibers in the adventitial layer of coronary arteries, whereas these fibers have a more complicated three-dimensional structure without a preferred orientation in the medial layer [74].

## 4. Fundamental Pathogenic Processes in Reconstructed Arteries

### 4.1. Inflammation

Inflammation is a series of processes activated in response to surgical and mechanical injury in reconstructed arteries. Inflammation can be divided into three phases: acute, sub-acute, and chronic phases [1,22–26]. The acute phase starts immediately following an injury and lasts several days. This phase is characterized by the activation of inflammatory mediators, elevation in the endothelial permeability, and leukocyte activation and adhesion. Inflammatory mediators include bradykinin, histamine, and cytokines. Bradykinin is a peptide generated by kallikrein protease-mediated cleavage of plasma kininogen expressed primarily in hepatic cells [75,76]. In blood vessels, bradykinin can cause vascular smooth muscle cell relaxation, resulting in vasodilation and elevation in blood flow to injured areas, and can induce an increase in endothelial permeability, facilitating inflammatory mediator transport across the endothelium and leukocyte adhesion [77]. Bradykinin also causes pain, swelling, and diuresis [78]. Histamine is an amino acid derivative from histidine under the action of histidine decarboxylase in primarily mast cells and basophils [79] and is primarily generated and stored in mast cells and basophils [80]. Upon inflammatory stimulation, histamine can be released to act on vascular endothelial cells to open the tight junction, resulting in an increase in endothelial permeability, a change causing edema. Cytokines are a superfamily of small proteins, expressed and released from primarily leukocytes in response to injury [1,81]. The majority of cytokines, such as interleukin 1α (IL1α), IL2, IL3, IL6, IL12, and chemokines, induce leukocyte activation, adhesion, and extravasation, although several cytokines such as IL10, IL27, and IL35 exert an opposite effect [1]. Overall, during the acute phase, the inflammation-stimulating cytokines are dominant to accelerate inflammatory responses.

The acute inflammatory phase is followed by the sub-acute phase, which lasts for several weeks. This phase is characterized by growth factor upregulation, endothelial cell proliferation and angiogenesis, vascular smooth muscle cell proliferation and migration from the host artery to the reconstructed artery, and over-generation of extracellular matrix components, primarily including the collagen matrix and proteoglycans. Several growth factors, including vascular endothelial growth factors (VEGFs), platelet-derived growth factors (PDGFs), and fibroblast growth factors (FGFs), are commonly expressed and released from vascular cells in response to injury [1]. These growth factors regulate vascular cell proliferation and migration via autocrine and paracrine mechanisms. VEGFs can induce vascular endothelial cell proliferation, an essential process for the repair of lost endothelial cells and angiogenesis. PDGFs and FGFs promote vascular smooth muscle cell proliferation and migration from the host artery to the reconstructed artery [1,82]. In autologous vein-based arterial constructs, smooth muscle cells can also migrate from the venous media to the sub-endothelial space, contributing to neointima formation [32,35]. The growth-factor-activated vascular cells can express and release extracellular matrix components, especially collagen, causing matrix accumulation and fibrosis. The chronic phase of inflammation is characterized by the continuous generation of extracellular matrix from the activated smooth muscle cells and fibroblasts, contributing to the advancement of fibrosis.

### 4.2. Thrombosis

Thrombosis is an acute process initiated in response to endothelial injury, hemorrhage, and exposure to biomaterial- and matrix-based arterial constructs. Thrombosis can start with blood coagulation caused by the formation of insoluble fibrin gels from its soluble precursor fibrinogen. This process requires the formation and action of thrombin, a proteinase that can cleave fibrinogen to generate fibrin. Thrombin arises from its precursor prothrombin, an inactive soluble plasma protein expressed and released from the liver [83], under the action of the proteinase prothrombinase. Injured endothelial cells and fibroblasts can express and release this proteinase [84], thus inducing blood coagulation.

Although blood coagulation is a process necessary to stop hemorrhage in the event of trauma, it causes the formation of thrombi, which are pathological structures composed of fibrin and blood cells, including erythrocytes, leukocytes, and platelets, and are found on the surface of reconstructed arteries [1,85,86]. The fibrin gel established during coagulation can attract and entrap blood cells. The entrapped leukocytes can upregulate and release cytokines that can continuously activate and attract leukocytes from the circulatory system to the fibrin gel [87,88]. This process, together with continuous fibrin gel development and blood cell entrapment, contributes to thrombus development [1,87,88]. An extreme case of endothelial injury in the host artery is endothelial denudation, resulting in the exposure of the supporting extracellular matrix. Platelets can adhere to selected matrix components, facilitating blood coagulation and thrombosis [89,90]. In mild injury, a thrombus grows slowly and can be covered by endothelial cells that are regenerated from surrounding endothelial cells. This endothelialization process prevents fibrin formation, blood cell entrapment, and enlargement of the thrombus. A small thrombus usually does not significantly interfere with blood flow. However, in severe injury, rapid fibrin gel formation and blood cell entrapment can occur, resulting in the formation of massive thrombi that can partially or completely obstruct blood flow and cause acute failure of reconstructed arteries. Furthermore, thrombi are not stable and can detach from the base to form emboli, resulting in blockade of distal arteries and ischemic injury [1].

### 4.3. Intimal Hyperplasia

Intimal hyperplasia in reconstructed arteries is cell proliferation to increase the cell density within the intima, primarily involving smooth muscle cells, a process resulting in the formation of neointima [30,91,92]. Neointima is focal in nature, often localized to the junction of the host artery with the reconstructed artery, where anastomosis causes injury, and regions exposed to vortex blood flow, where the level of fluid shear stress is low [32,35,93]. In structure, neointima is composed of leukocytes, platelets, smooth muscle cells, and extracellular matrix (primarily collagen and proteoglycans) with endothelial cells on the surface [1,32,57]. In reconstructed arteries, neointima develops from thrombi, involving smooth muscle cell migration from the host artery (and the vascular wall in the case of vein-based arterial constructs) and endothelialization. The consequence of neointima formation is restenosis and failure of reconstructed arteries.

### 5. Anti-Inflammatory and Anti-Thrombogenic Activities of Elastic Laminae

*5.1. Elastic Laminae-Based Protection against Arterial Inflammation*

The arterial media experience much reduced inflammatory activity compared with the arterial intima in response to a given level of injury. The arterial intima is susceptible to leukocyte infiltration, erythrocyte and platelet deposition, and smooth muscle cell hyperplasia, resulting in inflammation, thrombosis, and intimal hyperplasia, whereas the arterial media are rarely inflicted by these pathological processes. The arterial medial resistance to inflammation, thrombosis, and cell hyperplasia can possibly be attributed to the presence of elastic laminae. Key evidence that supports this concept is the capability of the elastic laminae to suppress leukocyte activities.

The arterial elastic laminae resist leukocyte adhesion. In a cell-culture-based test, the elastic-lamina-rich medial matrix and the collagen-rich adventitial matrix were prepared from the mouse aorta, and the elastic lamina surface was exposed by NaOH treatment [8,9]. The prepared matrix specimens were placed on separate culture dishes. Mouse leukocytes were isolated and cultured on the elastic-lamina-rich and collagen-rich matrix substrates. At selected time points (3, 6, 12, and 24 h), the matrix specimens were removed from the culture dishes and used for counting leukocytes. The density of leukocytes on the surface of the collagen-rich adventitial matrix was about 50 to 105 times higher than that on the elastic-lamina-rich medial matrix from 3 to 24 h of culture. These observations support the concept that the elastic laminae prevent leukocyte adhesion.

The arterial elastic laminae prevent leukocyte migration. In a rat arterial reconstruction model in vivo, allogenic aortic matrix scaffolds were prepared by removing cells and grafted into the aorta [9]. Whereas dense leukocytes were found in the collagen-rich adventitia of the aortic matrix scaffold at 1, 10, and 30 days following aortic grafting, leukocytes were rarely present in the elastic-lamina-rich media of the aortic matrix scaffold. The density of leukocytes within the collagen-rich adventitia was about 2000 to 4000 times higher than that within the elastic-lamina-rich media of the aortic matrix scaffold. An interesting observation was that, even at the end of the aortic matrix scaffold, leukocytes were unable to migrate into the inter-elastic lamina gaps, which were considerably larger than the leukocytes. These observations demonstrate the capability of the elastic laminae to inhibit leukocyte migration.

It is important to note that the arterial elastic laminae and their degradation products, elastin-derived peptides, may behave differently in the regulation of inflammatory responses. Whereas the elastic laminae prevent leukocyte adhesion and infiltration, elastin-derived peptides exert the opposite effect [94]. Selected elastin-derived peptides may induce inflammation-stimulating processes by activating the elastin receptor complex and cathepsin A-neuraminidase 1 complex signaling systems, which contribute to inflammatory responses and atherogenesis [94]. It is possible that the exposure of selected domains of a complete 3D-folded elastin molecule is required for the anti-inflammatory action of the arterial elastic laminae. Selected elastin-derived peptides, on the other hand, may exhibit a pro-inflammatory action when the inhibitory domains are disassembled during elastin degradation.

Another point to address is why the arterial media are more resistant to leukocyte infiltration than the arterial intima. In addition to the difference in extracellular matrix composition as discussed above, different cell types—endothelial cells in the intima and smooth muscle cells in the media—may play distinct roles in the control of inflammatory responses. Injured endothelial cells may facilitate leukocyte adhesion and infiltration, whereas smooth muscle cells may hypothetically inhibit these inflammatory activities. However, the following evidence does not support the anti-inflammatory role of smooth muscle cells. First, when the internal elastic lamina was damaged mechanically, leukocytes were able to migrate into the arterial media in the presence of smooth muscle cells in vivo [9]. Second, in decellularized arterial scaffolds, leukocytes were not able to migrate into the gaps between the elastic laminae, even though the gap width was larger than the leukocyte diameter [9]. Thus, the arterial elastic laminae, but not the smooth muscle cells, resist leukocyte adhesion and infiltration. It should be noted that elastic lamina fragmentation occurs in aged arteries. This change often leads to smooth muscle cell migration from the arterial media to the intima, contributing to neointima formation [95,96]. It remains to be demonstrated whether leukocytes can migrate into the media of aged arteries.

### 5.2. Elastic Lamina-Mediated Prevention of Vascular Smooth Muscle Cell Proliferation and Neointima Formation

The arterial elastic laminae suppress vascular smooth muscle cell proliferation because of the inhibitory action of elastin [13,15,16]. In mice with elastin gene deficiency ($Eln-/-$), the arterial smooth muscle cells exhibit an over-proliferative phenotype [15]. Humans with elastin gene mutation and elastin deficiency, found in supravalvular aortic stenosis and Williams–Beuren syndrome, express a similar phenotype in large arteries, often associated with excessive smooth muscle cell proliferation and arterial hypertrophy and stenosis, resulting in blood flow obstruction [12,14–16,19,21]. In experimental coronary artery restenosis, administration of elastin peptides to the injured artery results in a significant reduction in the rate of neointima formation [13].

Arterial elastic laminae could effectively prevent neointima formation in an arterial matrix implantation model [8]. In this investigation, aortic extracellular matrix scaffolds were harvested from donor rats and prepared to remove cells and expose the basal lamina, internal elastic lamina, or adventitial collagen by NaOH treatment. Matrix scaffolds with the three different surface components were implanted into the aortas of recipient rats and examined at 5, 10, and 20 days. The rate of smooth muscle cell proliferation, evaluated by the BrdU incorporation assay, differed substantially across the three allogenic aortic matrix scaffolds with distinct surface components. The elastic lamina surface exhibited the lowest BrdU index compared with the basal lamina and collagen surfaces at a selected time point. The highest BrdU index was found at the collagen surface. In the same allogenic aortic matrix scaffold implantation model, the elastic lamina surface was associated with the lowest level of neointima compared with the basal lamina and collagen surfaces, whereas the collagen surface was associated with the highest level of neointima. These observations support the concept that the arterial elastic laminae inhibit smooth muscle cell proliferation and neointima formation.

### 5.3. Elastin-Enhanced Actin Filament Generation in Vascular Smooth Muscle Cells

The coexistence of elastic laminae in the arterial media suggests a role for elastin in the development and maintenance of the contractile phenotype of the vascular smooth muscle cells. The building block of elastin, tropoelastin, has been demonstrated to cause the generation of actin filaments in vascular smooth muscle cells [7,97]. One peptide from tropoelastin, VGVAPG, has been suggested as a key element responsible for the myofibrillogenesis of smooth muscle cells [97]. Mechanistically, this short peptide, as well as tropoelastin, can induce actin polymerization by activating the G protein-coupled receptor, i.e., the RhoA signaling pathway [97]. These observations support the concept that the elastic laminae serve as a regulatory factor for the development and maintenance of the contractile phenotype of vascular smooth muscle cells.

The arterial elastic laminae play a role in the development of vascular smooth muscle $\alpha$ actin filaments in bone-marrow-derived CD34-positive cells [10]. Organ- and tissue-specific environmental conditions have long been considered cues that induce stem cell differentiation into specified functional cells. The arterial elastic laminae can create a condition in favor of developing the smooth muscle cell contractile phenotype, characterized by the presence of orderly aligned smooth muscle $\alpha$ actin filaments, which contrasts with the proliferative and synthetic phenotype found in the neointima. In an in vitro test, bone-marrow-derived CD34-positive cells developed smooth muscle $\alpha$ actin filaments when cultured on the surface of the mouse arterial elastic laminae, but these cells did not form smooth muscle $\alpha$ actin filaments when cultured on the adventitial collagen matrix [4]. These observations are consistent with the coexistence of smooth muscle cells with the elastic laminae in the arterial media, but not with the collagen matrix in the adventitia, supporting a role for the elastic laminae in the induction of smooth muscle cells from stem cells as well as in the maintenance of the contractile phenotype of the smooth muscle cells.

### 5.4. Mechanisms of the Inhibitory Action of Elastic Laminae

A fundamental question is how the arterial elastic laminae exert an inhibitory effect on the adhesion, proliferation, and migration of vascular cells and leukocytes. A prior investigation demonstrated that an inhibitory signaling pathway, involving signal regulatory protein α (SIRPα) and Src homology 2 domain-containing protein tyrosine phosphatase-1 (SHP1), potentially mediates the inhibitory action of the arterial elastic laminae [9]. The elastic lamina component, elastin, can bind to and activate SIRPα, a transmembrane receptor that can recruit and activate SHP1, an enzyme capable of dephosphorylating selected substrate proteins, including growth factor receptor protein tyrosine kinases, focal adhesion kinase, and Src homology 2 domain-containing protein tyrosine kinase (Figure 1). Growth factor receptor protein tyrosine kinases transmit growth factor signals to cause cell proliferation and migration, and focal adhesion kinase and Src homology 2 domain-containing protein tyrosine kinase relay matrix-dependent integrin signals to stimulate cell adhesion. Dephosphorylation of protein tyrosine kinases usually suppresses the activity of these kinases as well as the kinase-induced cell activities. Given such actions of the protein tyrosine kinases, the deactivation of these kinases in response to SHP1 results in the inhibition of vascular cell and leukocyte adhesion, proliferation, and migration [9]. In summary, the inhibition of the growth factor protein tyrosine kinase activity by the SHP1 action may promote the development of the contractile phenotype of vascular smooth muscle cells. This concept is consistent with the observation that growth factor-activated proliferative smooth muscle cells in injury-induced neointima (a structure without elastic laminae, but with elevated growth factor signaling actions) exhibit much reduced and more irregularly organized α actin filaments compared with healthy arterial medial smooth muscle cells that reside within the gaps between elastic laminae and are subject to a minimal level of growth factor activity [32,35,36]. However, the molecular regulatory mechanisms downstream to SHP1 need further investigation.

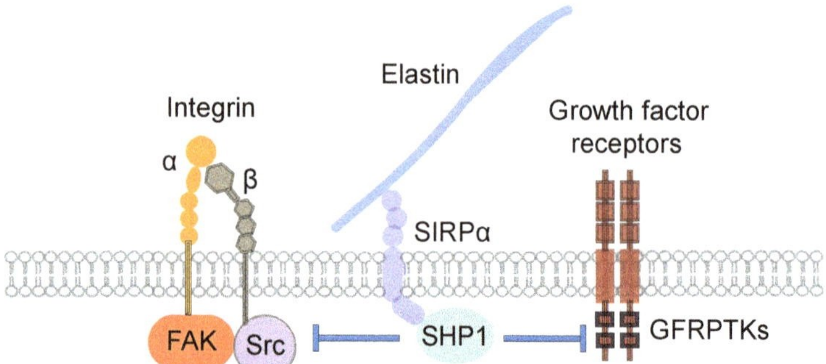


**Figure 1.** Mechanisms of the inhibitory action of elastin. SIRPα: Signal regulatory protein α. GFRPTK: Growth factor receptor protein tyrosine kinases. FAK: Focal adhesion kinase. Src: Src homology 2 domain-containing protein tyrosine kinase. SHP1: Src homology 2 domain-containing protein tyrosine phosphatase-1.

## 6. Application of Elastic Laminae and Elastin-Based Materials to Arterial Reconstruction

Autogenous vein- and arterial graft-based arterial reconstruction is an effective approach for the treatment of occlusive arterial disorders. Autogenous venous and arterial grafts have been considered the most reliable graft types because of their natural properties and performance, although these grafts are prone to thrombosis, inflammation, and intimal hyperplasia, resulting in graft stenosis and failure [98–103]. However, due to limited quantity, vascular disease, or prior harvests, suitable autogenous grafting materials are often unavailable. Consequently, researchers have been working intensively over the past several decades to develop reliable tissue-engineered and synthetic-material-based grafts for vascular reconstruction.

A major challenge in the development of effective tissue-engineered and synthetic vascular grafts is the prevention of inflammatory and thrombogenic responses [104–107]. Numerous synthetic vascular graft types have been developed, but are not suitable for small diameter artery reconstruction due to host inflammatory and thrombogenic responses. Synthetic materials possess poor blood compatibility, often activating leukocytes and causing blood coagulation, which are fundamental processes that lead to inflammation and thrombosis [108].

Tissue-engineering approaches have been used to develop patient-matching cellularized vascular constructs with natural properties [107,109], which are efficacious for replacing failed small diameter arteries. Although few tissue-engineered vascular grafts have attained widespread clinical acceptance, several vascular tissue-engineering strategies have been promoted for developing arterial constructs. In particular, cell-sheet-based approaches have been used to produce natural blood-vessel-like constructs, some of which have been tested in clinical trials with excellent results [108]. For example, long-term results from a recent human study with a cell-sheet-derived graft for hemodialysis access showed great promise as an alternative to synthetic grafts [109,110]. However, these approaches require tedious and complex culture and maturation processing. They also require a patient-matching cell harvest and expansion phase. The need for patient-matching cells extends the preparation time, increases the cost, and shortens the shelf-life, all of which precludes the feasibility of meeting the urgent needs of patients who require off-the-shelf grafts. Furthermore, it may not be possible to reproduce the exquisite material properties inherent to native arterial tissue.

Decellularized arterial matrix scaffolds hold greater promise as implant materials for small-diameter arterial reconstruction [111,112]. These scaffolds possess ideal mechanical properties, porosity, and cell adhesive and regenerative properties. However, native arteries that are decellularized to avoid immunogenic responses no longer have an endothelium, displaying an exposed thin collagen-rich matrix layer (from the basal lamina) to the circulating blood, promoting potent thrombosis and inflammation upon engraftment. Decellularization approaches therefore typically require a patient-matching endothelial layer on the luminal surface of arterial constructs before engraftment [107]. However, endothelial cells grown on arterial constructs are not stable and can detach rapidly when exposed to the arterial blood flow. The field of vascular engineering has not developed an effective arterial construct with an intact native endothelium despite many decades of intensive effort [113]. Furthermore, this approach requires cell harvesting, seeding, and expansion, precluding urgent uses.

To date, materials used for vascular graft engineering, including synthetic polymers and collagen matrices, can cause profound inflammatory and thrombogenic responses when placed in blood contact [104,107,114], contributing to intimal hyperplasia and graft stenosis. Since the arterial elastic-lamina-dominant matrix can effectively prevent leukocyte adhesion and transmigration in matrix-based arterial reconstruction and reduce inflammation and thrombosis [115,116], this matrix may be used to establish a blood contacting layer, as a substitute for an intact endothelium, to alleviate inflammatory and thrombogenic responses in engineered vascular grafts. Several vascular graft-engineering approaches have been developed to exploit the anti-inflammatory and anti-thrombosis properties of elastin [117]. One approach incorporates synthetic elastin or the tropoelastin precursor into engineered materials [118–125]. This approach can produce a synthetic graft with thrombogenic and immunogenic suppressive properties; however, the elastin-containing material lacks the mechanical properties of the native artery that has been optimized through evolution for cell and tissue regeneration/remodeling.

Another approach relies on decellularized arterial elastic laminae. In this approach, the elastic laminae are either completely purified from a donor artery or left interconnected with a portion of the native collagen matrix. In the case of complete elastic lamina purification, a synthetic scaffold is required to support the purified native elastic lamina layer due to the loss of the mechanical strength from the removal of the collagen matrix [126]. In the other case, if the elastic laminae remain interconnected with the native collagen matrix, it is necessary to deplete the basal lamina to expose the internal elastic lamina as the blood-contacting surface while maximizing the preservation of the mural collagen matrix and other native constituents to retain the natural regenerative properties. One approach is to incubate the native artery in an acidic or basic solution for varying times to completely deplete the basal lamina from the blood-contacting surface while reducing the dissolution of mural collagen matrix [8].

All materials developed so far for vascular graft engineering have critical limitations that preclude their widespread clinical acceptance. There is a pressing need to develop clinically feasible, universally applicable, and shelf-stable vascular grafting materials. An ideal vascular graft should match the mechanical characteristics of the native artery. Concurrently, viable vascular grafts must re-establish the properties of a native endothelium at the blood-contacting surface, avoid provoking inflammatory responses, remain shelf-stable over extended time periods prior to use, be easy to handle, and be commercially and clinically feasible. The main challenges have been mimicking a confluent endothelium on the luminal surface of grafting materials and difficulties in generating biocompatible arterial constructs. A future direction is to use decellularized native matrix materials with a uniform elastic lamina blood-contacting surface while retaining the outstanding mechanical properties, porosity, and cell-interaction capability of native arteries. We have made considerable progress in engineering such a vascular grafting material by selective surface collagen depletion in decellularized porcine internal mammary arteries (Figure 2). Once the approach is optimized, the elastin-rich grafts may be used for arterial reconstruction without the need for endothelialization. Acellular grafting materials with a luminal elastic lamina surface are shelf-stable and can be stocked and used to meet the urgent needs of patients.

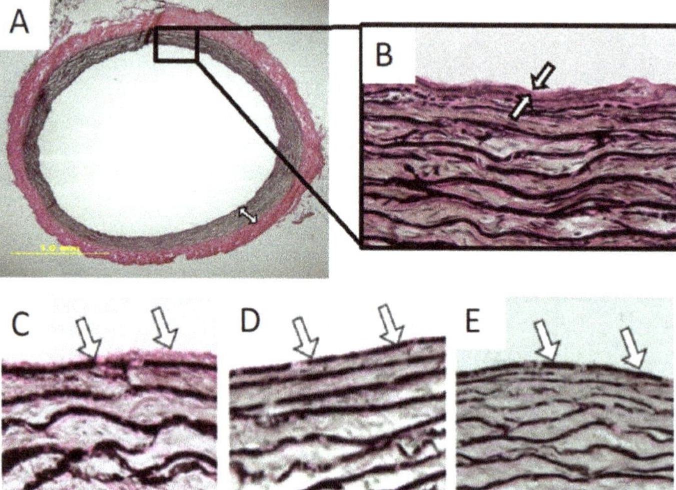

**Figure 2.** Preparation of arterial matrix scaffolds with a luminal elastic lamina surface. (**A,B**) Porcine mammary artery cross section (**A**) and close-up (**B**) showing collagen matrix (red) and abundant elastic laminae (black) in the arterial wall. The thin layer of native luminal surface collagen is identified by arrows. (**C–E**) Arterial matrix specimens processed by using high-viscosity acidic gels ((**C**), control gel; (**D**), 2 h treatment in glycolic acid gel; (**E**), 2 h treatment in lactic acid gel). These treatments can selectively remove the internal collagens while preserving the mural collagens. Additional work is required to optimize this approach.

It is important to address the technical aspect of preparing arterial matrix scaffolds with a luminal elastic lamina surface. We have used a high viscosity acidic gel to restrict matrix depletion activity to the artery luminal surface. The gel (made from polylactic acid or polyglycolic acid) is of sufficient molecular weight to reduce acid diffusion into the arterial wall. Thus, the natural matrix composition and structure of native arteries can be preserved, including mechanical and remodeling properties. Future work along this direction could produce commercially and clinically feasible small-diameter vascular grafts that can be mass-produced at low cost by using donor specimens. These strategies will potentially establish a basis for the development of matrix material-based grafts for small artery reconstruction.

Another potential application for elastin-based materials is to coat arterial stents for preventing in-stent thrombosis and neointima formation. Similar to biomaterial-based arterial constructs, stents can cause thrombosis rapidly following placement, resulting in neointima formation and arterial restenosis [127,128]. This has long been a challenging problem. Coating stents with various agents, such as anticoagulants, biomaterials, antimitotic substances, and corticosteroids [127], has been considered a potentially effective approach for reducing stent-induced neointima formation. However, the current coating agents may cause substantial inflammatory responses (biomaterials) or may only exert anti-thrombogenic and anti-hyperplastic effects for a short period (anticoagulants, antimitotic substances, and corticosteroids). Elastin is an insoluble stable polymer that, if able to firmly adhere to stents, may exert anti-inflammatory, anti-thrombogenic, and anti-hyperplastic effects for a longer time.

### 7. Concluding Remarks

The elastin-based extracellular matrix exhibits an inhibitory effect on leukocyte and vascular smooth muscle cell proliferation, adhesion, and migration. Given that these cell activities contribute to neointima formation in reconstructed arteries, this feature renders the elastin matrix a potential material for constructing the blood-contacting surface of arterial constructs. Preliminary experimental investigations have provided promising results. However, the mechanisms of the inhibitory elastin action remain to be investigated, and it is challenging to coat the luminal surface of an arterial construct with an elastin-based matrix.

**Author Contributions:** Conceptualization, S.Q.L.; writing—original draft preparation, J.G., S.Q.L. and B.J.T.; writing—review and editing, J.G., S.Q.L. and B.J.T. All authors have read and agreed to the published version of the manuscript.

**Funding:** This work was supported by NIH (JG, R01HL157642 to BJT) and NSF (SQL).

**Institutional Review Board Statement:** The animal study protocol was approved by the Ethics Committee (IACUC) of Michigan Technological University (380482, 2015).

**Informed Consent Statement:** Not applicable.

**Data Availability Statement:** The data presented in this study are available in Figure 2.

**Conflicts of Interest:** The authors declare no conflict of interest.

## References

1. Liu, S.Q. *Cardiovascular Engineering: A Protective Approach*, 1st ed.; McGraw-Hill: New York, NY, USA, 2020.
2. Galkina, E.; Ley, K. Immune and Inflammatory Mechanisms of Atherosclerosis. *Annu. Rev. Immunol.* **2009**, *27*, 165–197. [CrossRef] [PubMed]
3. Hansson, G.K.; Robertson, A.-L.; Söderberg-Nauclér, C. Inflammation and atherosclerosis. *Annu. Rev. Pathol. Mech. Dis.* **2006**, *1*, 297–329. [CrossRef] [PubMed]
4. Ross, R. Cell biology of atherosclerosis. *Annu. Rev. Physiol.* **1995**, *57*, 791–804. [CrossRef] [PubMed]
5. de Fougerolles, A.R.; Sprague, A.G.; Nickerson-Nutter, C.L.; Chi-Rosso, G.; Rennert, P.D.; Gardner, H.; Gotwals, P.J.; Lobb, R.R.; Koteliansky, V.E. Regulation of inflammation by collagen-binding integrins $\alpha 1\beta 1$ and $\alpha 2\beta 1$ in models of hypersensitivity and arthritis. *J. Clin. Investig.* **2000**, *105*, 721–729. [CrossRef] [PubMed]

6.  Goldman, R.; Harvey, J.; Hogg, N. VLA-2 is the integrin used as a collagen receptor by leukocytes. *Eur. J. Immunol.* **1992**, *22*, 1109–1114. [CrossRef]
7.  Lahti, M.; Heino, J.; Käpylä, J. Leukocyte integrins $\alpha L\beta 2$, $\alpha M\beta 2$ and $\alpha X\beta 2$ as collagen receptors—Receptor activation and recognition of GFOGER motif. *Int. J. Biochem. Cell Biol.* **2013**, *45*, 1204–1211. [CrossRef]
8.  Liu, S.Q.; Tieche, C.; Alkema, P.K. Neointima Formation on Elastic Lamina and Collagen Matrix Scaffolds Implanted in the Rat Aorta. *Biomaterials* **2004**, *25*, 1869–1882. [CrossRef]
9.  Liu, S.Q.; Alkema, P.K.; Tieche, C.; Tefft, B.J.; Liu, D.Z.; Sumpio, B.E.; Caprini, J.A.; Li, Y.C.; Paniagua, M. Negative Regulation of Monocyte Adhesion to Arterial Elastic Laminae by Signal-Regulatory Protein Alpha and SH2 Domain-Containing Protein Tyrosine Phosphatase-1. *J. Biolog. Chem.* **2005**, *280*, 39294–39301. [CrossRef]
10. Liu, S.Q.; Tefft, B.J.; Zhang, A.; Zhang, L.-Q.; Wu, Y.H. Formation of smooth muscle $\alpha$ actin filaments in CD34-positive bone marrow cells in elastic lamina-dominantmatrix ofarteries. *Matrix Biol.* **2008**, *27*, 282–294. [CrossRef]
11. Cocciolone, A.J.; Hawes, J.Z.; Staiculescu, M.C.; Johnson, E.O.; Murshed, M.; Wagenseil, J.E. Elastin, arterial mechanics, and cardiovascular disease. *Am. J. Physiol. Heart Circ. Physiol.* **2018**, *315*, H189–H205. [CrossRef]
12. Collins, R.T., 2nd. Cardiovascular disease in Williams syndrome. *Circulation* **2013**, *127*, 2125–2134. [PubMed]
13. Karnik, S.K.; Brooke, B.S.; Bayes-Genis, A.; Sorensen, L.; Wythe, J.D.; Schwartz, R.S.; Keating, M.T.; Li, D.Y. A critical role for elastin signaling in vascular morphogenesis and disease. *Development* **2003**, *130*, 411–423. [CrossRef] [PubMed]
14. Lasio, M.L.D.; Kozel, B.A. Elastin-Driven Genetic Diseases. *Matrix Biol.* **2018**, *71–72*, 144–160. [CrossRef]
15. Li, D.Y.; Brooke, B.; Davis, E.C.; Mecham, R.P.; Sorensen, L.K.; Boak, B.B.; Eichwald, E.; Keating, M.T. Elastin is an essential determinant of arterial morphogenesis. *Nature* **1998**, *393*, 276–280. [CrossRef] [PubMed]
16. Li, D.Y.; Faury, G.; Taylor, D.G.; Davis, E.C.; Boyle, W.A.; Mecham, R.P.; Stenzel, P.; Boak, B.; Keating, M.T. Novel arterial pathology in mice and humans hemizygous for elastin. *J. Clin. Investig.* **1998**, *102*, 1783–1787. [CrossRef] [PubMed]
17. Lin, C.-J.; Cocciolone, A.J.; Wagenseil, J.E. Elastin, arterial mechanics, and stenosis. *Am. J. Physiol. Cell Physiol.* **2022**, *322*, C875–C886. [CrossRef]
18. Liu, S.Q. *Bioregenerative Engineering: Principles and Applications*; Wiley Interscience: Hoboken, NJ, USA, 2007.
19. Pober, B.R.; Johnson, M.; Urban, Z. Mechanisms and treatment of cardiovascular disease in Williams-Beuren syndrome. *J. Clin. Investig.* **2008**, *118*, 1606–1615. [CrossRef]
20. Tieche, C.; Alkema, P.K.; Liu, S.Q. Vascular elastic laminae: Anti-inflammatory properties and potential applications to arterial reconstruction. *Front. Biosci.* **2004**, *9*, 2205–2217. [CrossRef]
21. Urban, Z.; Riazi, S.; Seidl, T.L.; Katahira, J.; Smoot, L.B.; Chitayat, D.; Boyd, C.D.; Hinek, A. Connection between elastin haploinsufficiency increased cell proliferation in patients with supravalvular aortic stenosis Williams-Beuren syndrome. *Am. J. Hum. Genet.* **2002**, *71*, 30–44. [CrossRef]
22. Ascione, R.; Lloyd, C.T.; Underwood, M.J.; Lotto, A.A.; Pitsis, A.A.; Angelini, G.D. Inflammatory response after coronary revascularization with or without cardiopulmonary bypass. *Ann. Thorac. Surg.* **2000**, *69*, 1198–1204. [CrossRef]
23. de Vries, M.R.; Quax, P.H.A. Inflammation in Vein Graft Disease. *Front. Cardiovasc. Med.* **2018**, *5*, 3. [CrossRef] [PubMed]
24. Li, N.; Astudillo, R.; Ivert, T.; Hjemdahl, P. Biphasic pro-thrombotic and inflammatory responses after coronary artery bypass surgery. *J. Thromb. Haemost.* **2003**, *1*, 470–476. [CrossRef] [PubMed]
25. Tang, L.; Eaton, J.W. Inflammatory Responses to Biomaterials. *Am. J. Clin. Pathol.* **1995**, *103*, 466–471. [CrossRef]
26. Ward, A.O.; Angelini, G.D.; Caputo, M.; Evans, P.C.; Johnson, J.L.; Suleiman, M.S.; Tulloh, R.M.; George, S.J.; Zakkar, M. NF-κB inhibition prevents acute shear stress-induced inflammation in the saphenous vein graft endothelium. *Sci. Rep.* **2020**, *10*, 15133. [CrossRef] [PubMed]
27. Lin, P.H.; Chen, C.; Bush, R.L.; Yao, Q.; Lumsden, A.; Hanson, S.R. Small-caliber heparin-coated ePTFE grafts reduce platelet deposition and neointimal hyperplasia in a baboon model. *J. Vasc. Surg.* **2004**, *39*, 1322–1328. [CrossRef] [PubMed]
28. Sarkar, S.; Sales, K.M.; Hamilton, G.; Seifalian, A.M. Addressing thrombogenicity in vascular graft construction. *J. Biomed. Mater. Res. B Appl. Biomater.* **2007**, *82*, 100–108. [CrossRef] [PubMed]
29. Tucker, E.I.; Marzec, U.M.; White, T.C.; Hurst, S.; Rugonyi, S.; McCarty, O.J.T.; Gailani, D.; Gruber, A.; Hanson, S.R. Prevention of vascular graft occlusion and thrombus-associated thrombin generation by inhibition of factor XI. *Blood* **2009**, *113*, 936–944. [CrossRef]
30. Cooley, B.C. Murine Model of Neointimal Formation and Stenosis in Vein Grafts. *Arterioscler. Thromb. Vasc. Biol.* **2004**, *24*, 1180–1185. [CrossRef]
31. Davies, M.G.; Owens, E.L.; Mason, D.P.; Lea, H.; Tran, P.K.; Vergel, S.; Hawkins, S.A.; Hart, C.E.; Clowes, A.W. Effect of Platelet-Derived Growth Factor Receptor-$\alpha$ and -$\beta$ Blockade on Flow-Induced Neointimal Formation in Endothelialized Baboon Vascular Grafts. *Circ. Res.* **2000**, *86*, 779–786. [CrossRef]
32. Liu, S.Q. Prevention of focal intimal hyperplasia in rat vein grafts by using a tissue engineering approach. *Atherosclerosis* **1998**, *140*, 365–377. [CrossRef]
33. Liu, S.Q. Biomechanical basis of vascular tissue engineering. *Crit. Rev. Biomed. Eng.* **1999**, *27*, 75–148. [CrossRef] [PubMed]
34. Liu, S.Q.; Moore, M.M.; Glucksberg, M.R.; Mockros, L.F.; Grotberg, J.B.; Mok, A.P. Partial prevention of monocyte and granulocyte activation in experimental vein grafts by using a biomechanical engineering approach. *J. Biomech.* **1999**, *32*, 1165–1175. [CrossRef]
35. Liu, S.Q. Focal expression of angiotensin II type 1 receptor and smooth muscle cell proliferation in the neointima of experimental vein grafts: Relation to eddy blood flow. *Arterioscler. Thromb. Vasc. Biol.* **1999**, *19*, 2630–2639. [CrossRef] [PubMed]

36. Liu, S.Q.; Moore, M.M.; Yap, C. Prevention of mechanical stretch-induced endothelial and smooth muscle cell injury in experimental vein grafts. *J. Biomech. Eng.* **2000**, *122*, 31–38. [CrossRef]
37. Moore, M.M.; Goldman, J.; Patel, A.; Chien, S.; Liu, S.Q. Role of tensile stress and strain in the induction of cell death in experimental vein grafts. *J. Biomech.* **2001**, *34*, 289–297. [CrossRef]
38. Xu, Q. Mouse Models of Arteriosclerosis: From Arterial Injuries to Vascular Grafts. *Am. J. Pathol.* **2004**, *165*, 1–10. [CrossRef] [PubMed]
39. Rosenbloom, J.; Abrams, W.R.; Indik, Z.; Yeh, H.; Ornstein-Goldstein, N.; Bashir, M.M. Structure of the elastin gene. *Ciba Found. Symp.* **1995**, *192*, 59–74. [PubMed]
40. Bashir, M.M.; Indik, Z.; Yeh, H.; Ornstein-Goldstein, N.; Rosenbloom, J.C.; Abrams, W.; Fazio, M.; Uitto, J.; Rosenbloom, J. Characterization of the complete human elastin gene. Delineation of unusual features in the 5′-flanking region. *J. Biol. Chem.* **1989**, *264*, 8887–8891. [CrossRef]
41. Davis, E.C. Stability of elastin in the developing mouse aorta: A quantitative radioautographic study. *Histochemistry* **1993**, *100*, 17–26. [CrossRef]
42. Lefevre, M.; Rucker, R.B. Aorta elastin turnover in normal and hypercholesterolemic Japanese quail. *Biochim. Biophys. Acta* **1980**, *630*, 519–529. [CrossRef]
43. Keeley, F.W.; Hussain, R.A.; Johnson, D.J. Pattern of accumulation of elastin and the level of mRNA for elastin in aortic tissue of growing chickens. *Arch. Biochem. Biophys.* **1990**, *282*, 226–232. [CrossRef]
44. Hew, Y.; Grzelczak, Z.; Lau, C.; Keeley, F.W. Identification of a large region of secondary structure in the 3′-untranslated region of chicken elastin mRNA with implications for the regulation of mRNA stability. *J. Biol. Chem.* **1999**, *14*, 14415–14421. [CrossRef] [PubMed]
45. Johnson, D.J.; Robson, P.; Hew, Y.; Keeley, F.W. Decreased elastin synthesis in normal development and in long-term aortic organ and cell cultures is related to rapid and selective destabilization of mRNA for elastin. *Circ. Res.* **1995**, *77*, 1107–1113. [CrossRef]
46. Indik, Z.; Yeh, H.; Ornstein-Goldstein, N.; Sheppard, P.; Anderson, N.; Rosenbloom, J.C.; Peltonen, L.; Rosenbloom, J. Alternative splicing of human elastin mRNA indicated by sequence analysis of cloned genomic complementary, DNA. *Proc. Natl. Acad. Sci. USA* **1987**, *84*, 5680–5684. [CrossRef]
47. Indik, Z.; Yeh, H.; Ornstein-Goldstein, N.; Kucich, U.; Abrams, W.; Rosenbloom, J.C.; Rosenbloom, J. Structure of the elastin gene and alternative splicing of elastin mRNA: Implications for human disease. *Am. J. Med. Genet.* **1989**, *34*, 81–90. [CrossRef]
48. Miao, M.; Reichheld, S.E.; Muiznieks, L.D.; Sitarz, E.E.; Sharpe, S.; Keeley, F.W. Single nucleotide polymorphisms and domain/splice variants modulate assembly and elastomeric properties of human elastin. Implications for tissue specificity and durability of elastic tissue. *Biopolymers* **2017**, *107*, e23007. [CrossRef]
49. Gray, W.R.; Sandberg, L.B.; Foster, J.A. Molecular model for elastin structure and function. *Nature* **1973**, *246*, 461–466. [CrossRef]
50. Foster, J.A.; Bruenger, E.; Gray, W.R.; Sandberg, L.B. Isolation and amino acid sequences of tropoelastin peptides. *J. Biol. Chem.* **1973**, *248*, 2876–2879. [CrossRef] [PubMed]
51. Sandberg, L.B.; Weissman, N.; Gray, W.R. Structural features of tropoelastin related to the sites of cross-links in aortic elastin. *Biochemistry* **1971**, *10*, 52–56. [CrossRef] [PubMed]
52. Cox, B.A.; Starcher, B.C.; Urry, D.W. Communication: Coacervation of tropoelastin results in fiber formation. *J. Biol. Chem.* **1974**, *249*, 997–998. [CrossRef]
53. Vrhovski, B.; Weiss, A.S. Biochemistry of tropoelastin. *Eur. J. Biochem.* **1998**, *258*, 1–18. [CrossRef] [PubMed]
54. Yeo, G.C.; Keeley, F.W.; Weiss, A.S. Coacervation of tropoelastin. *Adv. Colloid Interface Sci.* **2011**, *167*, 94–103. [CrossRef]
55. Lucero, H.A.; Kagan, H.M. Lysyl oxidase: An oxidative enzyme and effector of cell function. *Cell. Mol. Life Sci.* **2006**, *63*, 2304–2316. [CrossRef] [PubMed]
56. Baldock, C.; Oberhauser, A.F.; Ma, L.; Lammie, D.; Siegler, V.; Mithieux, S.M.; Tu, Y.; Chow, J.Y.H.; Suleman, F.; Malfois, M.; et al. Shape of tropoelastin, the highly extensible protein that controls human tissue elasticity. *Proc. Natl. Acad. Sci. USA* **2011**, *108*, 4322–4327. [CrossRef] [PubMed]
57. Wagenseil, J.E.; Mecham, R.P. New insights into elastic fiber assembly. *Birth Defects Res. C Embryo Today* **2007**, *81*, 229–240. [CrossRef]
58. Ozsvar, J.; Yang, C.; Cain, S.A.; Baldock, C.; Tarakanova, A.; Weiss, A.S. Tropoelastin and Elastin Assembly. *Front. Bioeng. Biotechnol.* **2021**, *9*, 643110. [CrossRef]
59. Shapiro, S.D.; Endicott, S.K.; Province, M.A.; Pierce, J.A.; Campbell, E.J. Marked longevity of human lung parenchymal elastic fibers deduced from prevalence of D-aspartate and nuclear weapons-related radiocarbon. *J. Clin. Investig.* **1991**, *87*, 1828–1834. [CrossRef]
60. Hungerford, J.E.; Owens, G.K.; Argraves, W.S.; Little, C.D. Development of the aortic vessel wall as defined by vascular smooth muscle and extracellular matrix markers. *Dev. Biol.* **1996**, *178*, 375–392. [CrossRef]
61. Gibbons, C.A.; Shadwick, R.E. Functional similarities in the mechanical design of the aorta in lower vertebrates and mammals. *Experientia* **1989**, *45*, 1083–1088. [CrossRef]
62. Aaron, B.B.; Gosline, J.M. Elastin as a random-network elastomer: A mechanical and optical analysis of single elastin fibers. *Biopolymers* **1981**, *20*, 1247–1260. [CrossRef]

63. Koenders, M.M.; Yang, L.; Wismans, R.G.; van der Werf, K.O.; Reinhardt, D.P.; Daamen, W.; Bennink, M.L.; Dijkstra, P.J.; van Kuppevelt, T.H.; Feijen, J. Microscale mechanical properties of single elastic fibers: The role of fibrillin-microfibrils. *Biomaterials* **2009**, *30*, 2425–2432. [CrossRef] [PubMed]
64. Lillie, M.A.; David, G.J.; Gosline, J.M. Mechanical role of elastin-associated microfibrils in pig aortic elastic tissue. *Connect. Tissue Res.* **1998**, *37*, 121–141. [CrossRef] [PubMed]
65. Sherebrin, M.H.; Song, S.H.; Roach, M.R. Mechanical anisotropy of purified elastin from the thoracic aorta of dog and sheep. *Can. J. Physiol. Pharmacol.* **1983**, *61*, 539–545. [CrossRef]
66. Gautieri, A.; Vesentini, S.; Redaelli, A.; Buehler, M.J. Hierarchical structure and nanomechanics of collagen microfibrils from the atomistic scale up. *Nano Lett.* **2011**, *11*, 757–766. [CrossRef] [PubMed]
67. Heim, A.J.; Matthews, W.G.; Koob, T.J. Determination of the elastic modulus of native collagen fibrils via radial indentation. *Appl. Phys. Lett.* **2006**, *89*, 181902. [CrossRef]
68. Farand, P.; Garon, A.; Plante, G.E. Structure of large arteries: Orientation of elastin in rabbit aortic internal elastic lamina and in the elastic lamellae of aortic media. *Microvasc. Res.* **2007**, *73*, 95–99. [CrossRef] [PubMed]
69. Lillie, M.A.; Gosline, J.M. Mechanical properties of elastin along the thoracic aorta in the pig. *J. Biomech.* **2007**, *40*, 2214–2221. [CrossRef]
70. Yu, X.; Turcotte, R.; Seta, F.; Zhang, Y. Micromechanics of elastic lamellae: Unravelling the role of structural inhomogeneity in multi-scale arterial mechanics. *J. R. Soc. Interface* **2018**, *15*, 20180492. [CrossRef]
71. Trachet, B.; Ferraro, M.; Lovric, G.; Aslanidou, L.; Logghe, G.; Segers, P.; Stergiopulos, N. Synchrotron-based visualization and segmentation of elastic lamellae in the mouse carotid artery during quasi-static pressure inflation. *J. R. Soc. Interface* **2019**, *16*, 20190179. [CrossRef]
72. Kamenskiy, A.V.; Dzenis, Y.A.; Kazmi, S.; Pemberton, M.A.; Pipinos, I.I.; Phillips, N.Y.; Herber, K.; Woodford, T.; Bowen, R.E.; Lomneth, C.S.; et al. Biaxial mechanical properties of the human thoracic and abdominal aorta, common carotid, subclavian, renal and common iliac arteries. *Biomech. Model. Mechanobiol.* **2014**, *13*, 1341–1359. [CrossRef]
73. Holzapfel, G.A.; Sommer, G.; Gasser, C.T.; Regitnig, P. Determination of layer-specific mechanical properties of human coronary arteries with nonatherosclerotic intimal thickening and related constitutive modeling. *Am. J. Physiol. Heart Circ. Physiol.* **2005**, *289*, H2048–H2058. [CrossRef]
74. Chen, H.; Kassab, G.S. Microstructure-based biomechanics of coronary arteries in health and disease. *J. Biomech.* **2016**, *49*, 2548–2559. [CrossRef]
75. Chen, H.M.; Liao, W.S. Molecular analysis of the differential hepatic expression of rat kininogen family genes. *Mol Cell Biol.* **1993**, *13*, 6766–6777. [PubMed]
76. Zhao, A.; Lew, J.-L.; Huang, L.; Yu, J.; Zhang, T.; Hrywna, Y.; Thompson, J.R.; de Pedro, N.; Blevins, R.A.; Peláez, F.; et al. Human Kininogen Gene Is Transactivated by the Farnesoid X Receptor. *J. Biol. Chem.* **2003**, *278*, 28765–28770. [CrossRef] [PubMed]
77. Kaplan, A.P.; Joseph, K.; Silverberg, M. Molecular mechanisms in allergy and clinical immunology. *Pathw. Bradykinin Form. Inflamm. Dis.* **2002**, *109*, 195–209.
78. Wong, M.K.S. Bradykinin. In *Handbook of Hormones, Comparative Endocrinology for Basic and Clinical Research*; Takei, Y., Ando, H., Tsutsui, K., Eds.; Academic Press: Tokyo, Japan, 2015.
79. Borriello, F.; Iannone, R.; Marone, G. Histamine Release from Mast Cells and Basophils. *Handb. Exp. Pharmacol.* **2017**, *241*, 121–139. [PubMed]
80. Moriguchi, T.; Takai, J. Histamine and histidine decarboxylase: Immunomodulatory functions and regulatory mechanisms. *Genes Cells* **2020**, *25*, 443–449. [CrossRef]
81. Commins, S.P.; Borish, L.; Steinke, J.W. Immunologic Messenger Molecules: Cytokines, Interferons, and Chemokines. *J. Allergy Clin. Immunol.* **2010**, *125*, S53–S72. [CrossRef] [PubMed]
82. Liu, S.Q.; Tang, D.; Tieche, C.; Alkema, P. Pattern formation of vascular smooth muscle cells subject to non-uniform fluid shear stress: Mediation by cell density gradients. *Am. J. Physiol. Heart Circ. Physiol.* **2003**, *285*, H1071–H1080.
83. Davie, E.W.; Kulman, J.D. An overview of the structure and function of thrombin. *Semin. Thromb. Hemost.* **2006**, *32* (Suppl. 1), 3–15. [CrossRef]
84. Cohen, C.T.; Turner, N.A.; Moake, J.L. Human endothelial cells and fibroblasts express and produce the coagulation proteins necessary for thrombin generation. *Sci. Rep.* **2021**, *11*, 21852. [CrossRef] [PubMed]
85. Li, X.; Sim, M.M.S.; Wood, J.P. Recent Insights Into the Regulation of Coagulation and Thrombosis. *Arterioscler. Thromb. Vasc. Biol.* **2020**, *40*, e119–e125. [CrossRef] [PubMed]
86. Tomaiuolo, M., Sr.; Brass, L.F.; Stalker, T.J. Regulation of platelet activation and coagulation and its role in vascular injury and arterial thrombosis. *Interv. Cardiol. Clin.* **2017**, *6*, 1–12. [CrossRef]
87. Kattula, S.; Byrnes, J.R.; Wolberg, A.S. Fibrinogen and Fibrin in Hemostasis and Thrombosis. *Arterioscler. Thromb. Vasc. Biol.* **2017**, *37*, e13–e21. [CrossRef]
88. Luyendyk, J.P.; Schoenecker, J.G.; Flick, M.J. The multifaceted role of fibrinogen in tissue injury and inflammation. *Blood* **2019**, *133*, 511–520. [CrossRef]
89. Groves, P.H.; Penny, W.J.; Cheadle, H.A.; Lewis, M.J. Exogenous nitric oxide inhibits in vivo platelet adhesion following balloon angioplasty Get access Arrow. *Cardiovasc. Res.* **1992**, *26*, 615–619. [CrossRef]

90. Konishi, H.; Katoh, Y.; Takaya, N.; Kashiwakura, Y.; Itoh, S.; Ra, C.; Daida, H. Platelets Activated by Collagen Through Immunoreceptor Tyrosine-Based Activation Motif Play Pivotal Role in Initiation and Generation of Neointimal Hyperplasia after Vascular Injury. *Circulation* **2002**, *105*, 912–916. [CrossRef] [PubMed]
91. Meng, Q.-H.; Irvine, S.; Tagalakis, A.D.; McAnulty, R.J.; McEwan, J.R.; Hart, S.L. Inhibition of neointimal hyperplasia in a rabbit vein graft model following non-viral transfection with human iNOS cDNA. *Gene Ther.* **2013**, *20*, 979–986. [CrossRef] [PubMed]
92. Torsney, E.; Mayr, U.; Zou, Y.; Thompson, W.D.; Hu, Y.; Xu, Q. Thrombosis and Neointima Formation in Vein Grafts Are Inhibited by Locally Applied Aspirin Through Endothelial Protection. *Circ. Res.* **2004**, *94*, 1466–1473. [CrossRef]
93. Kohler, T.R.; Kirkman, T.R.; Kraiss, L.W.; Zierler, B.K.; Clowes, A.W. Increased blood flow inhibits neointimal hyperplasia in endothelialized vascular grafts. *Circ. Res.* **1991**, *69*, 1557–1565. [CrossRef]
94. Gayral, S.; Garnotel, R.; Castaing-Berthou, A.; Blaise, S.; Fougerat, A.; Berge, E.; Montheil, A.; Malet, N.; Wymann, M.P.; Maurice, P.; et al. Elastin-derived peptides potentiate atherosclerosis through the immune Neu1-PI3Kγ pathway. *Cardiovasc. Res.* **2014**, *102*, 118–127. [CrossRef]
95. Nagai, Y.; Metter, E.J.; Earley, C.J.; Kemper, M.K.; Becker, L.C.; Lakatta, E.G.; Fleg, J.L. Increased Carotid Artery Intimal-Medial Thickness in Asymptomatic Older Subjects With Exercise-Induced Myocardial Ischemia. *Circulation* **1998**, *98*, 1504–1509. [CrossRef] [PubMed]
96. Virmani, R.; Avolio, A.P.; Mergner, W.J.; Robinowitz, M.; Herderick, E.E.; Cornhill, J.F.; Guo, S.Y.; Liu, T.H.; Ou, D.Y.; O'Rourke, M. Effect of aging on aortic morphology in populations with high and low prevalence of hypertension and atherosclerosis. Comparison between occidental and Chinese communities. *Am. J. Pathol.* **1991**, *139*, 1119–1129. [PubMed]
97. Karnik, S.K.; Wythe, J.D.; Sorensen, L.; Brooke, B.J.; Urness, L.D.; Li, D.Y. Elastin induces myofibrillogenesis via a specific domain, VGVAPG. *Matrix Biol.* **2003**, *22*, 409–425. [CrossRef] [PubMed]
98. Papageorgopoulou, C.P.; Nikolakopoulos, K.M.; Ntouvas, I.G.; Papadoulas, S. Chronic Limb Ischemia due to Thrombosis of an Aneurysmal Degeneration of the Autogenous Vein Graft: A Case Report. *Aorta* **2022**, *10*, 77–79. [CrossRef]
99. Kulik, A. Commentary: Yin and Yang: Antiplatelet and lipid-lowering therapies to prevent vein graft thrombosis and atherosclerosis after coronary artery bypass graft surgery. *J. Thorac. Cardiovasc. Surg.* **2022**, *163*, 1042–1043. [CrossRef]
100. Yeo, J.W.; Law, M.S.N.; Lim, J.C.L.; Ng, C.H.; Tan, D.J.H.; Tay, P.W.L.; Syn, N.; Tham, H.Y.; Huang, D.Q.; Siddiqui, M.S.; et al. Meta-analysis and systematic review: Prevalence, graft failure, mortality, and post-operative thrombosis in liver transplant recipients with pre-operative portal vein thrombosis. *Clin. Transplant.* **2022**, *36*, e14520. [CrossRef]
101. Kato, T.; Fuke, M.; Nagai, F.; Nomi, H.; Kanzaki, Y.; Yui, H.; Maruyama, S.; Nagae, A.; Sakai, T.; Saigusa, T.; et al. Successful endovascular treatment with a stent graft for chronic deep vein thrombosis with multiple arteriovenous fistulas: A case report. *J. Med. Case Rep.* **2022**, *16*, 257. [CrossRef]
102. Esenboga, K.; Baskovski, E.; Ates, B.N.; Ozyuncu, N.; Turhan, S.; Tutar, E. Thrombotic Complication of COVID-19: A Case Report of Acute Saphenous Vein Graft Thrombosis in a Newly Diagnosed Patient. *Turk Kardiyol. Dern. Ars.* **2022**, *50*, 228–230. [CrossRef]
103. Tatterton, M.; Wilshaw, S.P.; Ingham, E.; Homer-Vanniasinkam, S. The use of antithrombotic therapies in reducing synthetic small-diameter vascular graft thrombosis. *Vasc. Endovasc. Surg.* **2012**, *46*, 212–222. [CrossRef]
104. De Visscher, G.; Mesure, L.; Meuris, B.; Ivanova, A.; Flameng, W. Improved endothelialization and reduced thrombosis by coating a synthetic vascular graft with fibronectin and stem cell homing factor SDF-1alpha. *Acta Biomater.* **2012**, *8*, 1330–1338. [CrossRef] [PubMed]
105. Mallis, P.; Kostakis, A.; Stavropoulos-Giokas, C.; Michalopoulos, E. Future Perspectives in Small-Diameter Vascular Graft Engineering. *Bioengineering* **2020**, *7*, 160. [CrossRef] [PubMed]
106. Radke, D.; Jia, W.; Sharma, D.; Fena, K.; Wang, G.; Goldman, J.; Zhao, F. Tissue Engineering at the Blood-Contacting Surface: A Review of Challenges and Strategies in Vascular Graft Development. *Adv. Healthc. Mater.* **2018**, *7*, e1701461. [CrossRef] [PubMed]
107. Fang, S.; Ellman, D.G.; Andersen, D.C. Review: Tissue Engineering of Small-Diameter Vascular Grafts and Their In Vivo Evaluation in Large Animals and Humans. *Cells* **2021**, *10*, 713. [CrossRef]
108. Imashiro, C.; Shimizu, T. Fundamental Technologies and Recent Advances of Cell-Sheet-Based Tissue Engineering. *Int. J. Mol. Sci.* **2021**, *22*, 425. [CrossRef]
109. McAllister, T.N.; Maruszewski, M.; Garrido, S.A.; Wystrychowski, W.; Dusserre, N.; Marini, A.; Zagalski, K.; Fiorillo, A.; Avila, H.; Manglano, X.; et al. Effectiveness of haemo-dialysis access with an autologous tissue-engineered vascular graft: A multicentre co-hort study. *Lancet* **2009**, *373*, 1440–1446. [CrossRef]
110. Wystrychowski, W.; Garrido, S.A.; Marini, A.; Dusserre, N.; Radochonski, S.; Zagalski, K.; Antonelli, J.; Canalis, M.; Sammartino, A.; Darocha, Z.; et al. Long-term results of autologous scaffold-free tissue-engineered vascular graft for hemodialysis access. *J. Vasc. Access.* **2022**. [CrossRef]
111. Kristofik, N.J.; Qin, L.; Calabro, N.E.; Dimitrievska, S.; Li, G.; Tellides, G.; Niklason, L.E.; Kyriakides, T.R. Improving in vivo outcomes of decellularized vascular grafts via incorporation of a novel extracellular matrix. *Biomaterials* **2017**, *141*, 63–73. [CrossRef]
112. Gui, L.; Muto, A.; Chan, S.A.; Breuer, C.K.; Niklason, L.E. Development of Decellularized Human Umbilical Arteries as Small-Diameter Vascular Grafts. *Tissue Eng. Part A* **2009**, *15*, 9. [CrossRef]
113. Wolfe, J.T.; Shradhanjali, A.; Tefft, B.J. Strategies for Improving Endothelial Cell Adhesion to Blood-Contacting Medical Devices. *Tissue Eng. Part B Rev.* **2022**, *28*, 1067–1092. [CrossRef]
114. Cooley, B.C. Collagen-induced thrombosis in murine arteries and veins. *Thromb. Res.* **2013**, *131*, 49–54. [CrossRef]

115. Sa, Q.; Hoover-Plow, J.L. EMILIN2 (Elastin microfibril interface located protein), potential modifier of thrombosis. *Thromb. J.* **2011**, *9*, 9. [CrossRef] [PubMed]
116. Kawecki, C.; Hezard, N.; Bocquet, O.; Poitevin, G.; Rabenoelina, F.; Kauskot, A.; Duca, L.; Blaise, S.; Romier, B.; Martiny, L.; et al. Elastin-derived peptides are new regulators of thrombosis. *Arterioscler. Thromb. Vasc. Biol.* **2014**, *34*, 2570–2578. [CrossRef]
117. Wang, Z.; Liu, L.; Mithieux, S.M.; Weiss, A.S. Fabricating Organized Elastin in Vascular Grafts. *Trends Biotechnol* **2021**, *39*, 505–518. [CrossRef] [PubMed]
118. Sugiura, T.; Agarwal, R.; Tara, S.; Yi, T.; Lee, Y.U.; Breuer, C.K.; Weiss, A.S.; Shinoka, T. Tropoelastin inhibits intimal hyperplasia of mouse bioresorbable arterial vascular grafts. *Acta Biomater.* **2017**, *52*, 74–80. [CrossRef]
119. Ryan, A.J.; O'Brien, F.J. Insoluble elastin reduces collagen scaffold stiffness, improves viscoelastic properties, and induces a contractile phenotype in smooth muscle cells. *Biomaterials* **2015**, *73*, 296–307. [CrossRef]
120. Jordan, S.W.; Haller, C.A.; Sallach, R.E.; Apkarian, R.P.; Hanson, S.R.; Chaikof, E.L. The effect of a recombinant elastin-mimetic coating of an ePTFE prosthesis on acute thrombogenicity in a baboon arteriovenous shunt. *Biomaterials* **2007**, *28*, 1191–1197. [CrossRef] [PubMed]
121. Koens, M.J.; Krasznai, A.G.; Hanssen, A.E.; Hendriks, T.; Praster, R.; Daamen, W.F.; van der Vliet, J.A.; van Kuppevelt, T.H. Vascular replacement using a layered elastin-collagen vascular graft in a porcine model: One week patency versus one month occlusion. *Organogenesis* **2015**, *11*, 105–121. [CrossRef]
122. McKenna, K.A.; Hinds, M.T.; Sarao, R.C.; Wu, P.C.; Maslen, C.L.; Glanville, R.W.; Babcock, D.; Gregory, K.W. Mechanical property characterization of electrospun recombinant human tropoelastin for vascular graft biomaterials. *Acta Biomater.* **2012**, *8*, 225–233. [CrossRef]
123. Ibanez-Fonseca, A.; Flora, T.; Acosta, S.; Rodriguez-Cabello, J.C. Trends in the design and use of elastin-like recombinamers as biomaterials. *Matrix Biol.* **2019**, *84*, 111–126. [CrossRef]
124. Vassalli, M.; Sbrana, F.; Laurita, A.; Papi, M.; Bloise, N.; Visai, L.; Bochicchio, B. Biological and structural characterization of a naturally inspired material engineered from elastin as a candidate for tissue engineering applications. *Langmuir* **2013**, *29*, 15898–15906. [CrossRef] [PubMed]
125. Wang, Z.; Mithieux, S.M.; Vindin, H.; Wang, Y.; Zhang, M.; Liu, L.; Zbinden, J.; Blum, K.M.; Yi, T.; Matsuzaki, Y.; et al. Rapid Regeneration of a Neoartery with Elastic Lamellae. *Adv. Mater.* **2022**, *34*, e2205614. [CrossRef] [PubMed]
126. McCarthy, C.W.; Ahrens, D.C.; Joda, D.; Curtis, T.E.; Bowen, P.K.; Guillory, R.J., 2nd; Liu, S.Q.; Zhao, F.; Frost, M.C.; Goldman, J. Fabrication and Short-Term in Vivo Performance of a Natural Elastic Lamina-Polymeric Hybrid Vascular Graft. *ACS Appl. Mater. Interfaces* **2015**, *7*, 16202–16212. [CrossRef] [PubMed]
127. Babapulle, M.N.; Eisenberg, M.J. Coated Stents for the Prevention of Restenosis: Part, I. *Circulation* **2002**, *106*, 2734–2740. [CrossRef] [PubMed]
128. Nusca, A.; Viscusi, M.M.; Piccirillo, F.; De Filippis, A.; Nenna, A.; Spadaccio, C.; Nappi, F.; Chello, C.; Mangiacapra, F.; Grigioni, F.; et al. In Stent Neo-Atherosclerosis: Pathophysiology, Clinical Implications, Prevention, and Therapeutic Approaches. *Life* **2022**, *12*, 393. [CrossRef]

 *bioengineering*

*Review*

# Engineering Spatiotemporal Control in Vascularized Tissues

Astha Khanna [1], Beu P. Oropeza [2,3,4] and Ngan F. Huang [2,3,4,5,*]

1 Graver Technologies, Newark, NJ 07105, USA
2 Stanford Cardiovascular Institute, Stanford University, Stanford, CA 94305, USA
3 Department of Cardiothoracic Surgery, Stanford University, Stanford, CA 94305, USA
4 Center for Tissue Regeneration, Veterans Affairs Palo Alto Health Care System, Palo Alto, CA 94304, USA
5 Department of Chemical Engineering, Stanford University, Stanford, CA 94305, USA
* Correspondence: ngantina@stanford.edu

**Abstract:** A major challenge in engineering scalable three-dimensional tissues is the generation of a functional and developed microvascular network for adequate perfusion of oxygen and growth factors. Current biological approaches to creating vascularized tissues include the use of vascular cells, soluble factors, and instructive biomaterials. Angiogenesis and the subsequent generation of a functional vascular bed within engineered tissues has gained attention and is actively being studied through combinations of physical and chemical signals, specifically through the presentation of topographical growth factor signals. The spatiotemporal control of angiogenic signals can generate vascular networks in large and dense engineered tissues. This review highlights the developments and studies in the spatiotemporal control of these biological approaches through the coordinated orchestration of angiogenic factors, differentiation of vascular cells, and microfabrication of complex vascular networks. Fabrication strategies to achieve spatiotemporal control of vascularization involves the incorporation or encapsulation of growth factors, topographical engineering approaches, and 3D bioprinting techniques. In this article, we highlight the vascularization of engineered tissues, with a focus on vascularized cardiac patches that are clinically scalable for myocardial repair. Finally, we discuss the present challenges for successful clinical translation of engineered tissues and biomaterials.

**Keywords:** vascularization; tissue engineering; 3d bioprinting; biomaterials; cardiac engineering; extracellular matrix

**Citation:** Khanna, A.; Oropeza, B.P.; Huang, N.F. Engineering Spatiotemporal Control in Vascularized Tissues. *Bioengineering* 2022, 9, 555. https://doi.org/10.3390/bioengineering9100555

Academic Editors: Gary Chinga Carrasco and Mario Petretta

Received: 2 September 2022
Accepted: 11 October 2022
Published: 14 October 2022



## 1. Introduction

Cardiovascular disease (CVD) remains a major cause of morbidity and mortality all over the world, causing approximately 17.9 million deaths per year in the United States [1]. Heart failure, the inability of heart to provide sufficient blood flow to the body, is largely attributed to the non-regenerative cardiomyocytes that provide contractility to the heart [2]. Myocardial injury results in cardiac remodeling and adverse fibrosis, resulting in fibrotic scar tissue [3]. Patients with advanced failure resort to autologous or allogenic graft transplantation, often limited by factors such as donor site morbidity, non-availability of appropriate donor tissue in autologous transplantation, and the high probability of disease transmission associated with immunosuppression in allogenic transplantation [4]. However, failure of graft integration by inadequate vascularization post implantation can result in failure of graft. Tissue engineering to replace diseased tissues and organs is an alternative to autologous or allogenic graft transplantation. In the last two decades, tissue engineering has advanced the restoration and replacement of tissues and organs by using cells and biomolecules in three-dimensional (3D) biomaterials [4]. Although different avascular tissues such as bladder, cartilage, and epidermis have been successfully fabricated and progressed to clinical translation [5], tissue engineering of complex tissues that are thicker and vascularized have been limited in success due to the lack of a robust microvascular network [5]. Living cells typically reside within 200 μm of blood circulation for efficient oxygen and nutrient exchange for survival and functional bioactivity long-term. Due to the constraint of oxygen diffusion limit, most 3D engineered tissues of physiological architecture require adequate vascularization of tissue or perfusion source [5].

Strategies to induce vasculature formation in engineered tissues seek to mimic the re-vascularization process known as angiogenesis in response to tissue ischemia [6]. Understanding the mechanisms regulating angiogenesis can help discover clues to fabricate engineered tissues embedded with functional and mature vascular network [7]. The process of angiogenesis involves stimulation of quiescent vascular endothelial cells (ECs) from an existing vessel and their activation caused by high concentration of pro-angiogenic factors released by inflammatory cellular population as a signaling response to injury, hypoxia, or a combination of both. The ECs activate by proliferating and sensing the chemical gradients of soluble factors, resulting in stimulus directed elongation of new vessels via migration and secretion of soluble factors molecules that recruit perivascular support cells [7]. Perivascular cells, composed of capillary pericytes and smooth muscle cells in larger vessels, move towards the new vessels formed to cover the endothelium, imparting stability, inducing cell differentiation, and regulating permeability of vessel [8]. The critical processes include temporal regulation, spatial arrangement of the stimuli, crosstalk between cells and molecules, active remodeling, and organization of extracellular matrix (ECM). Dysregulation of any of the processes can result in abnormal development of new vasculature can occur due to disruption of the tightly regulated factors needed for angiogenesis.

Numerous engineering techniques have been developed to generate biomaterial constructs with unique spatial modifications and complex microarchitectures. In this review, we highlight the recent developments and state-of-the-art approaches in biofabrication techniques, including, but not limited to, electrospinning, micropatterning, and 3D bioprinting techniques for the generation of spatially defined biomaterials of optimal geometrical and topographical characteristics for modulating proliferation, migration, and differentiation of cells in contact with engineered scaffolds (Figure 1). We also review the temporal control advancements focused on the controlled release of growth factors and drugs to recapitulate the unique dynamic features of the ECM to direct biological processes, such as stem cell differentiation and functional tissue regeneration. The advancements and limitations of current biofabrication techniques for spatiotemporal regulation in multiscale materials and biomimetic microenvironments are also discussed. As an example, the spatiotemporal regulation strategies for the generation of vascularized engineered cardiac patches are discussed. Finally, we conclude by providing a perspective on the future challenges and opportunities in the development of biomaterials for tissue engineering applications.

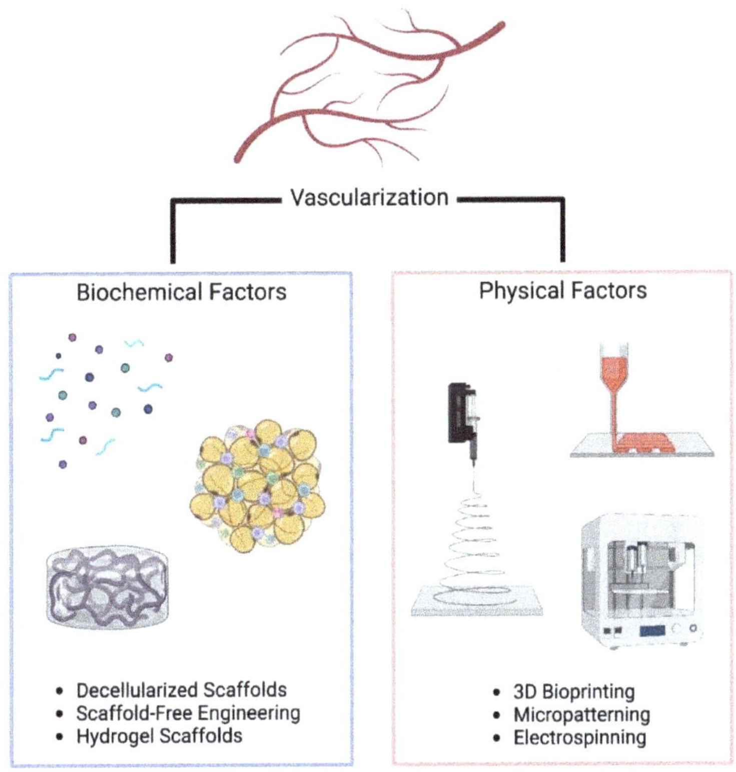

**Figure 1.** Overview of vascularization strategies using biochemical and biophysical factors.

## 2. Temporal Biology of Angiogenesis

During the process of angiogenesis, novel capillaries are formed from existing vasculature via the process of sprouting and intussusception. Several cellular functions occur during angiogenesis categorized into a phase of activation involving initiation and progression and a phase of resolution that includes processes of termination and maturation of vessel [9]. Biochemical processes such as ischemia or inflammation causes sprouting that promotes stimulatory autocrine and paracrine cytokines release, including the very potent vascular endothelial growth factor (VEGF) [10]. Leaky vasculature results from local basement membrane degradation and vessel endothelial cell's migration, governed by a specialized tip cell [11]. Dll4-Notch lateral inhibition between neighboring endothelial cells in a feedback loop with VEGF–VEGF receptor (VEGFR) signaling is characterized to be the 'central pattern generating' (CPG) mechanism that results in the selection of migratory tip cells. The tip cells inhibit their neighboring cells, which are termed 'stalk cells' [11]. Several studies have demonstrated the part of ECM in the process of angiogenesis. The ECM creates physical scaffold necessary to maintain blood vessel organization and participates in biochemical and biophysical signaling transduction during angiogenesis. Specifically, collagen and fibronectin stimulate EC tubular morphogenic events [12]. Laminin facilitates endothelial cell tip formation and sprouting and are also critical for maintaining vascular homeostasis, and proteoglycans regulate endothelial cell migration to form new vessels [12]. During angiogenesis, matrix metalloproteinases (MMPs) degrade collagen and other ECM components, facilitating endothelial cell migration from pre-existing vessels towards angiogenic stimuli. Polymerization of plasma-derived proteins such as fibrinogen and fibronectin cause generation of a provisional ECM that facilitates endothelial cell extension in the environment. Stalk cells proliferation behind the tip cell, and mural cell populations recruitment induce the sprout's elongation and formation of lumen [13]. The process of intussusception encompasses the existing vasculature remodeling through the protrusion and fusion of opposite vessel walls via split of an existing vessel to form a branched structure.

In comparison to sprouting, intussusceptive angiogenesis is rapid in capillary network expansion as it relies on reorganizing existing endothelial cells and not cellular division. Triggers such as hypoxia and injury facilitate angiogenesis resulting in immature microvascular networks through processes such as sprouting and intussusception, and pericyte invasion and subsequent cytokine-mediated cell induction that stabilizes the vessel walls and basement membrane followed by remodeling [13]. The premature vessel network is further remodeled to form an efficiently perfused vascular bed for oxygenation of the tissue. Vessel ablation of this nature is regulated by hemodynamic signals from local vasculature as endothelial cells integrate with nearby perfused network or undergo apoptosis causing vascular regression without pro-survival stimuli including shear flow and growth factor gradients [14].

Tissue engineering aims at inducing an angiogenic response from host ECs to utilize the natural capacity of our system for tissue vascularization through angiogenesis and regenerative techniques employed in specific applications, such as an infarcted heart. At the cellular level, mechanical and chemical signals govern the internal signaling effect and consequent biological responses including migration, proliferation, and differentiation. Hence, temporal conjugation of biomolecules in a tissue engineered scaffold plays a major role in development of mature vascular network. Commonly, non-covalent adsorption of growth factors and other molecules incorporated into engineered scaffolds is employed. Thus, growth factor release depends on the affinity of the molecules with the scaffold material or regulated through molecular diffusion kinetics. This approach is advantageous when scaffolds are employed for the controlled release of molecules to organs or tissues.

### 3. Growth Factors Regulation in Angiogenesis

Angiogenesis is controlled by cell–cell and cell-ECM interactions through crosstalk between VEGF and Notch signaling mechanisms. Novel vascular structures are regulated by the surrounding cells and modulated by secretion of platelet-derived growth factor (PDGF) and VEGF secreted by ECs and vascular smooth muscle cells [15]. Among the factors that impact EC activation status are proteins called cytokines. A tissue can dictate the cellular response to a given cytokine. Hence, cytokines are considered as specialized symbols in intercellular interaction. This interaction is influenced by three factors, including the concentration of other cytokines in the environment; chemical and biological interactions between ECM, cells, and cytokines; and the cytoskeleton [16].

The signal protein most commonly studied to influence angiogenesis are VEGF, acidic fibroblast growth factor (aFGF), and basic fibroblast growth factor (bFGF) [17]. VEGF and FGF-2 have been studied in vitro to positively regulate several endothelial cell functions, that includes cellular proliferation, migration, extracellular proteolytic activity, and tube formation [17]. In addition, although a myriad of factors have been demonstrated to be active in the experimental setting, they are not all relevant to the endogenous regulation of new blood vessel formation. On the list of molecules that are active during the phase of activation, VEGF meets most of the criteria of a vasculogenic or angiogenic factor.

Angiogenesis regulators may act either directly on ECs or indirectly by inducing the production of direct-acting regulators by inflammatory and other non-EC populations. Thus, in contrast to VEGF and FGF-2, which are direct endothelial cell mitogens, the cytokines transforming growth factor-$\beta$ (TGF-$\beta$) and tumor necrosis factor-$\alpha$ (TNF-$\alpha$) have been studied to inhibit EC growth in vitro and are therefore direct-acting negative regulators [18]. However, both TGF-$\beta$ and TNF-$\alpha$ are angiogenic in vivo, and it has been demonstrated to induce angiogenesis indirectly by stimulating the production of direct-acting positive regulators from stromal and chemoattracted inflammatory cells; hence, TGF-$\beta$ and TNF-$\alpha$ are considered to be indirect positive regulators [19]. TGF-$\beta$ has also been proposed to be a potential mediator of the phase of resolution due to its capacity to inhibit endothelial cell proliferation and migration directly, reduce extracellular proteolysis, and promote matrix deposition in vitro. In vitro, TGF-β has also been studied to promote the organization of single endothelial cells embedded in three-dimensional collagen gels into tubelike structures, further signifying its role in the phase of resolution [18].

Other cytokines that have been studied to regulate angiogenesis in vivo include HGF, EGF/TGF-α, PDGF-BB, interleukins (IL-1, IL-6, and IL-12), interferons, GM-CSF, PlGF, proliferin, and proliferin-related protein. Angiogenesis can also be regulated by a variety of noncytokine or nonchemokine factors, including enzymes (angiogenin and PD-ECGF/TP), inhibitors of matrix-degrading proteolytic enzymes (TIMPs) and of PAs (PAIs), extracellular matrix components/coagulation factors or fragments (thrombospondin, angiostatin, hyaluronan, and its oligosaccharides), soluble cytokine receptors, prostaglandins, adipocyte lipids, and copper ions. The roles of these bioactive molecules are summarized in Table 1.

**Table 1.** Bioactive molecules and their effects on vascularization of tissue engineered constructs.

| Bioactive Molecules | Angiogenic Effects | Ref |
|---|---|---|
| VEGF | Facilitates EC migration and proliferation<br>Regulates EC proliferation, migration, and survival; allows mobilization of BM-derived cells such as HSCs, and recruit SMCs for stabilization of vessel. | [20] |
| FGF | FGF-2 Enhances EC proliferation.<br>bFGF facilitates the activation, proliferation, and migration of EPC; regulate vasculogenesis and the formation of immature primary vascular networks.<br>FGF-2 Interacts with ECM molecules such as heparin, heparan sulfate proteoglycans (HSPGs); promotes EC response and neovascularization process.<br>FGF-2 facilitates proliferation of ECs, SMCs; endothelial capillary formation | [20] |
| IGF-1 | Facilitates formation of neovasculature from the endothelium of pre-existing vessels andInduces endothelial cell migration for vascularization<br>Induces the activation of the PI3-kinase/Akt signaling pathway and expression of growth factors | [21] |
| PDGF | Promotes vessel maturation by recruitment of MSCs, pericytes, and SMCs.<br>Facilitates remodeling by inducing collagenases secretion by fibroblasts.<br>Increases VEGF production and promote angiogenesis<br>Regulates the production of ECM molecules for basement membrane and blood vessel stabilization | [22] |
| TGF-β | Promotes EC migration, proliferation, and differentiation.<br>Increases VEGF secretion by ECs; and PGF and bFGF expression by SMCs. Enhances angiogenesis.<br>Facilitates vessel stabilization and maturation<br>Stimulates ECM deposition | [23] |
| HGF | Induces VEGF secretion<br>Promotes angiogenesis by ECs expression of VEGF. | [24] |
| TNF-α | Inhibits proliferation of endothelial cells; promotes angiogenesis | [23] |
| Angiopoietin | Facilitates TGF-β-induced differentiation of MSCs.<br>Promotes vessel maturation<br>Inhibits VEGF activity and facilitates EC-SMC interactions<br>Enhnaces type IV collagen deposition<br>Promotes EC proliferation<br>Induces VEGF mediated angiogenic sprouting. | [22] |
| SDF-1 | Facilitates vessel stabilization by recruitment of progenitors of SMCs<br>Initiate vascular remodeling; upregulate metalloproteinases and downregulate angiostatin | [25] |

Abbreviations: VEGF (Vascular Endothelial growth factor); FGF (Fibroblast growth factor); IGF-1 (insulin-like growth factor); PDGF (Platelet-derived growth factor); TGF-ß (Transforming growth factor-beta); HGF (Hepatocyte-growth factor); SDF-1 (Stromal cell derived growth factor); MSCs (Mesenchymal stem cells); VSMCs (Vascular Smooth muscle cells); TNF-α (Tumor necrosis factor α); EC (Endothelial cells); PGF (Placental growth factor); IGF-1 (Insulin growth factor -1); SMCs (Smooth muscle cells); EC (Endothelial cells); HSPGs (Heparan sulfate proteoglycans); FGF-2 (Fibroblast growth factor -2), bFGF (basic Fibroblast growth factor); HSCs (Hematopoietic stem cells); BM (Bone marrow); PGF (Placental growth factor); ECM (Extracellular matrix).

*Combinatorial Regulation Chemical Factors in Engineering Vascularized Tissues*

Biomaterial matrices functionalized with angiogenic growth factors have been extensively researched to promote vascularization. A myriad of angiogenic growth factors such as VEGF, PDGF-BB, bFGF, hepatocyte growth factor (HGF), insulin-like growth factor (IGF), and TGF-$\beta$ have been widely studied to promote vascularization in pathological disease models [26]. All key angiogenic growth factors (VEGF, FGF-2, IGF, HGF, PDGF-BB, and TGF-$\beta$1) bind to specific sites in the ECM; their release kinetics are based on their binding affinity and the proteases action to cleave the ECM or the ECM-binding growth factor domain [26]. It has been demonstrated through in vitro and in vivo studies that insufficient angiogenic growth factor exposure can inhibit angiogenesis, and subsequently, overexpression of growth factor can inhibit the function of vascular smooth muscle cells and pericytes population and form immature and unstable vessels [27]. The dose and duration of growth factor release has been studied to play a critical role in therapeutic applications. Several strategies have been employed to control the release of growth factors from biodegradable scaffolds. For example, angiogenesis has been enhanced using heparin or heparan sulfate-mimetic molecules covalently crosslinked with the collagen type I scaffold via 1-ethyl-3-dimethyl aminopropyl carbodiimide (EDC) and N-hydroxysuccinimide (NHS) for release of heparin-binding growth factors [28]. Further, angiogenesis has been studied to be enhanced by the combination of VEGF and FGF with a heparin-immobilized scaffold compared with a single growth factor molecule. Biomaterials have also been functionalized using surface modification strategies or heparin-binding ECM domain addition. For example, sequestration of multiple growth factors (VEGF-A165, PDGF-BB, and BMP-2) can be achieved using a fibrin matrix covalently crosslinked with multifunctional recombinant fibronectin (FN) fragments, including both its 12th and 14th type III repeats (FN III12-14) and FN III9-10 for enhanced angiogenic effects [29]. Angiogenic growth factors can also be altered for enhanced binding affinity to biomaterials for enhanced affinity with growth factors. Sacchi et al. achieved covalently crosslinking of fibrin hydrogels with VEGF fused to a sequence derived from $\alpha_2$-plasmin inhibitor ($\alpha_2$-PI$_{1-8}$) for controlled VEGF release by enzymatic cleavage, that resulted in stable and functional angiogenesis [30].

Incorporation of short bioactive peptides onto 3D scaffolds has gained interest as an effective method to achieve vascularization. Several approaches have been employed to study the effects of the immobilized bioactive peptides on vascular network formation. Increased EC attachment, growth, and migration were achieved by incorporation of integrin to ECM derived short peptide adhesive sequences such as collagen (Arg-Gly-Asp (RGD)), laminin (e.g., Tyr-Ile-Gly-Ser-Arg (YIGSR) and Ser-Ile-Lys-Val-Ala-Val (SIKVAV)), and FN (e.g., RGD and Arg-Glu-Asp-Val (REDV)) that increased angiogenesis [31]. Hydrogel activation by functional RGD and REDV sequences in an elastin-like recombinamer-based hydrogel caused improved EC adhesion and in vivo angiogenic potential via general cell adhesion and specific endothelial cell adhesion.

Several strategies have been taken to deliver bioactive molecules from tissue engineered scaffolds that mimic those associated with angiogenesis. Although the delivery of single-factor soluble factors such as bFGF can induce EC proliferation [32], the delivery of combinatorial growth factors better mimic the complexity of the angiogenic process. Various studies have demonstrated controlled dose and duration of growth factor release from biodegradable materials. Heparin-binding growth factors, VEGF and FGF-2 delivered from heparin-immobilized scaffolds exhibited an increased degree of angiogenesis in comparison to individual growth factor response [33]. Multiple growth factors (VEGF-A165, PDGF-BB, and BMP-2) were sequestered using fibrin matrix covalently crosslinked with multifunctional recombinant fibronectin (FN) fragments (12th and 14th type III repeats (FN III12-14) and FN III9-10) and exhibited enhanced angiogenic effects in a mouse model of chronic wound healing [33]. In another example, Kuttappan et al. functionalized a nanocomposite fibrous scaffold with combinations of VEGF, FGF-2, and BMP2 for differential growth factor release [34] that resulted in increased tissue vascularization. Furthermore, since growth factor delivery based on scaffold degradation can lead to an initial burst release, it was shown that increasing the crosslinking density of gelatin could improve growth factor retention. Turner et al. achieved controlled release of VEGF or BMP2 based on the progressive proteolytic degradation of the scaffold using crosslinked gelatin microspheres. However, the non-specific degradation of the scaffold and non-uniform growth factor release necessitates the need for development of more advanced systems for optimal release [35].

Incorporation of different bioactive molecules in sequential layers of polymers; a technique called layer-by-layer (LBL), can be employed for sequential delivery of growth factors. The incorporation of bioactive molecules and action of matrix-degrading enzymes causes sequential delivery of growth factors. A Polycaprolactone (PCL) scaffold was developed with sequential layers of heparin and VEGF was developed. Long-term anti-thrombogenic effect of tissue engineered graft was achieved by initial burst release of VEGF, facilitated by ECM degrading enzyme metallopeptidase-2 (MMP-2) and controlled release of heparin [36].

An enzyme-sensitive linker to link pro-angiogenic molecules covalently to the scaffold has been studied to promote angiogenesis. Linking the linker sequence to a specific enzyme (e.g., MMPs, serine, or cysteine proteinases) regulates time-bound release as enzymes are produced by cells at specific times during differentiation or angiogenesis. Light, an external stimulus for smart drug-delivery platforms, has been studied in various biomedical applications including image-guided surgery, and the photopolymerization and -degradation of tissue engineering scaffolds, for the advantages of its noninvasive properties, high spatial resolution, temporal control, and simple to use [37]. Light-sensitive linkers have been used to covalently bind molecules, with UV or near infrared (NIR) light used to release "incorporated" biomolecules. Light-responsive delivery systems must possess high spatial and temporal regulation over drug release, employ nonionizing radiation, formed of biocompatible materials, and flexible to be tailored to the needed application [38].

Encapsulation is another technique for controlled release of bioactive molecules. It has the advantage of providing protection for growth factors, increasing their half-life. Studies have been performed to design a scaffold patterned with composite microspheres, with the spatiotemporal release of proteins [39]. Lai et al. employed the technique of encapsulation via nanofibers and gelatin nanoparticles to form a scaffold for sequential release of VEGF, PDGF, FGF, and EGF (epithelial growth factor). This resulted in enhanced endothelial cell proliferation and development of vascular networks [40]. Various approaches have been employed for controlled local delivery of angiogenic growth factors, however, the limitation of their inherent inability to control the geometric architecture of vascular networks needs to be addressed for optimal 3D tissue construction. These techniques can be employed to control growth factor delivery recapitulating the temporal pattern observed in physiological angiogenesis, but with limited complexity. Despite the promise growth factor delivery or bioactive-peptide-guided vascular network formation, these approaches still lack the control network geometry, for generation of a spatially controllable 3D mature vascular network. Therefore, advancements in fabrication technologies below aim to fabricate spatially controllable 3D vascular networks using scaffolds.

## 4. Spatial Control in Engineering Vascularized Tissues

Besides temporal regulation of growth factors, angiogenesis is strictly regulated by spatial signals to govern vessel sprouting and maturation. Physiological cues including inflammation and ischemia induce release of growth factor molecules, cytokines to create a gradient within the extracellular matrix domain that results in generation of a spatially controlled rearrangement of neovessels. Tissue engineering strategies seek to develop systems with spatial control, such as direct cellular patterning using 3D bioprinting, electrospinning and soft lithography for more precisely controllable neo-vessel formation. These systems have been highlighted below.

### 4.1. Three-Dimensiona; Bioprinting

Three-dimensional bioprinting is a multidisciplinary approach to spatially pattern cellular and biological components by employing a layer-by-layer process to deposit and generate 3D organ analogs and tissues platforms. Researchers have widely employed 3D bioprinting to generate 3D constructs with vascular networks for creation of more geometrically complex tissue structures. Bioprinting utilizes two manufacturing approaches of direct and indirect printing to create tissue constructs. Direct printing includes printing of bio-ink droplets (cell-laden hydrogels) containing cellular and extracellular components into designed vascular network structures. In contrast, indirect printing involves a cell-free scaffold or component bioprinted with cell-laden hydrogel layers. Using these methods, combination of cells, biomaterials, and growth factors can produce complex constructs with spatial organization with dimensional micron-sized channels and pore sizes to direct angiogenesis.

The most commonly employed bioprinting techniques are based on inkjet, extrusion, and lasers methods (Table 2). Inkjet bioprinting involves layer-by-layer dispersion of bio-ink droplets on a construct using a thermal or piezoelectric actuator. In inkjet bioprinting, the printhead is placed over the printing bed, followed by generation of a 3D tissue using bioink droplets created by thermal, electrostatic, or piezoelectric inkjet bioprinters [41]. This approach has the advantage of generating picoliter-scale drops with a ~30 to 60 µm printing resolution. It utilizes crosslinking molecules with hydrogels having rapid gelation characteristics for generation of organized networks. For example, alginate-based bio-inks can be printed into a calcium chloride solution as they rapidly crosslink. This technique has been used to create 200-µm diameter vessels. Cui et al. demonstrated effective simultaneous printing of ECs and fibrin-based vascular networks. The aligned ECs proliferated to form a confluent tubular form within the printed channels within the fibrin scaffold after 28 days of culture [42]. This approach has the advantage of low cost due to the potential to adapt regular printers and for printing multiple cell types. The thickness of constructs printed using inkjet methods are limited by weak structural support due to low concentration of hydrogel.

Extrusion-based bioprinting encompasses layer-by-layer printing of bio-ink by employing a syringe and piston for dispensing through nozzles on a microscale level. Extrusion-based approaches employ relatively higher amounts of hydrogels such as alginate and Pluronic F-127 for the generation of stable 3D cellular constructs. Millik et al. optimized an advanced extrusion system and bio-ink blend for generation of highly organized and perfusable cell-loaded microvasculature [43]. Tubes produced with a wide range of diameters (500–1500 µm) and wall thicknesses (60–280 µm) using the co-axial system. Gao et al. recently constructed a coaxial extrusion system for concurrent flow of calcium solution (interior) and alginate solution (exterior) [44]. Hollow and high strength calcium alginate filaments of cell-laden, 3D hydrogel structures were successfully fabricated with microchannels that were perfusable. One of the limitations is the sub-optimal mechanical stability and structural integrity, for printing clinically scalable tissue constructs. In a study conducted by Kim et al., an advanced 3D printer known as an integrated tissue-organ printer (ITOP), was employed to generate stable and multiform human-scale tissue constructs. [45]. The ITOP patterns multi- cell-laden composite hydrogels composed of gelatin, fibrinogen, hyaluronic acid, and glycerol and present a PCL polymer (of high strength) and a sacrificial Pluronic F-127 hydrogel with strong mechanical characteristics. The application of ITOP to generate a human-scale mandible, calvarial bone, cartilage, and skeletal muscle was successfully performed in vivo with mature and perfusable tissue formation [46]. The ITOP has been demonstrated to allow advanced 3D bioprinting and generation of clinically translatable tissues.

Lastly, laser-assisted bioprinting is an effective method for printing precise microvasculature, although fewer studies have been discussed on this. Laser-based bioprinting can be conducted using photopolymerization or laser-induced forward transfer method. Although this technique is costly, it can print cells at very high resolution obviating the exposure to high shear stress Wu and Ringeisen [47] employed laser bioprinting for generation of branch/stem structures with HUVECs followed by culture of human umbilical vein smooth muscle cells on the printed HUVEC constructs. The formed microvasculature possessed two stems and a stable lumina recapitulating the microvascular network.

**Table 2.** Comparison of bioprinting methods for fabrication of vascularized engineered tissues.

| Bioprinting Technique | Bioprinted Cellular Types | Vascularization Application | Limitations | Ref |
|---|---|---|---|---|
| Inkjet Based Bioprinting | Human umbilical vein endothelial cells (HUVECs) Rat Smooth muscle cells (SMCs) | • Heterocellular tissue patterned with <100 µm droplets. • Employs thermal, electromagnetic, or piezoelectric strategy for deposition of "ink" droplets • Rapid printing speeds with high resolution. • Potential to print biomaterials with low viscosity. • Availability and ease of using multiple bioinks. High-cellular viability and relatively less expensive | • Low material viscosity (<10 Pa.s). • Lack of precision with respect to droplet size. Requirement for low viscosity bioink. • Nozzle clogging and cellular distortion due to high-cell density. • Low mechanical strength. Inability to provide continuous stream of material | [48] |
| Extrusion Based Bioprinting | Human umbilical vein endothelial cells (HUVECs) Human umbilical vein smooth muscle cells (HUVSMCs), human bone marrow derived mesenchymal stem cells (hMSCs) Mouse embryonic fibroblasts (MEF) | • Coaxial extrusion enables >80 cm long vascular conduits of lumen diameter 1520 µm to be developed. Heterogenous tissues constructs can be created (>1 cm in thickness and 10 cm$^3$ volume). Multicellular spheroids (>400µm diameter) are bioprinted and double layered small diameter conduits of diameter 2.5 mm. • Potential for printing biomaterials with high cellular densities (higher than $1 \times 10^6$ cells/mL). • Continuous stream of material can be generated High viscosity bioinks such as polymers, clay-based substrates can be printed. | • Low printing resolution (>100 µm) and slow printing speeds. • Loss of cellular viability and distortion of cellular structure due to the pressure to expel the bioink. | [49] |
| | | • Microchannels of width > 100µm can be obtained. Fast printing speeds and potential to print biomaterials with broader viscosity gradient (1–300 mPa/s). • High precision and resolution (1 cell/droplet) can be achieved. • High density of cells can be printed- 108/mL | • Takes longer to generate—need to prepare reservoirs/ribbons. Low cellular viability compared to other techniques. • Thermal damage can cause loss of cells. • Intense UV radiation needed for SLA for crosslinking process. • Large amount of materials needed and high cost. • Longer post processing time and fewer materials have been found SLA-compatible. | [50,51] |

Abbreviations: HUVEC (Human Umbilical Vein Endothelial Cells); hMSCs (Human Mesenchymal Stem Cells); MEF (Mouse Embryonic fibroblasts); NIH ST3 (Murine Fibroblasts); SMCs (Smooth Muscle cells); SLA (Stereolithography) bioprinting.

Multi-Material Bioprinting

For printing heterogenous and complex tissues, traditional bioprinters have the limitation of deposition of single bio-ink formulation from the single nozzle or postprocessing through layer-by-layer deposition. Multi-material bioprinting integrating multi-material platforms for bioprinting heterogenous, multicellular and functional tissue constructs has recently gained attention. It is advantageous for concurrent or sequentially depositing different materials such as cell-laden hydrogels or extracellular matrix structures, sacrificial materials, and polymers for scaffolding with hierarchical microstructure. These have been very promising for native tissue biomimicry. A myriad of multiple-head multi-material bioprinting strategies have been established for the constructing multicellular and zonally stratified organization of blood vessels for cell depositing on exogenous or other biomaterials. Tan et al. employed multi-head multi materials bioprinting for generation of generated concentric and self-supporting tubular structures. The group used two extrusion-based bioprinting printheads; first printhead was employed to extrude the alginate-xanthan gum hydrogel blend bio-ink in a circular pattern and, second printhead was designed for extruding crosslinker solution into inner-side of the printed circular pattern for high mechanical stability of tube wall [52]. Another group, Campbell et al., extruded several hydrogels using integration of a single printhead equipped with a selector valve to switch between separate syringe pumps, to allow sequential and controlled biofabrication of heterogeneous, multilayered and multicellular complex vascular tissue structures [53]. Pre-crosslinked cell-laden alginate-collagen blends of specific viscosity used as bio-ink with EC-laden bio-ink. This was successfully deposited and sequentially surrounded by extrusion of SMC-laden bioink.

Besides using bio-ink-based bioprinting, a scaffold-free approach using multicellular spheroids and cylinders have been employed using for generating vascular constructs [54]. Forgacs et al. demonstrated the application of scaffold-free multi-material bioprinting multilayered and multicellular vascular tubes, as one printhead was employed for deposition of agarose rods (molding template) and other was a pre-set extruded multicellular spheroid or cylinder [55] The fabrication of vascular tubes with linear and bifurcated geometries has been employed. Furthermore, a double-layered vascular construct composed of inner layer (HUVSMCs cylinders) and outer layer (human dermal fibroblast (HDF) cylinders) was created to recapitulate layers of native blood vessels –tunica media and tunica adventitia, respectively. Kucukgul et al. generated a biomimetic macrovascular construct using an algorithmic model and successfully performed scaffold-free bioprinting of aortic tissue constructs on capillary-based extrusion. Human aorta of mouse embryonic fibroblast (MEF) aggregates and agarose structures from two separate printheads were imaged [56].

Microfluidic multi-material bioprinting techniques have also been employed for the fabrication of vascular structures [57,58]. Attalla et al. deposited several viscous hydrogels from a multiaxial microfluidic printhead to engineer tubular constructs with cell-laden bioinks and the crosslinker solution added using needles in the microfluidic chip and concentrically dispensed from the nozzle [57]. Zhou et al. bioprinted a vessel-like tubular construct by employing a capillary-based microfluidic printhead where cell-laden alginate was released from six outer channels of the multi-barrel capillary nozzle, and $CaCl_2$ was released from the central channel for crosslinking the bioink solution into a lumen [58]. Feng et al. developed a multicomponent bioprinting platform for biofabrication of artificial vessels where two alginate-based bioinks encapsulated with HUVECs and embryonic rat cardiomyocytes were extruded from the coaxial microfluidic printhead on a rotating material resulting in layer-by-layer fabrication of concentric ring structure [59]. Table 3 provides a comparison of various multi-material bioprinting approaches for the fabrication of vascular tissues.

**Table 3.** Multi-material bioprinting strategies for generation of vascularized tissues.

| Bioprinting Approach | Targeted Vascularized Tissue | Bioprinter Used | Bioink | Vascularization Impact | Ref |
|---|---|---|---|---|---|
| Multi-material bioprinting | Vascularized liver | Double nozzle printing system | ADSC-laden gelatin/alginate/fibrinogen Hepatocytes-laden gelatin/alginate/chitosan | Functional hepatocytes were formed with endothelial like structures in tissue construct. | [60] |
| | Vascularized bone | 3D-bioprinter with two controllable printheads | hMSCs laden gelatin-fibrinogen HUVEC laden gelatin-fibrinogen hydrogel | Osteogenic differentiation factors perfusion through vascular network resulted in osteogenic tissue formation. | [61] |
| | Vascularized cardiac patch | Multi-head extrusion-based 3D bioprinting | ECs within sacrificial gelatin CMs laden ECM bioink | Heart structure with mechanically stable and robust perfusable vessels | [62] |
| | Vascularized tissue model | 3D bioprinter with more than two controllable printheads | Fibroblast-cell laden GelMA EC injection through microchannels | Fabrication of vascularized tissue constructs. | [63] |
| Dual 3D bioprinting | SLA-based and extrusion-based bioprinting | Vascularized bone | ECs and hMSCs laden VEGF modified Gel MA-based bioink | Spatial controlled localization of growth factors and perfusion lead to interconnected vascularized bone construct. | [64] |
| | Extrusion and inkjet bioprinting | Vascularized skin | Adipose-derived dECM and fibrinogen bioink encapsulated human adipocytes Fibroblast cells laden skin dECM and fibrinogen | Formation of vascularized channels between dermis and hypodermis leads to maturation of epidermis with human like structure. | [65] |
| | Extrusion-based and SLA-based bioprinting platform | Multiphasic hybrid construct vascular conduit model | Cells encapsulated within PEGDA | Diffusion of media into cells resulted in a thick construct | [66] |
| | Co-axial and extrusion bioprinting platform | Vascular model | Human coronary artery SMCs laden modified Gel MA | Bioprinted vascular construct with biomechanics, perfusion ablility and permeability. | [67] |
| Co-axial Bioprinting | Coaxial nozzle bioprinting | Vascularized muscle | Endothelial cell-laden vascular dECM | Formation of pre-vascularized muscle with integration into the host tissue and functional recovery. | [68] |
| | Coaxial nozzle bioprinting | Perfusable renal tissue | Hybrid hydrogel bioink incorporated with kidney dECM and alginate | Renal proximal tube integrated into the host tissues in vivo | [69] |
| | Coaxial nozzle bioprinting | Vascularized intestinal villi | HUVEC extruded from core region of coaxial nozzle | Human intestine regeneration and organ-on-a-chip system | [70] |
| | Coaxial bioprinting platform | Vascularized tissue > 1 cm | Cell-laden GelMA Endothelial cell laden gelatin | Generation of tissue models | [71] |
| Light-based bioprinting | LIFT-Based bioprinting | Vascularized cardiac patch | Deposition of MSCs on a cardiac patch within ECs mesh structure | Pre-vascularized patches with enhanced angiogenesis | [72] |
| | DLP based-bioprinting | Vascularized thick tissue | Photopolymerizable glycidyl methacrylate-hyaluronic acid and GelMA | Fabrication of vascularized tissue constructs with high resolution. | [73] |

Abbreviations: EC (Endothelial cells); dECM (decellularized extracellular matrix); MSCs (Mesenchymal stem cells); Gel MA (Gelatin methacryloyl); HUVEC (Human vascular endothelial cells); PEGDA (Polyethylene glycol diacrylate); DLP (Digital Light Processing); SMCs (Smooth muscle cells); VEGF (Vascular endothelial growth factor); LIFT (Laser-induced forward transfer); CMs (Cardiomyocytes); ADSC (Adipose derived stem cells); SLA (Stereolithography).

Despite its capability for extruding meter-long vascular-like constructs, this approach is limited in the recapitulation of branched vascular tissues. Microfluidic multi-material bioprinting approaches, particularly the ones combined with coaxial nozzles, have the capability to recapitulate the native vascular tissues. Therefore, microfluidic printheads should be studied further to support the creation of freeform, multiscale vascular constructs in integration with the embedded bioprinting technique. Multi-material bioprinting platforms emerge as a powerful tool for replication of heterocellular and hierarchical composition of living tissues and organs needed for successful translation of engineered tissues and organs for clinical applications.

*4.2. Electrospinning*

Electrospinning employs electrical forces to generate nanofibers of a wide range of materials [74]. Nanofibers are generated using a polymeric solution injected from the syringe to a center of high electric field. As electrostatic forces get higher than the surface tension of the polymeric solution, this results in the formation of a Taylor cone with rapid acceleration of narrow jet towards the target (collector), connected to the ground with opposite charge [74]. In the past decade, electrospinning has been explored for generation of nano-fiber based microvasculature. This approach allows for fine control over diameter, porosity, and degradation rate. In addition, this technique results in formation of fibers having diameter similar to native ECM (50–500 nm) and mimic natural topographical cues [75]. Bioink deposited remade nanofibers have been designed for macroscale hydrogel constructs with nanoscale spatial control. Integration of electrospinning with 3D fiber deposition has been used to generate multiscale scaffold of a PEG/poly (butylene terephthalate) (PBT) block copolymer of native microarchitecture. In vitro studies revealed long term viability and high metabolic rate of human mesenchymal stromal cells cultured on these scaffolds arranged aligned along the scaffold [76]. This study demonstrated the application of multiscale, multi-compound, and multifunctional engineered tissues to recapitulate the complex native tissues.

In another study, Kim et al. studied endothelial differentiation of induced pluripotent stem cells (iPSCs) with topographically aligned 3D electrospun PCL scaffolds to produce iPSC-derived ECs (iPSC-ECs) [77] (Figure 2). The group reported enhanced gene expression of EC phenotypic markers CD31, CD144, and nitric oxide synthase within 3D scaffolds, compared to on 2D PCL films. Parallel-aligned vascular-like networks cultured with iPSC-ECs displayed 70% longer branch length in comparison to randomly oriented scaffolds (Figure 2C) This study revealed ability of fiber topography mechanism for modulating vascular network-like formation and patterning of structures. Wanjare et.al engineered cardiovascular tissues and demonstrated a dominant role of scaffold anisotropy to maintain human induced pluripotent stem cell-derived organization of cardiomyocytes (iCMs) and contractile function [78]. In another example, Kenar et al. blended poly(L-lactide-co-ε-caprolactone) (PCL) with collagen and hyaluronic acid and designed a micro-fibrous composite scaffold with enhanced length of vasculature [79]. Furthermore, Cui et al. demonstrated improved pre-vascularization of constructs by pre-seeding HUVEC on LBL aligned (PCL)/cellulose nanofiber matrices pre- implantation [80]. In vivo, the aligned fiber matrices integrated with the host vasculature. Electrospinning has the potential to produce matrices capable of mimicking ECM for enhanced in vivo angiogenesis through the fiber spatial organization, further highlighting the scope of the technique to improve vascularization.

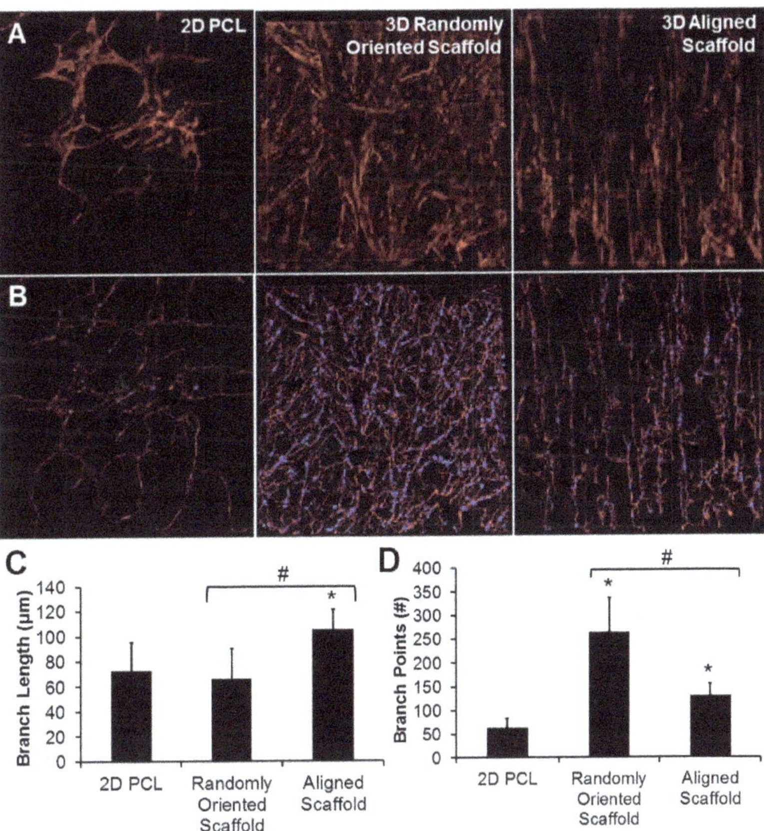



**Figure 2.** Vascular network-like formation in 3D microfibrous scaffolds. (**A**) 3D stacked confocal images of CD31 staining in 2D PCL film, 3D randomly oriented scaffold, and 3D aligned scaffold. (**B**) Transformation of CD31 expression into skeletonized filaments; (**C,D**) Quantification of branch length (**C**) and branch points (**D**). * indicates statistically significant relationship to 2D polycaprolactone (PCL) film, and # indicates statistically significant comparison between 3D groups. $p < 0.05$ (n = 5). Reproduced with permission from [77].

### 4.3. Patterning of Bioactive Molecules

Biofabrication methods can be employed to pattern bioactive molecules within scaffolds to emulate biochemical gradients present in natural angiogenesis and promote in situ vascularization via integration with the host vascular network. Owing to its potent effect on angiogenesis, VEGF is the most commonly employed growth factor for the patterning of scaffolds. For example, Alsop et al. printed VEGF in a spatially defined manner onto a collagen-glycosaminoglycan scaffold by employing photolithography. The group reported high cell infiltration and immature vascular networks [81]. Directional vessel growth via pre-defined release of VEGF have been induced by using hydrogels [82]. Promotion of aligned vasculature as the hydrogel is printed parallel to existing vasculature and not perpendicular depicts the precise control. Studies have also incorporated growth factor combinations into the scaffold material for recapitulating the sequential stages of angiogenesis. Combinations of VEGF, FGF, and BMP2, and with VEGF and Angiopoietin have been studied to improve angiogenesis [83]. Even with the use of multiple growth factors, the platforms are still relatively basic compared to the complex process of angiogenesis. Incorporation of bioactive molecules in the scaffold material does not adequately ensure its spatial localization due to diffusion and burst release. Therefore, for development of functionalized decellularized scaffolds, several strategies have been employed to use heparin via endpoint attachment. This approach has been successful for binding and controlled release of heparin-binding growth factors, for example, VEGF, for enhanced angiogenesis [84]. Another group printed biodegradable polymer scaffold with concurrent zones of VEGF and VEGF inhibitors and achieved spatially restricted signaling [85].

Patterning bioactive molecules within scaffolds is another way for mimicking biochemical gradients or promoting in situ vascularization as a result off integration with the host vascular network. This spatial micropatterning approach to formation of vascular network can achieve a spatial resolution of less than 10 μm. This technique can engineer spatially organized ECs using microfabrication technologies, such as soft lithography and photopolymerization [86]. The steps to soft lithography include (1) design of pattern; (2) photomask and master fabrication; (3) fabrication of polydimethylsiloxane (PDMS) stamp; and (4) micro- and nano-structure fabrication employing the stamp. Raghavan et al. employed soft lithography techniques on microfabricated PDMS templates with intended geometries [87]. Spatially arranged endothelial cords were formed using a suspension of ECs in collagen gel introduced into the channel and stimulated with VEGF and bFGF. Baranski et al. studied micropatterned EC cords with human hepatocytes implanted into nude mice. Implanted cords directed rapid vascularization, and anastomosis of the cords with the host vasculature. [88] Another strategy for the controlled release of bioactive molecules is encapsulation. Encapsulation also provides a layer of protection for growth factors, improving their short half-life. Minardi et al. achieved the spatiotemporal release of proteins by patterning the scaffold with composite microspheres [89]. In addition, Nazarnezhad et al. produced a scaffold to achieve concurrent release of VEGF, PDGF, FGF, and EGF by encapsulation via nanofibers and gelatin nanoparticles [90]. Due to gradual release of these growth factors sustained for 45 days, enhanced endothelial cell proliferation and development of vascular-like structures was observed.

Along with growth factors, scaffolds have been functionalized with peptides to induce vascular growth and network formation. Covalent binding of peptides to the scaffold helps to pattern peptides on to surfaces and scaffolds with relative resilience to processing. These peptides possess greater flexibility than growth factors and can incorporate angiogenic domains. Lei et al. micropatterned SVVYGLR peptide strips on polymer surfaces using photolithography. Directional regulation and morphogenesis of ECs cultured onto 10, and 50 μm aligned peptide strips generated tubular structures [91]. Chow et al. employed peptide-PCL conjugates with selective affinity for glycosaminoglycans (GAGs), in combination with sequential electrospinning techniques, to direct the spatial patterning of GAGs all through the scaffold [92]. This helps to protect the bioactivity of the GAGs and native ECM, for a more clinically translatable tissue structure. These studies underline the importance of spatial organization in vascularization strategies. Angiogenesis has been studied to be modulated by micropatterning strong mechanical forces, as convex part of micropatterned vessel walls allows for preferential blood vessel formation. Huang's group demonstrated parallel-aligned micropatterned channels as well as nanopatterned collagen scaffolds to promote the organization and migration of ECs [93]. Laminar flow applied to EC-seeded parallel-aligned nanofibrillar collagen scaffolds and orthogonal to the direction of collagen patterning, the cellular population preferentially remained organized along the spatial patterning direction.

Patterning of biomolecules shows great promise for promoting vascularization in engineered scaffolds and tissues owing to the high spatial precision of micropatterning techniques, Biomolecules such as peptides can be incorporated into micropatterned or nanopatterned scaffolds (Table 4). The limitations of random distribution of growth factors have been addressed by biomolecules encapsulation pre-patterning of scaffolds. The use of nanoparticles is being explored for strict control on biomolecule patterning and release. In addition, as the size-scale of spatial micropatterned substrates is limited, hence, strategies that allow generation of larger-scale vascular networks are actively studied.

**Table 4.** Fabrication techniques/structures to promote vascularization in tissue engineered constructs.

| Technique/Structures | Application | Limitations | Ref |
|---|---|---|---|
| 3D Bioprinting | • Developed tissue constructs mimic the spatial, mechanochemical, and temporal characteristics of native tissues<br>• Microchannels of width > 100 μm can be obtained<br>• Heterogenous tissues constructs can be created (>1 cm in thickness and 10 cm3 volume). Multicellular spheroids (>400 μm diameter) are bioprinted and double layered small diameter conduits of diameter 2.5 mm.<br>• High accuracy and reproducibility<br>• High precision in 3D structure<br>• Modularity of bio-inks | • Print resolution is limited<br>• Print size limited to diffusion | [94] |
| Micropatterning | • Promotes cell alignment and cell density<br>• High reproducibility<br>• Can be integrated with other techniques. | • Limited complexity of organized tissue.<br>• Constructs unable to be implanted<br>• Size scale is limited | [95] |
| Hydrogel | • Biocompatible<br>• Can match tissue stiffness | • Limited cell directionality<br>• Fragile construct | [96] |
| Electrospinning | • High reproducibility<br>• Relatively low cost<br>• Cellular alignment maintained | • Low biocompatibility<br>• Limited tissue complexity. | [97] |
| Decellularized Scaffolds | • Recapitulate 3D organ specific architecture<br>• Native vascular network is largely preserved<br>• Low cytotoxicity | • Limited efficiency<br>• Limited tissue/organ donor availability<br>• Antigenicity from xenogenic tissues | [98] |
| Tissue Engineered Heart | • Constructs have native myocardial structure<br>• Cardiomyocyte contractility is maintained | • Low apparatus modularity<br>• Restricted applications | [99] |
| Scaffold-free Engineering | • High reproducibility and efficiency<br>• Physiological Cell–cell interaction<br>• Controlled growth factor release | • Limited accessibility<br>• Restricted applications.<br>• Lack of precision in network architecture | [100] |

Abbreviations: 3D (Three dimensional).

## 5. Spatiotemporal Regulation of Engineering Vascularized Cardiac Patches

*5.1. Vascularized Cardiac Patch with Temporal Regulation*

5.1.1. Engineering Vascularized Patch with Temporal Regulation In Vitro

Tissue engineering encompasses principles of engineering and biology for the generation of living tissues studied for drug screening, disease modeling, and therapeutic regeneration. Techniques to reprogram human somatic cells into iPSCs and differentiation into cardiomyocytes and other cardiac cells have been extensively studied to be efficient, which has led to accelerated progress towards the generation of engineered human cardiac muscle patch (hCMP) and heart tissue constructs [101,102]. Traditional methods for hCMP fabrication involve suspending cells within biocompatible material scaffolds or culture of two-dimensional sheets to form multilayered constructs. Recently, spatiotemporal techniques such as micropatterning and three-dimensional bioprinting have been employed to generate hCMP architectures at unprecedented spatiotemporal resolution [102]. One limitation of hCMP-based strategies for in vivo tissue repair is inadequate scalability, poor integration and engraftment rate, and the lack of functional vascular networks. Therefore, cardiac patches must be designed to allow assimilation with the host myocardium and synchronization. Porous scaffolds have been widely studied as a promising biomaterial as the architecture provides appropriate directions for cells for matrix penetration. This strategy can induce adequate rate of biomaterial degradation for new tissue reconstruction with improved nutrient supply and electrophysiological mediated integration. For instance, improved in vivo vascularization within the patch was achieved using VEGF-containing scaffolds [103]. Pre-vascularization of cardiac patch scaffolds has also been studied to improve mass transport. MSCs, through the release of angiogenic factors have been shown to support the formation of microvessels and their structure. ECM nanofibers and MSCs were shown to promote vascular constructs as a sheet when co-cultured with ECs. Shevach et al. suggested decorating decellularized matrices with gold nanoparticles and nanowires for improved electrical coupling, presenting stronger contractile force and lower excitation frequency [104].

Induced pluripotent stem cells (iPSCs) have gained attention as a key component of cardiac tissue engineering for understanding of cardiovascular disease mechanisms, drug responses, and developmental processes in human 3D tissue models. A wide range of iPSC-derived cardiac spheroids, organoids, and heart-on-a-chip models have been developed since the very first engineered tissue was fabricated more than two decades ago. The iPSC-derived cardiovascular cells can be differentiated by soluble factors (e.g., small molecules), extracellular matrix scaffolds, and exogenous biophysical maturation cues [105]. Efficient cardiomyocyte (CM) differentiation protocols, in combination with advancements in engineered biomaterials and organ-on-a-chip technology, have led to a variety of in vitro cardiac tissue models, ranging from spheroids and organoids to transplantable cardiac patches and 3D-bioprinted hearts. Protocols for differentiation into other major cardiovascular cell types (iPSC-derived endothelial cells, iPSC-ECs and iPSC-derived vascular smooth muscle cells) have been extensively studied in the past decade [106–109]. The major advantage of incorporating various iPSC derived cardiovascular cell types is to generate a more physiological construct, as demonstrated by the improved structural and functional maturity of multi-cell type microtissues. Multi-cellularity also enables researchers to study pathogenic mechanisms and drug responses to a specific cell that provides a versatile tool to study intercellular communication mechanisms (e.g., paracrine or contact-mediated). Despite the promise of CMs derived from human induced pluripotent stem cells (hiPSCs), these cells are found to be functionally immature and exhibit fetal-like features. To improve CM maturation, Mummery et al. showed that the tri-cellular combination of hiPSC-derived CMs, cardiac fibroblasts, and iPSC-ECs could enhance CM maturation in scaffold-free, three-dimensional microtissues [105,106]. Integration of engineered biomaterials with various microfabricated devices, stretch, and electrical circuits with conventional 2D approaches is being studied to overcome these limitations [110,111].

Biological materials have been extensively used as drug delivery vehicles. Various polymeric materials have been successfully studied to encapsulate or entrap biomolecular components resulting in low-dimension particles (microns to sub-nano scale). Such particles enable the delivery of soluble and insoluble bioactive molecules to the target site, providing enhanced stability, drug half-life, pharmacokinetics, and drug specificity. Drug delivery platforms could also be fabricated from biomaterials incorporated with delivery agents. Different fabrication methods and chemical formulations could be designed for tailoring the mechanical properties of biomaterials. These matrices are used for the controlled release of drugs, depending on factors such as degradation/erosion rate, triggers, or environment factors. Neighboring cells in the natural microenvironment communicate with each other via paracrine pathways and factors mediated by proteins, small RNA molecules, and extracellular vesicles (EVs). ECM acts as a reservoir of signaling components, the incorporation of proteins and protein-binding features into biomaterials could mimic ECM function and induce cellular responses, such as cell proliferation, migration, and differentiation [112]. Therefore, ECM-based biomaterials are potential candidates as advanced drug delivery systems for spatial–temporal presentation and delivery of therapeutic drugs, imperative for endogenous cardiac tissue regeneration and to restore functionality post cardiac injuries.

### 5.1.2. Engineering Vascularized Patch with Temporal Regulation In Vivo

For localized and temporary delivery of bioactive molecules, controlled and sustained release systems have been designed. Polymeric materials, forming 2D and 3D matrices, have been extensively studied. As endogenous cardiac regeneration strategies usually target localized action, delivery systems impregnated within hydrogels and cardiac patches have gained interest. Injectable hydrogels have been shown to enhance cell survival and attenuate fibrotic responses immediately after myocardial infarction. Ruvinov et al. demonstrated improved cardiac regeneration by facilitating the release of two GFs-IGF-1 (considered cardioprotective) and hepatocyte growth factor (HGF; considered anti-fibrotic) [113]. The group used injectable alginate hydrogel capable of binding to growth factors with affinity-binding AlgS [114]. The injection of the proposed hydrogel into infarcted rat hearts resulted in reduced myocyte apoptosis and fibrosis, and cardiomyocyte proliferation attenuated infarct tissue. Cell-free heart patches have also been extensively studied to effectively induce endogenous cardiac regeneration such as the "paracrine effect". Cell-based therapies to improve cardiac regeneration involve secretion of cardioprotective factors that signal cells in the infarcted area. In one study, Jeske et al. hypothesized that iPSC- or iPSC-CM-derived EVs (i.e., microvesicles and exosomes) revealed similar effect [115]. The group studied and isolated these EVs in vitro for studying the function of miRNA content. The isolated EVs were encapsulated in a cell-free collagen hydrogel for their tendency to get rapidly consumed by recipient cells, to increase treatment efficacy. Prolonged release of EVs was achieved using a collagen hydrogel for up to 1 week in vivo. In a rat myocardial infarction model, the collagen hydrogel patch reduced scar formation and apoptosis of CMs and enhanced recovery of contractile functions. This system has the advantages of being independent of any cellular component, low risk of immunogenicity and optimal cellular viability and retention. In the following section, spatial regulation is discussed for engineered cardiac patches.

### *5.2. Engineering Vascularized Patch with Spatial Regulation*

### 5.2.1. Engineering Vascularized Patch with Spatial Regulation In Vitro

Multiscale architectures can be precisely engineered using capillary force lithography to recapitulate the topographical and ECM cues. Cellular morphology and directionality can be regulated by nanopatterning of materials through UV-assisted capillary force lithography technique that can impart ECM cues needed. Structures and directions of fibroblast cells have been studied to be predominantly influenced by nano-topography instead of microtopography [116]. Further, micro/nanopatterned transplantable patches composed of PLGA have been fabricated using capillary force lithography/wrinkling combinatory technique. The multiscale PLGA patches displayed augmented tissue adhesion to the underlying native tissue and optimal mechanical strength compared to the merely nanopatterned counterparts [117]. This combinatory strategy can greatly benefit by manipulating the substrate of cell culture to govern cell fate and functionality. Clinically relevant size constructs of hCMPs have been generated. However, patches with relatively larger surface areas (e.g., 8 cm$^2$) are comparatively thin (1.25 mm), leading to limited direct perfusion and limited thickness of hCMP to 1–2 mm. Hence, larger and thicker hCMP constructs have been optimized and improved. Engineering thick and viable hCMP generation is limited by inadequate recapitulation of characteristics of native myocardium, such as generation of optimum forces and action potentials. hCMP thickness is limited by the oxygen and nutrients from the vascular network post-transplantation, necessitating the cardiomyocytes to be within 100–200 μm distance from the capillaries [118]. hCMPs of optimum thicknesses need the formation of a dense internal vascular network that integrates with native circulation post-transplantation. Vascularization can be enhanced by including a combination of vascular and other cell types (ECs, SMCs, fibroblasts) and also employing nanoparticle-mediated extended release of pro-angiogenic factors, e.g., vascular endothelial growth factor (VEGF), fibroblast growth factor (FGF), and the Wnt activator CHIR99021 for infiltration of the native circulatory loop [119]. In addition, advanced biofabrication methods (e.g., micropatterning and 3D bioprinting) can be used to control the spatial orientation of the vascular network to improve mass transport and perfusion. These examples demonstrate the various techniques that can be used to induce spatial patterned vascularized patches.

### 5.2.2. Engineering Vascularized Patch with Spatial Regulation In Vivo

In the beginning, tissue printing was employed for generation of constructs human cardiac-derived CPCs and alginate with cardiac lineage commitment and high viability for 7 days in culture [120]. In subsequent experiments, human cardiomyocyte progenitor cells were printed in six perpendicularly printed layers into a hyaluronic acid matrix, and gelatin formed a patch of 4 cm$^2$ surface area. Improved measures of cardiac function were observed with increased expression of cardiac and vascular differentiation markers within 4 weeks in a murine MI (myocardial infarction) model [121]. Scaffold-free bioprinted hCMPs have been generated using loaded spheroids onto an array of needles, fused and hCMP cultured as needle holes got filled with surrounding tissue [122]. The construct remained engrafted and displayed vascularization for 7 days after implantation into infarcted rat hearts. Recently, a customized device has been developed for simultaneous loading of layer of spheroids into the needle array, substantially improving print time for larger engineered constructs. An advanced technique called multiphoton-excited (MPE) 3D bioprinting has been employed to improve limitation of low resolution of traditional bioprinting strategies that limit printing structural details to facilitate cellular interactions. This technique controls the architecture of photoactive polymers to reproduce the structural features of the ECM with high fidelity. MPE 3D-printed hCMPs composed of iPSC-derived CMs, ECs, and SMCs in a gelatin scaffold generated calcium transients post fabrication and beat synchronously within 1 day [123]. The printed patches exhibited significant improvements in a murine MI model with enhanced cardiac function (left ventricular ejection fraction and fractional shortening), apoptosis, vascularity, and cell growth. Photoactivated 3D bioprinting has also been studied with a bioink incorporating both ECM proteins and hiPSCs for fabrication of two-chambered structures with both inlet and outlet vessels. Mature cardiac cells were formed from proliferation and differentiation of hiPSCs in situ with human cardiac muscle recapitulating the chambers and large vessels of a human heart [124].

Elaborate research has been conducted for replacement of infarcted cardiac tissue with tissue engineered cardiac patches generated from biocompatible and bioabsorbable components including purified ECM molecules and heterogeneous mixtures of ECM molecules. Jang et al. generated a 3D prevascularized stem cell patch using the spatial arrangement of cardiac progenitor/MSCs with decellularized ECM bio-ink [125]. The cardiac patch reduced fibrosis and cardiac remodeling with enhanced cardiomyogenesis and neovascularization at the injured myocardium post-transplantation. Gao et al. successfully 3D printed an EPC/atorvastatin-loaded PLGA microspheres laden bioink (vascular tissue-derived ECM and alginate) and bio-blood vessel [126]. The engineered tissue exhibited enhanced viability, proliferation, and differentiation of endothelial progenitor cells (EPCs) with improved in vitro endothelialization. The bioblood vessel (BBV)-based technique showed significantly improved EPC function and recovery of ischemic injury in a nude mice hind limb ischemia model. We have previously shown that electrospun aligned microfiber scaffolds could be used for the co-culture of iPSC-derived cardiomyocytes and endothelial cells [78] Upon implantation in vivo, the patches composed of aligned scaffolds induced the formation of microvasculature that were preferentially aligned along the direction of the scaffold microfibers [127] (Figure 3).

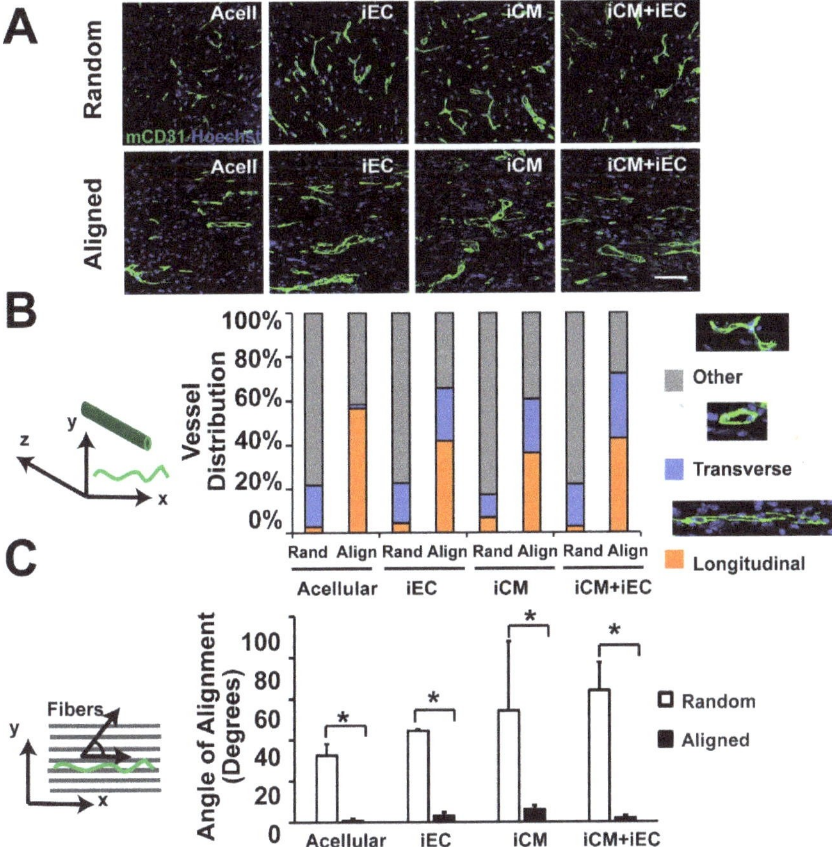


**Figure 3.** Vascularization of engineered myocardial tissue following subcutaneous implantation into mice. (**A**) Representative confocal microscopy images of CD31 staining (green) within en face sections of engineered myocardial tissues derived from randomly oriented or aligned nanofibrillar scaffolds containing iPSC-derived cardiomyocytes (iCMs), iPSC-derived endothelial cells (iECs), or iCM + iECs at 2 weeks after subcutaneous implantation. Acell denotes acellular scaffold. (**B**) Distribution of vessel orientations within explanted engineered myocardial tissue, relative to the axis of the aligned microfibers as longitudinal, transverse, or other. (**C**) Quantification of the global angle of vessel alignment within subcutaneously explanted engineered myocardial tissues, relative to the axis of the aligned microfibers. The global angle of vessel alignment is calculated as the angle formed by the direction of the longitudinally oriented vessel with respect to the axis of the aligned microfibers. For randomly oriented scaffolds, an arbitrary axis was selected ($n \geq 3$). Reprinted with permission from [127]. * Denotes statistically significant in comparison ($p < 0.05$).

In terms of vascularization of patches, apart from infiltration from the native vasculature, thicker hCMPs will likely need engineered vascularity pre-transplantation. Vascularization has been studied to be induced during the fabrication process by encapsulation of a sacrificial gelatin mesh in scaffold material, with subsequent melting of the gelatin mesh to produce a network of interconnected microfluidic channels. The sacrificial scaffolds generated a rudimentary endothelial network when seeded with human microvascular ECs. An alternative strategy using the sustained release of the angiogenic factor thymosin β to promote the outgrowth of vessels from explanted veins and arteries, forming a capillary bed within a hydrogel scaffold mimics the endogenous angiogenic process [128]. Vessel growth can also be induced with micropatterned polyglycerol sebacate scaffolds as they degrade post transplantation with infiltration of host blood cells into the microvessels [129]. Micropatterning has also been employed for organization of ECs into 'cords' that induce the capillary for integration with the host tissue post-transplantation. A recent study of 3D printed vessels using thermal inkjet printer for bioprinting human microvascular ECs and fibrin, resulting in generation of micro-sized fibrin channels lined with confluent cells [130]. Vasculature has also been bioprinted with an advanced extrusion platform that generated a sheath of photoactive, cell-laden bioink around an alginate core structure [131]. The alginate was dissolved with a $Ca^{2+}$ chelating agent post UV crosslinking, allowing the cellular population to proliferate and spread forming a perfusable biomimetic vasculature. However, due to limited penetration of UV radiation, in-depth polymerization can be induced via enzymatic reactions, such as the conversion of fibrinogen into fibrin with thrombin as a catalyst [132]. Thick (>1 cm) engineered osteogenic tissues [133] have been generated using the technique of co-printing of vascular and cellular inks in cast ECM material.

## 6. Challenges and Future Prospects

Despite the significant advancements in the field, the generation of thick tissues with functional and mature microvascular networks in vitro faces major challenges for successful clinical tissue/organ translation. Although successful vascularization attempts have made remarkable progress for implantable 3D constructs at clinically relevant scale, building vascular networks that mimic the complexity, microstructure, geometry, biochemical cues, and optimal organ cellular density remains a challenge. Furthermore, the appropriate and timely vascularization of the implanted 3D constructs needs more attention. Direct anastomosis of host microvascular network and preformed microvasculature as reperfusion process for clinically scalable 3D constructs is challenging. Therefore, newer techniques for rapid vascularization are required.

A prominent challenge in angiogenic therapy is the risk of undesired and uncontrolled tissue ingrowth. Significant challenges to fully utilize the potency of angiogenic growth factors therapy include precisely controlling in vivo distribution of growth factor dose and time-duration of bioactivity. Such optimizations are necessary to obviate unstable vessel formation and subsequent regression, hemangioma formation, or neointimal thickening.

Fabrication techniques for sustained release of growth factors from biomimetic scaffolds should render engineered scaffolds with optimal physical properties (e.g., pore size, water content, porosity, the interconnection of pores, etc.) for adequate vascularization. Cells cultured within matrices for generation of vascular structures by themselves result in unwanted architecture and patterns. Advanced techniques like electrospinning, patterning, and 3D printing can be combined to provide a scaffold-guided path for the cellular population [134]. Mimicking the natural hierarchical structure of tissue is imperative for engineered vascularization. Since the effectiveness of bioactive factors vary in vitro and in vivo environments, optimal properties of growth factor delivery from scaffolds should be carefully determined. Along with the need to improve in vitro vascularization strategies is the need to generate immune-evasive cells that can be genetically modified to prevent rejection. Successful implantation and anastomosis of vascularized tissues will further require advanced microsurgical expertise. Although technological advances have now led to the formation of functional and mature vasculature, some challenges remain. Currently, no single vascularization approach can produce a functional, bioactive, stable, and scalable vascular structure, although thin, simple vascular networks have been generated successfully. An optimized approach consisting of a tailored, synergistic combination of several tissue engineering techniques (cells, decellularized tissue, and growth molecules delivery) and inter-disciplinary systems (functionalized biomaterials and fabrication methods) will allow us to engineer improved vascular networks for the development of scalable vascularized 3D tissues.

With respect to the clinical translation of cardiac patches, successful translation will require tissue integration with the surrounding myocardium at three levels: physical and biochemical continuity, electrophysiological cues, and nutrient perfusion. Even though cardiac patches have been successful in improving cell viability and retention, there is a high risk of cells engrafted through such constructs or through intracardiac injection to provoke an immunogenic response, to result in immune rejection of the allograft [135]. In addition, cardiac patch transplantation not accompanied by immunosuppressants can significantly risk transplanted cell survival rate and can cause failing integration. Accessibility for cell migration from areas close to the infarcted zones needs to be allowed by optimal design for formation of blood vessels and nerves to integrate with the host. The presence of a fibrotic scar barrier results in failure of electrical integration with the host. Hence, these limitations must be addressed by cardiac patch designs to be clinically relevant for assimilation with the host myocardium and synchronization over large distances. Despite these challenges, it is envisioned that these limitations will be successfully overcome with time, and the successful implementation of scalable vascularized tissues will become reality in the future.

**Author Contributions:** Conceptualization, A.K. and N.F.H.; writing—original draft preparation, A.K.; writing—review and editing, A.K., B.P.O. and N.F.H. All authors have read and agreed to the published version of the manuscript.

**Funding:** This work was supported in part by grants to NFH from the US National Institutes of Health (R01 HL127113 and R01 HL142718), the US Department of Veterans Affairs (1I01BX002310, 1I01BX004259, RX001222), the National Science Foundation (1829534), California Institute for Regenerative Medicine (10603), and the American Heart Association (20IPA35360085 and 20IPA35310731). Schematic was created using BioRender.

**Conflicts of Interest:** The authors declare no conflict of interest.

## References


1. Tsao, C.W.; Aday, A.W.; Almarzooq, Z.I.; Alonso, A.; Beaton, A.Z.; Bittencourt, M.S.; Boehme, A.K.; Buxton, A.E.; Carson, A.P.; Commodore-Mensah, Y.; et al. Heart Disease and Stroke Statistics—2022 Update: A Report From the American Heart Association. *Circulation* **2022**, *145*, e153. [CrossRef] [PubMed]
2. Duan, X.; Liu, X.; Zhan, Z. Metabolic Regulation of Cardiac Regeneration. *Front. Cardiovasc. Med.* **2022**, *9*, 933060. [CrossRef] [PubMed]
3. Talman, V.; Ruskoaho, H. Cardiac fibrosis in myocardial infarction—From repair and remodeling to regeneration. *Cell Tissue Res.* **2016**, *365*, 563–581. [CrossRef] [PubMed]
4. Khanna, A. *Fabrication of Human Serum Albumin Film for Enhanced Hemocompatibility and Mitigation of Neointimal Hyperplasia Under Physiologically Relevant Flow Shear Conditions*; Clemson University: Clemson, SC, USA, 2017.
5. Khanna, A.; Luzinov, I.; Burtovvy, R.; Vatansever, F.; Langan, E., III; LaBerge, M. Fabrication of Human Serum Albumin film on expanded polytetrafluoroethylene (e-PTFE) for Enhanced Hemocompatibility and Adhesion Strength. In Proceedings of the Society for Biomaterials, Minneapolis, MN, USA, April 2017; p. 810.
6. Masson-Meyers, D.S.; Tayebi, L. Vascularization strategies in tissue engineering approaches for soft tissue repair. *J. Tissue Eng. Regen. Med.* **2021**, *15*, 747–762. [CrossRef] [PubMed]
7. Mastrullo, V.; Cathery, W.; Velliou, E.; Madeddu, P.; Campagnolo, P. Angiogenesis in Tissue Engineering: As Nature Intended? *Front. Bioeng. Biotechnol.* **2020**, *8*, 188. [CrossRef] [PubMed]
8. Cathery, W.; Faulkner, A.; Maselli, D.; Madeddu, P. Concise Review: The Regenerative Journey of Pericytes Toward Clinical Translation. *Stem Cells* **2018**, *36*, 1295–1310. [CrossRef] [PubMed]
9. Nitzsche, B.; Rong, W.W.; Goede, A.; Hoffmann, B.; Scarpa, F.; Kuebler, W.M.; Secomb, T.W.; Pries, A.R. Coalescent angiogenesis-evidence for a novel concept of vascular network maturation. *Angiogenesis* **2022**, *25*, 35. [CrossRef]
10. Peluzzo, A.M.; Autieri, M.V. Challenging the Paradigm: Anti-Inflammatory Interleukins and Angiogenesis. *Cells* **2022**, *11*, 587. [CrossRef]
11. Trindade, A.; Duarte, A. Notch Signaling Function in the Angiocrine Regulation of Tumor Development. *Cells* **2020**, *9*, 2467. [CrossRef]
12. Crosby, C.; Zoldan, J. Mimicking the physical cues of the ECM in angiogenic biomaterials. *Regen. Biomater.* **2019**, *6*, 61–73. [CrossRef]
13. Kant, R.J.; Coulombe, K.L. Integrated approaches to spatiotemporally directing angiogenesis in host and engineered tissues. *Acta Biomater.* **2018**, *69*, 42–62. [CrossRef] [PubMed]
14. Uccelli, A.; Wolff, T.; Valente, P.; Di Maggio, N.; Pellegrino, M.; Gürke, L.; Banfi, A.; Gianni-Barrera, R. Vascular endothelial growth factor biology for regenerative angiogenesis. *Swiss Med. Wkly.* **2019**, *149*, w20011. [CrossRef] [PubMed]
15. Campinho, P.; Vilfan, A.; Vermot, J. Blood Flow Forces in Shaping the Vascular System: A Focus on Endothelial Cell Behavior. *Front. Physiol.* **2020**, *11*, 552. [CrossRef]

16. Boyd, D.F.; Thomas, P.G. Towards integrating extracellular matrix and immunological pathways. *Cytokine* **2017**, *98*, 79–86. [CrossRef]
17. Ucuzian, A.A.; Gassman, A.A.; East, A.T.; Greisler, H.P. Molecular Mediators of Angiogenesis. *J. Burn Care Res.* **2010**, *31*, 158–175. [CrossRef] [PubMed]
18. Fallah, A.; Sadeghinia, A.; Kahroba, H.; Samadi, A.; Heidari, H.R.; Bradaran, B.; Zeinali, S.; Molavi, O. Therapeutic targeting of angiogenesis molecular pathways in angiogenesis-dependent diseases. *Biomed. Pharmacother.* **2018**, *110*, 775–785. [CrossRef]
19. Pepper, M.S. Manipulating Angiogenesis. *Arter. Thromb. Vasc. Biol.* **1997**, *17*, 605–619. [CrossRef]
20. Nazeer, M.A.; Karaoglu, I.C.; Ozer, O.; Albayrak, C.; Kizilel, S. Neovascularization of engineered tissues for clinical translation: Where we are, where we should be? *APL Bioeng.* **2021**, *5*, 021503. [CrossRef]
21. Jacobo, S.M.P.; Kazlauskas, A. Insulin-like Growth Factor 1 (IGF-1) Stabilizes Nascent Blood Vessels. *J. Biol. Chem.* **2015**, *290*, 6349–6360. [CrossRef]
22. Peters, E.B. Endothelial Progenitor Cells for the Vascularization of Engineered Tissues. *Tissue Eng. Part B Rev.* **2018**, *24*, 1–24. [CrossRef]
23. Zhang, Y.; Liu, J.; Zou, T.; Qi, Y.; Yi, B.; Dissanayaka, W.L.; Zhang, C. DPSCs treated by TGF-β1 regulate angiogenic sprouting of three-dimensionally co-cultured HUVECs and DPSCs through VEGF-Ang-Tie2 signaling. *Stem Cell Res. Ther.* **2021**, *12*, 281. [CrossRef] [PubMed]
24. Zhang, L.-L.; Xiong, Y.-Y.; Yang, Y.-J. The Vital Roles of Mesenchymal Stem Cells and the Derived Extracellular Vesicles in Promoting Angiogenesis After Acute Myocardial Infarction. *Stem Cells Dev.* **2021**, *30*, 561–577. [CrossRef]
25. Dierick, F.; Solinc, J.; Bignard, J.; Soubrier, F.; Nadaud, S. Progenitor/Stem Cells in Vascular Remodeling during Pulmonary Arterial Hypertension. *Cells* **2021**, *10*, 1338. [CrossRef]
26. Ren, X.; Zhao, M.; Lash, B.; Martino, M.M.; Julier, Z. Growth Factor Engineering Strategies for Regenerative Medicine Applications. *Front. Bioeng. Biotechnol.* **2020**, *7*, 469. [CrossRef] [PubMed]
27. Chu, H.; Wang, Y. Therapeutic angiogenesis: Controlled delivery of angiogenic factors. *Ther. Deliv.* **2012**, *3*, 693–714. [CrossRef] [PubMed]
28. Ikegami, Y.; Mizumachi, H.; Yoshida, K.; Ijima, H. Heparin-conjugated collagen as a potent growth factor-localizing and stabilizing scaffold for regenerative medicine. *Regen. Ther.* **2020**, *5*, 236–242. [CrossRef] [PubMed]
29. Wang, Z.; Wang, Z.; Lu, W.W.; Zhen, W.; Yang, D.; Peng, S. Novel biomaterial strategies for controlled growth factor delivery for biomedical applications. *NPG Asia Mater.* **2017**, *9*, e435. [CrossRef]
30. Martino, M.M.; Brkic, S.; Bovo, E.; Burger, M.; Schaefer, D.J.; Wolff, T.; Gürke, L.; Briquez, P.S.; Larsson, H.M.; Gianni-Barrera, R.; et al. Extracellular Matrix and Growth Factor Engineering for Controlled Angiogenesis in Regenerative Medicine. *Front. Bioeng. Biotechnol.* **2015**, *3*, 45. [CrossRef]
31. Tallawi, M.; Rosellini, E.; Barbani, N.; Cascone, M.G.; Rai, R.; Saint-Pierre, G.; Boccaccini, A.R. Strategies for the chemical and biological functionalization of scaffolds for cardiac tissue engineering: A review. *J. R. Soc. Interface* **2015**, *12*, 20150254. [CrossRef]
32. Hu, C.; Ayan, B.; Chiang, G.; Chan, A.H.P.; Rando, T.A.; Huang, N.F. Comparative Effects of Basic Fibroblast Growth Factor Delivery or Voluntary Exercise on Muscle Regeneration after Volumetric Muscle Loss. *Bioengineering* **2022**, *9*, 37. [CrossRef]
33. Sedlář, A.; Trávníčková, M.; Matějka, R.; Pražák, Š.; Mészáros, Z.; Bojarová, P.; Bačáková, L.; Křen, V.; Slámová, K. Growth Factors VEGF-A$_{165}$ and FGF-2 as Multifunctional Biomolecules Governing Cell Adhesion and Proliferation. *Int. J. Mol. Sci.* **2021**, *22*, 1843. [CrossRef] [PubMed]
34. Kuttappan, S.; Mathew, D.; Jo, J.-I.; Tanaka, R.; Menon, D.; Ishimoto, T.; Nakano, T.; Nair, S.V.; Nair, M.B.; Tabata, Y. Dual release of growth factor from nanocomposite fibrous scaffold promotes vascularisation and bone regeneration in rat critical sized calvarial defect. *Acta Biomater.* **2018**, *78*, 36–47. [CrossRef] [PubMed]
35. Turner, P.A.; Thiele, J.S.; Stegemann, J.P. Growth factor sequestration and enzyme-mediated release from genipin-crosslinked gelatin microspheres. *J. Biomater. Sci. Polym. Ed.* **2017**, *28*, 1826–1846. [CrossRef]
36. Cuenca, J.P.; Kang, H.-J.; Al Fahad, A.; Park, M.-K.; Choi, M.-J.; Lee, H.-Y.; Lee, B.-T. Physico-mechanical and biological evaluation of heparin/VEGF-loaded electrospun polycaprolactone/decellularized rat aorta extracellular matrix for small-diameter vascular grafts. *J. Biomater. Sci. Polym. Ed.* **2022**, *33*, 1664–1684. [CrossRef]
37. Damiri, F.; Kommineni, N.; Ebhodaghe, S.O.; Bulusu, R.; Jyothi, V.G.S.S.; Sayed, A.A.; Awaji, A.A.; Germoush, M.O.; Al-Malky, H.S.; Nasrullah, M.Z.; et al. Microneedle-Based Natural Polysaccharide for Drug Delivery Systems (DDS): Progress and Challenges. *Pharmaceuticals* **2022**, *15*, 190. [CrossRef]
38. Sagar, V.; Nair, M. Near-infrared biophotonics-based nanodrug release systems and their potential application for neuro-disorders. *Expert Opin. Drug Deliv.* **2017**, *15*, 137–152. [CrossRef] [PubMed]
39. Pandolfi, L.; Minardi, S.; Taraballi, F.; Liu, X.; Ferrari, M.; Tasciotti, E. Composite microsphere-functionalized scaffold for the controlled release of small molecules in tissue engineering. *J. Tissue Eng.* **2016**, *7*, 2041731415624668. [CrossRef]
40. Lai, H.-J.; Kuan, C.-H.; Wu, H.-C.; Tsai, J.-C.; Chen, T.-M.; Hsieh, D.-J.; Wang, T.-W. Tailored design of electrospun composite nanofibers with staged release of multiple angiogenic growth factors for chronic wound healing. *Acta Biomater.* **2014**, *10*, 4156–4166. [CrossRef]
41. Jamee, R.; Araf, Y.; Bin Naser, I.; Promon, S.K. The promising rise of bioprinting in revolutionizing medical science: Advances and possibilities. *Regen. Ther.* **2021**, *18*, 133–145. [CrossRef]

42.  Cui, H.; Zhu, W.; Huang, Y.; Liu, C.; Yu, Z.-X.; Nowicki, M.; Miao, S.; Cheng, Y.; Zhou, X.; Lee, S.-J.; et al. In vitro and in vivo evaluation of 3D bioprinted small-diameter vasculature with smooth muscle and endothelium. *Biofabrication* **2019**, *12*, 015004. [CrossRef]
43.  Millik, S.C.; Dostie, A.M.; Karis, D.G.; Smith, P.T.; McKenna, M.; Chan, N.; Curtis, C.D.; Nance, E.; Theberge, A.B.; Nelson, A. 3D printed coaxial nozzles for the extrusion of hydrogel tubes toward modeling vascular endothelium. *Biofabrication* **2019**, *11*, 045009. [CrossRef] [PubMed]
44.  Gao, Q.; He, Y.; Fu, J.-Z.; Liu, A.; Ma, L. Coaxial nozzle-assisted 3D bioprinting with built-in microchannels for nutrients delivery. *Biomaterials* **2015**, *61*, 203–215. [CrossRef] [PubMed]
45.  Min, S.; Ko, I.K.; Yoo, J.J. State-of-the-Art Strategies for the Vascularization of Three-Dimensional Engineered Organs. *Vasc. Spéc. Int.* **2019**, *35*, 77–89. [CrossRef] [PubMed]
46.  Kim, J.H.; Seol, Y.-J.; Ko, I.K.; Kang, H.-W.; Lee, Y.K.; Yoo, J.J.; Atala, A.; Lee, S.J. 3D Bioprinted Human Skeletal Muscle Constructs for Muscle Function Restoration. *Sci. Rep.* **2018**, *8*, 12307. [CrossRef] [PubMed]
47.  Wu, P.K.; Ringeisen, B.R. Development of human umbilical vein endothelial cell (HUVEC) and human umbilical vein smooth muscle cell (HUVSMC) branch/stem structures on hydrogel layers via biological laser printing (BioLP). *Biofabrication* **2010**, *2*, 014111. [CrossRef]
48.  Geckil, H.; Xu, F.; Zhang, X.; Moon, S.; Demirci, U. Engineering hydrogels as extracellular matrix mimics. *Nanomedicine* **2010**, *5*, 469–484. [CrossRef] [PubMed]
49.  Vega, S.; Kwon, M.; Burdick, J. Recent advances in hydrogels for cartilage tissue engineering. *Eur. Cells Mater.* **2017**, *33*, 59–75. [CrossRef] [PubMed]
50.  Guo, Y.; Yuan, T.; Xiao, Z.; Tang, P.; Xiao, Y.; Fan, Y.; Zhang, X. Hydrogels of collagen/chondroitin sulfate/hyaluronan interpenetrating polymer network for cartilage tissue engineering. *J. Mater. Sci. Mater. Med.* **2012**, *23*, 2267–2279. [CrossRef] [PubMed]
51.  Kérourédan, O.; Bourget, J.-M.; Rémy, M.; Crauste-Manciet, S.; Kalisky, J.; Catros, S.; Thébaud, N.B.; Devillard, R. Micropatterning of endothelial cells to create a capillary-like network with defined architecture by laser-assisted bioprinting. *J. Mater. Sci. Mater. Med.* **2019**, *30*, 28. [CrossRef]
52.  Tan, E.Y.S.; Yeong, W.Y. Concentric Bioprinting Of Alginate-Based Tubular Constructs Using Multi-Nozzle Extrusion-Based Technique. *Int. J. Bioprint.* **2015**, *201*, 49–56. [CrossRef]
53.  Campbell, J.; McGuinness, I.; Wirz, H.; Sharon, A.; Sauer-Budge, A.F. Multimaterial and Multiscale Three-Dimensional Bioprinter. *J. Nanotechnol. Eng. Med.* **2015**, *6*, 021005. [CrossRef]
54.  Ozler, S.B.; Bakirci, E.; Kucukgul, C.; Koc, B. Three-dimensional direct cell bioprinting for tissue engineering. *J. Biomed. Mater. Res. Part B Appl. Biomater.* **2016**, *105*, 2530–2544. [CrossRef] [PubMed]
55.  Norotte, C.; Marga, F.S.; Niklason, L.E.; Forgacs, G. Scaffold-free vascular tissue engineering using bioprinting. *Biomaterials* **2009**, *30*, 5910–5917. [CrossRef] [PubMed]
56.  Kucukgul, C.; Ozler, S.B.; Inci, I.; Karakas, E.; Irmak, S.; Gozuacik, D.; Taralp, A.; Koc, B. 3D bioprinting of biomimetic aortic vascular constructs with self-supporting cells. *Biotechnol. Bioeng.* **2014**, *112*, 811–821. [CrossRef]
57.  Attalla, R.; Puersten, E.; Jain, N.; Selvaganapathy, P.R. 3D bioprinting of heterogeneous bi- and tri-layered hollow channels within gel scaffolds using scalable multi-axial microfluidic extrusion nozzle. *Biofabrication* **2018**, *11*, 015012. [CrossRef]
58.  Zhou, Y.; Liao, S.; Tao, X.; Xu, X.-Q.; Hong, Q.; Wu, D.; Wang, Y. Spider-Inspired Multicomponent 3D Printing Technique for Next-Generation Complex Biofabrication. *ACS Appl. Bio Mater.* **2018**, *1*, 502–510. [CrossRef]
59.  Feng, F.; He, J.; Li, J.; Mao, M.; Li, D. Multicomponent bioprinting of heterogeneous hydrogel constructs based on microfluidic printheads. *Int. J. Bioprin.* **2019**, *5*, 39. [CrossRef]
60.  Li, S.; Xiong, Z.; Wang, X.; Yan, Y.; Liu, H.; Zhang, R. Direct Fabrication of a Hybrid Cell/Hydrogel Construct by a Double-nozzle Assembling Technology. *J. Bioact. Compat. Polym.* **2009**, *24*, 249–265.
61.  Kolesky, D.B.; Homan, K.A.; Skylar-Scott, M.A.; Lewis, J.A. Three-dimensional bioprinting of thick vascularized tissues. *Proc. Natl. Acad. Sci. USA* **2016**, *113*, 3179–3184. [CrossRef]
62.  Noor, N.; Shapira, A.; Edri, R.; Gal, I.; Wertheim, L.; Dvir, T. 3D Printing of Personalized Thick and Perfusable Cardiac Patches and Hearts. *Adv. Sci.* **2019**, *6*, 1900344. [CrossRef]
63.  Kolesky, D.B.; Truby, R.L.; Gladman, A.S.; Busbee, T.A.; Homan, K.A.; Lewis, J.A. 3D bioprinting of vascularized, heterogeneous cell-laden tissue constructs. *Adv. Mater.* **2014**, *26*, 3124. [CrossRef] [PubMed]
64.  Cui, H.; Zhu, W.; Nowicki, M.; Zhou, X.; Khademhosseini, A.; Zhang, L.G. Hierarchical Fabrication of Engineered Vascularized Bone Biphasic Constructs via Dual 3D Bioprinting: Integrating Regional Bioactive Factors into Architectural Design. *Adv. Healthc. Mater.* **2016**, *5*, 2174–2181. [CrossRef] [PubMed]
65.  Kim, B.S.; Gao, G.; Kim, J.Y.; Cho, D.W. 3D cell printing of perfusable vascularized human skin equivalent composed of epidermis, dermis, and hypodermis for better structural recapitulation of native skin. *Adv. Healthc. Mater.* **2019**, *8*, 1801019. [CrossRef] [PubMed]
66.  Shanjani, Y.; Pan, C.C.; Elomaa, L.; Yang, Y. A novel bioprinting method and system for forming hybrid tissue engineering constructs. *Biofabrication* **2015**, *7*, 045008. [CrossRef] [PubMed]
67.  Chen, E.P.; Toksoy, Z.; Davis, B.A.A.; Geibel, J.P. 3D Bioprinting of Vascularized Tissues for in vitro and in vivo Applications. *Front. Bioeng. Biotechnol.* **2021**, *9*, 664188. [CrossRef]

68. Choi, Y.-J.; Jun, Y.-J.; Kim, D.Y.; Yi, H.-G.; Chae, S.-H.; Kang, J.; Lee, J.; Gao, G.; Kong, J.-S.; Jang, J.; et al. A 3D cell printed muscle construct with tissue-derived bioink for the treatment of volumetric muscle loss. *Biomaterials* **2019**, *206*, 160–169. [CrossRef]

69. Singh, N.K.; Han, W.; Nam, S.A.; Kim, J.W.; Kim, J.Y.; Kim, Y.K.; Cho, D.-W. Three-dimensional cell-printing of advanced renal tubular tissue analogue. *Biomaterials* **2019**, *232*, 119734. [CrossRef]

70. Kim, W.; Kim, G. Intestinal Villi Model with Blood Capillaries Fabricated Using Collagen-Based Bioink and Dual-Cell-Printing Process. *ACS Appl. Mater. Interfaces* **2018**, *10*, 41185–41196. [CrossRef]

71. Shao, L.; Gao, Q.; Xie, C.; Fu, J.; Xiang, M.; He, Y. Directly coaxial 3D bioprinting of large-scale vascularized tissue constructs. *Biofabrication* **2020**, *12*, 035014. [CrossRef]

72. Gaebel, R.; Ma, N.; Liu, J.; Guan, J.; Koch, L.; Klopsch, C.; Gruene, M.; Toelk, A.; Wang, W.; Mark, P.; et al. Patterning human stem cells and endothelial cells with laser printing for cardiac regeneration. *Biomaterials* **2011**, *32*, 9218–9230. [CrossRef]

73. Zhu, W.; Qu, X.; Zhu, J.; Ma, X.; Patel, S.; Liu, J.; Wang, P.; Lai, C.S.E.; Gou, M.; Xu, Y.; et al. Direct 3D bioprinting of prevascularized tissue constructs with complex microarchitecture. *Biomaterials* **2017**, *124*, 106–115. [CrossRef] [PubMed]

74. Xue, J.; Wu, T.; Dai, Y.; Xia, Y. Electrospinning and Electrospun Nanofibers: Methods, Materials, and Applications. *Chem. Rev.* **2019**, *119*, 5298–5415. [CrossRef] [PubMed]

75. Wang, C.; Wang, J.; Zeng, L.; Qiao, Z.; Liu, X.; Liu, H.; Zhang, J.; Ding, J. Fabrication of Electrospun Polymer Nanofibers with Diverse Morphologies. *Molecules* **2019**, *24*, 834. [CrossRef]

76. Caddeo, S.; Boffito, M.; Sartori, S. Tissue Engineering Approaches in the Design of Healthy and Pathological In Vitro Tissue Models. *Front. Bioeng. Biotechnol.* **2017**, *5*, 40. [CrossRef]

77. Kim, J.J.; Hou, L.; Yang, G.; Mezak, N.P.; Wanjare, M.; Joubert, L.M.; Huang, N.F. Microfibrous Scaffolds Enhance Endothelial Differentiation and Organization of Induced Pluripotent Stem Cells. *Cell. Mol. Bioeng.* **2017**, *10*, 417–432. [CrossRef] [PubMed]

78. Wanjare, M.; Hou, L.; Nakayama, K.H.; Kim, J.J.; Mezak, N.P.; Abilez, O.J.; Tzatzalos, E.; Wu, J.C.; Huang, N.F. Anisotropic microfibrous scaffolds enhance the organization and function of cardiomyocytes derived from induced pluripotent stem cells. *Biomater. Sci.* **2017**, *5*, 1567–1578. [CrossRef]

79. Kenar, H.; Ozdogan, C.Y.; Dumlu, C.; Doger, E.; Kose, G.T.; Hasirci, V. Microfibrous scaffolds from poly(l-lactide-co-ε-caprolactone) blended with xeno-free collagen/hyaluronic acid for improvement of vascularization in tissue engineering applications. *Mater. Sci. Eng. C* **2018**, *97*, 31–44. [CrossRef]

80. Cui, L.; Li, J.; Long, Y.; Hu, M.; Li, J.; Lei, Z.; Wang, H.; Huang, R.; Li, X. Vascularization of LBL structured nanofibrous matrices with endothelial cells for tissue regeneration. *RSC Adv.* **2017**, *7*, 11462–11477. [CrossRef]

81. Alsop, A.T.; Pence, J.C.; Weisgerber, D.W.; Harley, B.A.; Bailey, R.C. Photopatterning of vascular endothelial growth factor within collagen-glycosaminoglycan scaffolds can induce a spatially confined response in human umbilical vein endothelial cells. *Acta Biomater.* **2014**, *10*, 4715–4722. [CrossRef]

82. O'Dwyer, J.; Murphy, R.; González-Vázquez, A.; Kovarova, L.; Pravda, M.; Velebny, V.; Heise, A.; Duffy, G.; Cryan, S. Translational Studies on the Potential of a VEGF Nanoparticle-Loaded Hyaluronic Acid Hydrogel. *Pharmaceutics* **2021**, *13*, 779. [CrossRef]

83. Bai, Y.; Leng, Y.; Yin, G.; Pu, X.; Huang, Z.; Liao, X.; Chen, X.; Yao, Y. Effects of combinations of BMP-2 with FGF-2 and/or VEGF on HUVECs angiogenesis in vitro and CAM angiogenesis in vivo. *Cell Tissue Res.* **2014**, *356*, 109–121. [CrossRef] [PubMed]

84. Wu, Q.; Li, Y.; Wang, Y.; Li, L.; Jiang, X.; Tang, J.; Yang, H.; Zhang, J.; Bao, J.; Bu, H. The effect of heparinized decellularized scaffolds on angiogenic capability. *J. Biomed. Mater. Res. Part A* **2016**, *104*, 3021–3030. [CrossRef] [PubMed]

85. Samorezov, J.E.; Alsberg, E. Spatial regulation of controlled bioactive factor delivery for bone tissue engineering. *Adv. Drug Deliv. Rev.* **2014**, *84*, 45–67. [CrossRef] [PubMed]

86. Qin, D.; Xia, Y.; Whitesides, G.M. Soft lithography for micro- and nanoscale patterning. *Nat. Protoc.* **2010**, *5*, 491–502. [CrossRef] [PubMed]

87. Raghavan, S.; Nelson, C.M.; Baranski, J.; Lim, E.; Chen, C. Geometrically Controlled Endothelial Tubulogenesis in Micropatterned Gels. *Tissue Eng. Part A* **2010**, *16*, 2255–2263. [CrossRef] [PubMed]

88. Baranski, J.D.; Chaturvedi, R.R.; Stevens, K.R.; Eyckmans, J.; Carvalho, B.; Solorzano, R.D.; Yang, M.T.; Miller, J.S.; Bhatia, S.N.; Chen, C.S. Geometric control of vascular networks to enhance engineered tissue integration and function. *Proc. Natl. Acad. Sci. USA* **2013**, *110*, 7586–7591. [CrossRef] [PubMed]

89. Minardi, S.; Taraballi, F.; Pandolfi, L.; Tasciotti, E. Patterning Biomaterials for the Spatiotemporal Delivery of Bioactive Molecules. *Front. Bioeng. Biotechnol.* **2016**, *4*, 45. [CrossRef]

90. Nazarnezhad, S.; Baino, F.; Kim, H.-W.; Webster, T.J.; Kargozar, S. Electrospun Nanofibers for Improved Angiogenesis: Promises for Tissue Engineering Applications. *Nanomaterials* **2020**, *10*, 1609. [CrossRef]

91. Lei, Y.; Zouani, O.F.; Rémy, M.; Ayela, C.; Durrieu, M.-C. Geometrical Microfeature Cues for Directing Tubulogenesis of Endothelial Cells. *PLoS ONE* **2012**, *7*, e41163. [CrossRef]

92. Chow, L.W.; Wang, L.-J.; Kaufman, D.B.; Stupp, S.I. Self-assembling nanostructures to deliver angiogenic factors to pancreatic islets. *Biomaterials* **2010**, *31*, 6154–6161. [CrossRef]

93. Huang, N.F.; Lai, E.S.; Ribeiro, A.J.; Pan, S.; Pruitt, B.L.; Fuller, G.G.; Cooke, J.P. Spatial patterning of endothelium modulates cell morphology, adhesiveness and transcriptional signature. *Biomaterials* **2013**, *34*, 2928–2937. [CrossRef] [PubMed]

94. Khanna, A.; Ayan, B.; Undieh, A.A.; Yang, Y.P.; Huang, N.F. Advances in three-dimensional bioprinted stem cell-based tissue engineering for cardiovascular regeneration. *J. Mol. Cell. Cardiol.* **2022**, *169*, 13–27. [CrossRef] [PubMed]

95. Khanna, A.; Zamani, M.; Huang, N.F. Extracellular Matrix-Based Biomaterials for Cardiovascular Tissue Engineering. *J. Cardiovasc. Dev. Dis.* **2021**, *8*, 137. [CrossRef] [PubMed]
96. Lu, X.; Khanna, A.; Luzinov, I.; Nagatomi, J.; Harman, M. Surface modification of polypropylene surgical meshes for improving adhesion with poloxamine hydrogel adhesive. *J. Biomed. Mater. Res. Part B Appl. Biomater.* **2018**, *107*, 1047–1055. [CrossRef]
97. Zaarour, B.; Zhu, L.; Jin, X. A Review on the Secondary Surface Morphology of Electrospun Nanofibers: Formation Mechanisms, Characterizations, and Applications. *Chem. Sel.* **2020**, *5*, 1335–1348. [CrossRef]
98. García-Gareta, E.; Abduldaiem, Y.; Sawadkar, P.; Kyriakidis, C.; Lali, F.; Greco, K.V. Decellularised scaffolds: Just a framework? Current knowledge and future directions. *J. Tissue Eng.* **2020**, *11*, 2041731420942903. [CrossRef]
99. White, K.A.; Olabisi, R.M. Spatiotemporal Control Strategies for Bone Formation through Tissue Engineering and Regenerative Medicine Approaches. *Adv. Healthc. Mater.* **2018**, *8*, e1801044. [CrossRef]
100. De Pieri, A.; Rochev, Y.; Zeugolis, D.I. Scaffold-free cell-based tissue engineering therapies: Advances, shortfalls and forecast. *NPJ Regen. Med.* **2021**, *6*, 68. [CrossRef]
101. Wanjare, M.; Huang, N.F. Regulation of the microenvironment for cardiac tissue engineering. *Regen. Med.* **2017**, *12*, 187–201. [CrossRef]
102. Huang, N.F.; Lee, R.J.; Li, S. Chemical and Physical Regulation of Stem Cells and Progenitor Cells: Potential for Cardiovascular Tissue Engineering. *Tissue Eng.* **2007**, *13*, 1809. [CrossRef]
103. Joshi, A.; Choudhury, S.; Gugulothu, S.B.; Visweswariah, S.S.; Chatterjee, K. Strategies to Promote Vascularization in 3D Printed Tissue Scaffolds: Trends and Challenges. *Biomacromolecules* **2022**, *23*, 2730–2751. [CrossRef]
104. Shevach, M.; Zax, R.; Abrahamov, A.; Fleischer, S.; Shapira, A.; Dvir, T. Omentum ECM-based hydrogel as a platform for cardiac cell delivery. *Biomed. Mater.* **2015**, *10*, 034106. [CrossRef]
105. Giacomelli, E.; Meraviglia, V.; Campostrini, G.; Cochrane, A.; Cao, X.; Van Helden, R.W.; Garcia, A.K.; Mircea, M.; Kostidis, S.; Davis, R.P.; et al. Human-iPSC-Derived Cardiac Stromal Cells Enhance Maturation in 3D Cardiac Microtissues and Reveal Non-cardiomyocyte Contributions to Heart Disease. *Cell Stem Cell* **2020**, *26*, 862. [CrossRef]
106. Zamani, M.; Karaca, E.; Huang, N.F. Multicellular Interactions in 3D Engineered Myocardial Tissue. *Front. Cardiovasc. Med.* **2018**, *5*, 147. [CrossRef]
107. Rufaihah, A.J.; Huang, N.F.; Jamé, S.; Lee, J.C.; Nguyen, H.N.; Byers, B.; De, A.; Okogbaa, J.; Rollins, M.; Reijo-Pera, R.; et al. Endothelial Cells Derived From Human iPSCS Increase Capillary Density and Improve Perfusion in a Mouse Model of Peripheral Arterial Disease. *Arter. Thromb. Vasc. Biol.* **2011**, *31*, e72–e79. [CrossRef]
108. Huang, N.F.; Dewi, R.E.; Okogbaa, J.; Lee, J.C.; Jalilrufaihah, A.; Heilshorn, S.C.; Cooke, J.P. Chemotaxis of human induced pluripotent stem cell-derived endothelial cells. *Am. J. Transl. Res.* **2013**, *5*, 510–520.
109. Wanjare, M.; Kuo, F.; Gerecht, S. Derivation and maturation of synthetic and contractile vascular smooth muscle cells from human pluripotent stem cells. *Cardiovasc. Res.* **2013**, *97*, 321. [CrossRef]
110. Ronaldson-Bouchard, K.; Ma, S.P.; Yeager, K.; Chen, T.; Song, L.; Sirabella, D.; Morikawa, K.; Teles, D.; Yazawa, M.; Vunjak-Novakovic, G. Advanced maturation of human cardiac tissue grown from pluripotent stem cells. *Nature* **2018**, *556*, 239–243. [CrossRef]
111. Ribeiro, A.J.S.; Ang, Y.-S.; Fu, J.-D.; Rivas, R.N.; Mohamed, T.M.A.; Higgs, G.C.; Srivastava, D.; Pruitt, B.L. Contractility of single cardiomyocytes differentiated from pluripotent stem cells depends on physiological shape and substrate stiffness. *Proc. Natl. Acad. Sci. USA* **2015**, *112*, 12705–12710. [CrossRef]
112. Rackov, G.; Garcia-Romero, N.; Esteban-Rubio, S.; Carrión-Navarro, J.; Belda-Iniesta, C.; Ayuso-Sacido, A. Vesicle-Mediated Control of Cell Function: The Role of Extracellular Matrix and Microenvironment. *Front. Physiol.* **2018**, *9*, 651. [CrossRef]
113. Ruvinov, E.; Leor, J.; Cohen, S. The promotion of myocardial repair by the sequential delivery of IGF-1 and HGF from an injectable alginate biomaterial in a model of acute myocardial infarction. *Biomaterials* **2011**, *32*, 565–578. [CrossRef] [PubMed]
114. Bar, A.; Cohen, S. Inducing Endogenous Cardiac Regeneration: Can Biomaterials Connect the Dots? *Front. Bioeng. Biotechnol.* **2020**, *8*, 126. [CrossRef] [PubMed]
115. Jeske, R.; Bejoy, J.; Marzano, M.; Li, Y. Human Pluripotent Stem Cell-Derived Extracellular Vesicles: Characteristics and Applications. *Tissue Eng. Part B Rev.* **2020**, *26*, 129–144. [CrossRef]
116. Bae, W.-G.; Kim, J.; Choung, Y.-H.; Chung, Y.; Suh, K.Y.; Pang, C.; Chung, J.H.; Jeong, H.E. Bio-inspired configurable multiscale extracellular matrix-like structures for functional alignment and guided orientation of cells. *Biomaterials* **2015**, *69*, 158–164. [CrossRef]
117. Leijten, J.; Seo, J.; Yue, K.; Santiago, G.T.-D.; Tamayol, A.; Ruiz-Esparza, G.U.; Shin, S.R.; Sharifi, R.; Noshadi, I.; Álvarez, M.M.; et al. Spatially and temporally controlled hydrogels for tissue engineering. *Mater. Sci. Eng. R Rep.* **2017**, *119*, 1–35. [CrossRef]
118. Wang, L.; Serpooshan, V.; Zhang, J. Engineering Human Cardiac Muscle Patch Constructs for Prevention of Post-infarction LV Remodeling. *Front. Cardiovasc. Med.* **2021**, *8*, 621781. [CrossRef]
119. Serbo, J.V.; Gerecht, S. Vascular tissue engineering: Biodegradable scaffold platforms to promote angiogenesis. *Stem Cell Res. Ther.* **2013**, *4*, 8. [CrossRef]
120. Gaetani, R.; Doevendans, P.A.; Metz, C.H.; Alblas, J.; Messina, E.; Giacomello, A.; Sluijter, J.P. Cardiac tissue engineering using tissue printing technology and human cardiac progenitor cells. *Biomaterials* **2012**, *33*, 1782–1790. [CrossRef]

121. Liu, J.; Miller, K.; Ma, X.; Dewan, S.; Lawrence, N.; Whang, G.; Chung, P.; McCulloch, A.D.; Chen, S. Direct 3D bioprinting of cardiac micro-tissues mimicking native myocardium. *Biomaterials* **2020**, *256*, 120204. [CrossRef]
122. Aguilar, I.N.; Olivos, D.J.; Brinker, A.; Alvarez, M.B.; Smith, L.J.; Chu, T.-M.G.; Kacena, M.A.; Wagner, D.R. Scaffold-free bioprinting of mesenchymal stem cells using the Regenova printer: Spheroid characterization and osteogenic differentiation. *Bioprinting* **2019**, *15*, e00050. [CrossRef]
123. Kupfer, M.E.; Lin, W.-H.; Ravikumar, V.; Qiu, K.; Wang, L.; Gao, L.; Bhuiyan, D.B.; Lenz, M.; Ai, J.; Mahutga, R.R.; et al. In Situ Expansion, Differentiation, and Electromechanical Coupling of Human Cardiac Muscle in a 3D Bioprinted, Chambered Organoid. *Circ. Res.* **2020**, *127*, 207–224. [CrossRef] [PubMed]
124. Gopinathan, J.; Noh, I. Recent trends in bioinks for 3D printing. *Biomater. Res.* **2018**, *22*, 11. [CrossRef] [PubMed]
125. Das, S.; Nam, H.; Jang, J. 3D bioprinting of stem cell-laden cardiac patch: A promising alternative for myocardial repair. *APL Bioeng.* **2021**, *5*, 031508. [CrossRef] [PubMed]
126. Gao, G.; Lee, J.H.; Jang, J.; Lee, D.H.; Kong, J.S.; Kim, B.S.; Choi, Y.J.; Jang, W.B.; Hong, Y.J.; Kwon, S.M.; et al. Tissue Engineered Bio-Blood-Vessels Constructed Using a Tissue-Specific Bioink and 3D Coaxial Cell Printing Technique: A Novel Therapy for Ischemic Disease. *Adv. Funct. Mater.* **2017**, *27*, 1700798. [CrossRef]
127. Wanjare, M.; Kawamura, M.; Hu, C.; Alcazar, C.; Wang, H.; Woo, Y.J.; Huang, N.F. Vascularization of Engineered Spatially Patterned Myocardial Tissue Derived From Human Pluripotent Stem Cells in vivo. *Front. Bioeng. Biotechnol.* **2019**, *7*, 208. [CrossRef]
128. Do, A.-V.; Khorsand, B.; Geary, S.M.; Salem, A.K. 3D Printing of Scaffolds for Tissue Regeneration Applications. *Adv. Healthc. Mater.* **2015**, *4*, 1742–1762. [CrossRef]
129. Reis, L.A.; Chiu, L.L.Y.; Feric, N.; Fu, L.; Radisic, M. Biomaterials in myocardial tissue engineering. *J. Tissue Eng. Regen. Med.* **2014**, *10*, 11–28. [CrossRef]
130. Barrs, R.; Jia, J.; Silver, S.E.; Yost, M.; Mei, Y. Biomaterials for Bioprinting Microvasculature. *Chem. Rev.* **2020**, *120*, 10887–10949. [CrossRef]
131. Tomasina, C.; Bodet, T.; Mota, C.; Moroni, L.; Camarero-Espinosa, S. Bioprinting Vasculature: Materials, Cells and Emergent Techniques. *Materials* **2019**, *12*, 2701. [CrossRef]
132. Zhao, P.; Huo, S.; Fan, J.; Chen, J.; Kiessling, F.; Boersma, A.J.; Göstl, R.; Herrmann, A. Activation of the Catalytic Activity of Thrombin for Fibrin Formation by Ultrasound. *Angew. Chem. Int. Ed.* **2021**, *60*, 14707–14714. [CrossRef]
133. Song, H.H.G.; Rumma, R.T.; Ozaki, C.K.; Edelman, E.R.; Chen, C.S. Vascular Tissue Engineering: Progress, Challenges, and Clinical Promise. *Cell Stem Cell* **2018**, *22*, 340. [CrossRef] [PubMed]
134. Xie, Z.; Gao, M.; Lobo, A.O.; Webster, T.J. 3D Bioprinting in Tissue Engineering for Medical Applications: The Classic and the Hybrid. *Polymers* **2020**, *12*, 1717. [CrossRef] [PubMed]
135. Madonna, R.; Van Laake, L.W.; Botker, H.E.; Davidson, S.M.; De Caterina, R.; Engel, F.; Eschenhagen, T.; Fernandez-Aviles, F.; Hausenloy, D.J.; Hulot, J.-S.; et al. ESC Working Group on Cellular Biology of the Heart: Position paper for Cardiovascular Research: Tissue engineering strategies combined with cell therapies for cardiac repair in ischaemic heart disease and heart failure. *Cardiovasc. Res.* **2019**, *115*, 488–500. [CrossRef] [PubMed]

 *bioengineering*

*Article*

# Hydrogel Coating Optimization to Augment Engineered Soft Tissue Mechanics in Tissue-Engineered Blood Vessels

Bryan T. Wonski [1], Bruce Fisher [2] and Mai T. Lam [1,*]

[1] Department of Biomedical Engineering, Wayne State University, Detroit, MI 48201, USA; wonski1bt@wayne.edu
[2] Plymouth Family Dentistry, Plymouth, MI 48170, USA
* Correspondence: mtlam@wayne.edu; Tel.: +1-(313)-577-0118; Fax: +1-(313)-577-8333

**Abstract:** Tissue engineering has the advantage of replicating soft tissue mechanics to better simulate and integrate into native soft tissue. However, soft tissue engineering has been fraught with issues of insufficient tissue strength to withstand physiological mechanical requirements. This factor is due to the lack of strength inherent in cell-only constructs and in the biomaterials used for soft tissue engineering and limited extracellular matrix (ECM) production possible in cell culture. To address this issue, we explored the use of an ECM-based hydrogel coating to serve as an adhesive tool, as demonstrated in vascular tissue engineering. The efficacy of cells to supplement mechanical strength in the coating was explored. Specifically, selected coatings were applied to an engineered artery tunica adventitia to accurately test their properties in a natural tissue support structure. Multiple iterations of three primary hydrogels with and without cells were tested: fibrin, collagen, and gelatin hydrogels with and without fibroblasts. The effectiveness of a natural crosslinker to further stabilize and strengthen the hydrogels was investigated, namely genipin extracted from the gardenia fruit. We found that gelatin crosslinked with genipin alone exhibited the highest tensile strength; however, fibrin gel supported cell viability the most. Overall, fibrin gel coating without genipin was deemed optimal for its balance in increasing mechanical strength while still supporting cell viability and was used in the final mechanical and hydrodynamic testing assessments. Engineered vessels coated in fibrin hydrogel with cells resulted in the highest tensile strength of all hydrogel-coated groups after 14 d in culture, demonstrating a tensile strength of 11.9 ± 2.91 kPa, compared to 5.67 ± 1.37 kPa for the next highest collagen hydrogel group. The effect of the fibrin hydrogel coating on burst pressure was tested on our strongest vessels composed of human aortic smooth muscle cells. A significant increase from our previously reported burst pressure of 51.3 ± 2.19 mmHg to 229 ± 23.8 mmHg was observed; however, more work is needed to render these vessels compliant with mechanical and biological criteria for blood vessel substitutes.

**Keywords:** tissue engineering; vascular graft; biomaterials; cells; hydrogels; fibrin; collagen; gelatin; genipin

**Citation:** Wonski, B.T.; Fisher, B.; Lam, M.T. Hydrogel Coating Optimization to Augment Engineered Soft Tissue Mechanics in Tissue-Engineered Blood Vessels. *Bioengineering* **2023**, *10*, 780. https://doi.org/10.3390/bioengineering10070780

Academic Editors: Gary Chinga Carrasco, Brandon J. Tefft and Ngan F. Huang

Received: 26 April 2023
Revised: 24 June 2023
Accepted: 27 June 2023
Published: 30 June 2023

## 1. Introduction

The engineering of soft tissues, despite its great potential to solve patient tissue supply issues, has yet to reach the clinic. The main issue is that engineered soft tissues continue to lack the strength needed for human application. Traditionally, stiff scaffolds were used to provide the needed strength; however, they create a significant difference in mechanical properties compared to the native tissue. In terms of blood vessel tissue engineering, commonly used stiff polymer scaffolds adversely result in compliance mismatch. Our laboratory established a scaffold-less technique to engineer blood vessels in which individual vascular ring segments are self-organized and then stacked into tubular structures to create an engineered vessel [1–3]. The advantage of a completely biologically engineered soft tissue, such as ours, is the biocompatibility in terms of mechanical property matching and cell matrix support. The disadvantage is the lack of extracellular matrix (ECM) resulting in insufficient strength.

In natural tissues, the ECM is the proverbial glue holding them together. The scientific question is whether the ECM can be sufficiently reproduced for engineered tissue construction. Replication of the ECM in cell culture continues to be significantly restricted due to the currently available tools. Growth factors have been the primary route to induce ECM production and deposition in cell culture with continued limited results [2,4,5]. Biomaterials are a viable option for exploration. Typically, polymers are the primary choice; however, issues of foreign body reaction and the discrepancy in mechanical properties between the polymer and natural tissue hinder soft tissue mechanics. Stiff polymers create compliance mismatch, interrupting hemostasis and inducing intimal hyperplasia, leading to occlusion [6,7]. Cell sheets formed into a tubular shape have offered a completely biological option that better matches the mechanical properties of blood vessels [8,9]. However, cell sheets lack the strength needed to function under blood pressure and are typically mechanically conditioned for months to promote sufficient ECM deposition to strengthen the vessel [9]. Hydrogels are often derived from ECM components and are more similar to soft tissue mechanics. Our hypothesis is that a hydrogel coating for a soft tissue-engineered blood vessel would serve as supplemental ECM, thus increasing vessel strength.

The goal of this research was to test a range of hydrogel coatings for their effects on the mechanical properties of engineered blood vessels. Fibrin, collagen, and gelatin hydrogels were chosen as viable candidates as they all serve naturally as matrix components. Fibrin forms a provisional scaffold during vessel injury repair as part of the coagulation cascade, encouraging cell adhesion, migration, proliferation, and differentiation [10]. Collagen is a major structural protein that constitutes the main support framework of the vascular ECM [11]. Gelatin is a sub-component of the collagen molecule, derived through collagen hydrolysis with beneficial properties of compliance and biodegradability [12,13]. Crosslinking agents facilitate a standard protocol for tuning hydrogel properties and for preventing the rapid biodegradation of hydrogels [14]. Glutaraldehyde and formaldehyde are commonly used crosslinkers for their effectiveness; however, they are highly cytotoxic and promote inflammation [15]. More recently, genipin was explored for its low toxicity as a new crosslinking agent [14]. Genipin is a natural substance derived from the gardenia fruit which bonds the free amino groups of lysine on the collagen molecule [14,15].

ECM-based fibrin, collagen, and gelatin hydrogels with and without genipin crosslinking were evaluated for their ability to serve as an effective outer surface coating to engineered vessels to improve overall vessel stability and strength. To test the coatings in a representative mechanical application, the coatings were applied to tissue-engineered tunica adventitia vessels as the adventitia is the primary strength element in a blood vessel. To further enhance the ECM properties, the effect of adding cells, i.e., fibroblasts, to the hydrogel coatings was explored with the rationale that the mechanical properties of the cells themselves would add structural integrity in addition to depositing additional ECM. The multiple coating iterations were applied to the engineered blood vessels. The coatings were evaluated for their mechanical strength. The coatings exhibiting the highest mechanical strength were then tested with cells. Burst pressure was evaluated with the optimal hydrogel coating. A hydrogel coating could effectively solve issues of the lack of ECM in engineered soft tissues, thus restoring mechanical strength and stability in a much simpler method than the current technique of weeks of mechanical conditioning. In addition, such a hydrogel coating has plausible applications for other engineered soft tissues.

## 2. Materials and Methods

*Patient Cell Harvest and Culture.* Patient cells were harvested from human abdominal skin tissues acquired with informed patient consent from abdominoplasty surgeries at Henry Ford Hospital (Detroit, MI, USA) and Henry Ford Medical Center Cottage (Grosse Pointe Farms, MI, USA) in accordance with both Wayne State University and Henry Ford Health System Institutional Review Board (IRB) guidelines. Patient dermal fibroblasts (PtFib) were extracted through an explant culture of full-thickness skin following adipose tissue removal. Skin tissues were cut into 3 × 3 mm sections and placed on 10% gelatin-coated Petri dishes. Following 1–2 weeks, PtFibs colonies populated the culture plate surface and the skin sections were discarded. PtFib cell cultures were maintained in growth media consisting of 89% Dulbecco's Modified Eagle Medium (DMEM) high glucose, 10% fetal bovine serum, and 1% antibiotic-antimycotic. Cells were expanded in 150 mm Petri dishes in an incubator at 37 °C and 5% $CO_2$ until their use in vascular tissue formation and in the coatings. Experiments were performed using cells between passages 3 to 8 to ensure a healthy morphology.

Human aortic smooth muscle cells (HASMCs; PCS-100-012, ATCC, Manassas, VA, USA) were used to assess vessel hemodynamics for burst pressure. HASMCs were expanded in smooth muscle growth media consisting of 88.5% DMEM; 5% of L-glutamine and fetal bovine serum; 1% antibiotic-antimycotic; and 0.1% of recombinant human insulin (rH-insulin), recombinant human epidermal growth factor (rH-EGF), recombinant human fibroblast growth factor (rH-FGF), and ascorbic acid. Cells were incorporated into engineered tissues in healthy morphologies of passages 3–6.

***Hydrogel Composition and Mechanics.*** Concentrations of fibrin, collagen, and gelatin hydrogel coatings alone were optimized based on maximizing tensile mechanics. To prepare the hydrogels for tensile testing, hydrogels were molded into strips. Rectangular molds were created by embedding a 3D-printed polylactic acid (PLA) rectangular shape (25 mm × 10 mm × 6 mm) into an uncured 10:1 base to curing reagent ratio of poly(dimethysiloxane) elastomer (PDMS) within a 60 mm polystyrene culture dish. Following overnight polymerization, the PLA was carefully removed and the PDMS culture dishes were sterilized by ethanol and UV prior to use. PLA filament (MP05823, MakerBot PLA Filament, Makerbot, New York, NY, USA) was printed on a MakerBot Replicator Mini (Makerbot).

Hydrogel strips were fabricated by casting 1 mL of the following hydrogel mixtures in a rectangular mold and incubated at 37 °C and 5% $CO_2$ overnight for 16 h. Two fibrin hydrogel compositions consisting of either 4.8 mg/mL or 9.6 mg/mL fibrinogen (0215112205, MP Biomedicals, Santa Ana, CA, USA) suspended in 880 μL of growth media and 120 μL of 100 U/mL thrombin (7592, BioVision, Milpitas, CA, USA) per gel were assessed. Additionally, fibrin–genipin gels were investigated by incorporating 2% weight genipin (G4796, Millepore Sigma, Burlington, MA, USA) to weight ECM ($w/w$) into the fibrinogen solution; however, the addition of genipin inhibited fibrin gel formation at this concentration. Genipin is derived from the gardenia fruit and is known for its crosslinking properties with low toxicity. Collagen hydrogels with and without 2% $w/w$ genipin crosslinking were created by combining 0.9 mL collagen solution and 0.1 mL neutralization solution (Rat Collagen Type I Acid Soluble Rat Tail Collagen, Advanced Biomatrix) for a final collagen concentration of 4 mg/mL. Gelatin 3D gels required the addition of genipin for thermal stability around 37 °C. Therefore, "gelapin" coatings consisting of 5 or 10% weight to volume ($w/v$) gelatin (G2500500G; Millepore Sigma, Burlington, MA, USA) solution in combination with 2, 5, or 10% $w/w$ genipin were tested. The composition of gelapin hydrogels were denoted by the percentage $w/v$ of gelatin and percentage $w/w$ of genipin. For example, 5:2% gelapin indicates gelapin gels composed of 5% $w/v$ gelatin and 2% $w/w$ genipin.

The mechanical properties of the hydrogel strips (n = 3–5 per group) were analyzed with tensile testing using a UStretch system equipped with 5 N load cell (CellScale, Waterloo, ON, Canada). Hydrogels were mounted to the actuators by BioRakes with 1.3 mm penetration depth and 0.9 mm spacing. Once attached to the system and under tension, digital calipers were used to measure the sample's initial length, width, and thickness. Uniaxial tensile testing was performed at a strain rate of 0.4 mm/min until failure. Stress–strain curves were produced from force–displacement data to determine the elastic modulus (E), ultimate tensile strength (UTS), max force, failure strength (FS), and elongation at failure of each sample. The optimized concentration determined for each hydrogel group was then applied to the tissue engineered blood vessel to test as a coating.

***Engineered Vessel Plates and Coating Mold Fabrication.*** Vascular tissue ring and vessel culture plates were created using a modified version of our lab's previously established methods [1–3]. For the ring culture plates, 35 mm diameter 6-well culture dishes were surface coated with PDMS consisting of a 10:1 base-to-curing reagent ratio and left to cure overnight. A 5 mm biopsy punch was used to create posts from a poured slab of cured PDMS. Once the 6-well plates' surface coatings had solidified, a 5 mm diameter PDMS post was adhered to the middle of each well using more PDMS. The vessel culture dishes consisted of polycarbonate tubing attached to a polycarbonate base using an acrylic solvent. PLA posts of 5 mm in diameter and post holders were 3D-printed and filed for smoothness. The posts were thinly coated with PDMS to reduce friction during ring stacking and vessel removal. Post holders were embedded in PDMS in the center of each vessel culture dish.

Molds for coating the vessels were created. A PLA model of a 5 mm diameter vessel with a 1.75 mm uniform wall thickness and 10 mm length fixed in the center of a 5 mm diameter and 20 mm long cylindrical post was 3D-printed. Negative impressions were created by placing a 3D-printed model horizontally on the PDMS surface coated 60 mm polystyrene culture plates followed by filling the dish with uncured PDMS until half of the model was submerged. After overnight polymerization, PLA models were removed from the PDMS leaving behind a negative cavity for vessel coating. Finally, all culture dishes were treated aseptically with ethanol and UV.

***Fabrication of Engineered Vascular Rings.*** Tissue-engineered vascular rings were formed using our lab's previously published methods [2]. Fibrin hydrogels were polymerized in custom 6-well culture dishes by depositing 0.5 mL of growth media followed by the addition of 40 µL of 100 U/mL thrombin followed by 160 µL of 20 mg/mL fibrinogen containing $1 \times 10^6$ HASMCs or PtFibs. Once polymerized, 2 mL of growth media containing an additional $1 \times 10^6$ HASMCs or PtFibs and supplemented with ascorbic acid and transforming growth factor-beta 1 (TGF-$\beta_1$) was seeded dropwise onto the hydrogel surface. Culture media was replaced 24 h after seeding and every 48 h thereafter. After seven days in culture, tissue rings were removed from the posts with sterile forceps and stacked to form vessels.

***Assembly of Engineered Vessels.*** Tissue-engineered vessels consisting of six or more rings were constructed; the rings adhered together by the optimized concentrations of each hydrogel: gelatin + genipin (termed "gelapin"), fibrin, collagen, or collagen–genipin. The culture time served as another variable in order to test the effects of matrix remodeling over time. Vessels were cultured for 1 d or 14 d and tested for circumferential and longitudinal strength (n = 4–7, 6-ring vessels) and for 4 or 16 weeks for hemodynamic analysis (n = 2–4, 12-ring vessels). After 7 days in culture, vascular tissue rings were transferred onto PLA vessel posts in the vessel plates and carefully pushed together. Next, 0.5 mL of solubilized hydrogel components containing $500 \times 10^3$ cells for every six rings stacked (i.e., 0.5 mL for 6-ring and 1 mL for 12-ring) were transferred into the vessel mold cavity followed by immediately submerging the tissue stack into the hydrogel solution (Figure 1). Vessel molds were placed in an incubator at 37 °C for 1 h after which the posts were rotated 180° and the hydrogel coating process was repeated to achieve complete coverage. After the vessels were fully coated, they were removed from the molds, inserted into the post holder of the vessel culture plates, and maintained in growth media until they were analyzed histologically and mechanically.

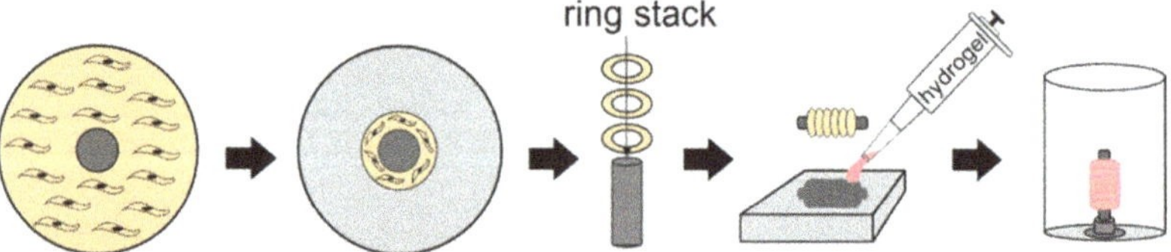


**Figure 1.** Protocol for hydrogel coating engineered vessels. Tissue rings were formed by inducing the self-organization of a vascular cell monolayer around a central post in a dish to form a ring. Rings were stacked to form the engineered vessel. The vessel was transferred to a mold containing a solution of an extracellular matrix hydrogel and cells (i.e., fibroblasts). After the first round of hydrogel coating polymerization, the vessel was rotated 180° and the coating process was repeated for complete coverage.

***Hydrogel Cell Viability Assay.*** The effect of the hydrogels alone on cellular viability was determined by a live/dead stain over a 7-d period. Gelapin, fibrin, collagen, or collagen–genipin hydrogels containing $200 \times 10^3$ PtFibs (n = 3 for each time point) were formed in custom-fabricated 6-well culture ring plates. On days 1, 3, and 7, the culture media was supplemented with green fluorescent calcein acetoxymethyl and red fluorescent ethidium homodimer-1 (Live/Dead Viability Cytotoxicity Kit, ThermoFisher, Waltham, MA, USA) demarcating live and dead cells, respectively. After 1 h, the hydrogels were transferred to fresh 35 mm polystyrene plates and five randomly selected areas on each gel were imaged by an EVOS Fluorescent Cell Imaging System for viable cell counts.

***Tissue Histology.*** Adventitia vessels cultured for 14 d in vitro were fixed in 10% formalin for 24 h and stored in 70% ethanol at 5 °C until dehydration. Tissue samples were dehydrated in graduations of 70%, 95%, and 100% ethanol over 8 h followed by xylene for 2 h. Following dehydration, samples were submerged in liquid paraffin wax for 2 h at 60 °C and then embedded in paraffin blocks for tissue sectioning. Cross-sectional tissue sections were cut at a thickness of 10 μm. The cellularity of adventitia vessels was assessed by hematoxylin and eosin (H and E) staining, while Picrosirius Red and Masson's Trichrome stains were used to visualize collagen.

***Mechanical Analysis of Vascular Tissues.*** Longitudinal and circumferential tensile mechanics of hydrogel-coated 6-ring adventitia vessels were obtained using a UStretch system equipped with a 5 N load cell (CellScale, Waterloo, ON, USA). Adventitia vessels cultured for 1 d and 14 d (n = 4–7 per hydrogel group at each time point) were tensile tested longitudinally. For longitudinal tensile tests, 3D-printed PLA stages were utilized to mount samples onto the UStretch actuators. Prior to mounting, the inner and outer diameter of each sample was measured with digital calipers. The vessels were adhered to the 3D-printed stage at each end by VetBond tissue adhesive (3M, St. Paul, MN, USA) and the initial length ($L_0$) was recorded. Samples were stretched at a strain rate of 0.4 mm/min until complete failure. The optimal hydrogel group was defined as the group with the highest longitudinal tensile strength. Once the optimal group was identified, in this case, the fibrin hydrogel group, the remaining mechanical analyses were focused on this group.

Circumferential tensile experiments were performed on fibrin-coated vessels following 1 d (n = 5) and 14 d (n = 7) culture periods. Briefly, vessels were mounted to the actuators by inserting metal hooks through the lumen. Under slight tension, wall thicknesses, length, and initial width measurements were recorded. Samples were stretched until failure at a strain rate of 0.4 mm/min. Following longitudinal and circumferential testing, stress–strain data were analyzed to determine the elastic modulus, ultimate tensile strength, maximum force, failure strength, and elongation at failure.

***Hemodynamic Analysis.*** The hemodynamic strength of fibrin-coated engineered tunica media vessels was evaluated by burst pressure testing following 16 weeks of culture. A custom bioreactor consisting of a peristaltic pump (WT600-2J, Longer Precision Pump Corporation, Boonton, NJ, USA) connected by silicone tubing to a media reservoir, vessel chamber with 3D-printed vessel tubing connectors, and pressure gauge was used to subject vessels to pulsatile flow. Vessels were perfused with water at a pulse rate of 60 pulses per minute for 30 s. Following the 30 s priming period, the tubing downstream of the pressure gauge and vessel chamber was clamped and pressure was monitored on the pressure gauge until the vessel ruptured.

***Statistical Analysis.*** All statistics were performed in SPSS (IBM, Armonk, NY, USA). Results are presented as means ± standard deviation. For preliminary ECM material concentration mechanical testing, statistical analysis within each group (fibrin, collagen, and gelapin) was performed by one-way ANOVA for gelapin strips, whereas independent *t*-tests were used to compare the two fibrin concentrations, collagen, and collagen–genipin hydrogels. Additionally, one-way ANOVAs were performed to compare the longitudinal mechanics between groups for the hydrogel and adventitia vessel mechanics. To compare the effects of cell incorporation into the hydrogel coatings on longitudinal and circumferential mechanics, vessels cultured for 1 d or 14 d were analyzed via independent *t*-tests for each coating group. Following the ANOVA test for multiple groups, Tukey's B post hoc test was performed to determine the significance between groups. The statistical significance was assessed using a *p*-value less than or equal to 0.05.

## 3. Results

*Genipin Crosslinker Increases the Mechanical Properties of Hydrogels.* Gelapin, fibrin, and collagen-based hydrogel mechanical properties were analyzed by tensile testing to identify the ideal concentration within each group based on the ultimate tensile strength (Table 1; Figure 2a). Pure gelatin gels (without genipin) remained in the liquid form at incubator temperatures (37 °C) and hence were not stable as hydrogels alone in this application. Increasing the gelatin concentration from 5% to 10% resulted in a higher average ultimate tensile strength when compared to gels with equivalent genipin concentration ratios. Interestingly, the genipin concentration was found to be inversely related to the average ultimate tensile strength, failure strength, and elongation of gels with constant gelatin concentration. Gelapin hydrogels consisting of 10% gelatin and 2% genipin (i.e., 10:2% gelapin) had a significantly higher average ultimate tensile strength of 6.46 ± 1.19 kPa ($p < 0.05$) relative to all other hydrogel combinations and thus was determined to be the optimized gelapin hydrogel for coating.

**Table 1.** Average tensile properties of hydrogels alone.

| Material | E (kPa) | UTS (kPa) | Max Force (N) | FS (kPa) | Elongation (%) |
|---|---|---|---|---|---|
| 5:2% Gelapin | 2.25 ± 0.597 | 4.26 ± 0.556 [c,d] | 0.125 ± 0.008 [c,d] | 3.76 ± 0.427 | 151 ± 22.9 |
| 5:5% Gelapin | 2.82 ± 0.310 | 2.68 ± 0.572 [d] | 0.080 ± 0.010 [d] | 2.63 ± 0.528 | 85.8 ± 11.9 |
| 5:10% Gelapin | 5.36 ± 1.85 | 1.81 ± 0.615 [a,d,e] | 0.063 ± 0.018 [a,d,e] | 1.35 ± 1.07 | 43.3 ± 11.9 |
| 10:2% Gelapin | 7.04 ± 1.79 [f,h,i] | 6.46 ± 1.19 [a,b,c,e,f,g] | 0.182 ± 0.037 [a,b,c,e,f] | 6.12 ± 1.03 | 99.1 ± 15.8 |
| 10:5% Gelapin | 12.6 ± 4.35 [a,b,c] | 4.24 ± 0.546 [c,d] | 0.141 ± 0.025 [c,d] | 3.79 ± 0.565 | 45.5 ± 13.2 |
| 10:10% Gelapin | 17.3 ± 4.50 [a,b,c,d] | 3.05 ± 0.644 [d] | 0.103 ± 0.027 [d] | 1.87 ± 0.700 | 26.8 ± 4.36 |
| Fibrin (4.8 mg/mL) | 2.68 ± 0.693 | 1.98 ± 0.413 | 0.057 ± 0.008 | 1.79 ± 0.229 | 70.5 ± 15.9 |
| Fibrin (9.6 mg/mL) | 2.54 ± 0.247 [h,i] | 2.31 ± 0.781 [d,i] | 0.061 ± 0.017 | 2.22 ± 0.680 | 81.1 ± 6.95 |
| Collagen (4 mg/mL) | 24.8 ± 6.77 [i] | 3.99 ± 0.808 [d] | 0.089 ± 0.016 | 0.78 ± 0.386 | 118 ± 40.2 |
| Collagen–Genipin (4 mg/mL—2%) | 38.2 ± 13.6 [d,g,h] | 5.15 ± 1.51 [g] | 0.102 ± 0.022 | 0.808 ± 0.566 | 102 ± 42.3 |

[a] Statistically significant difference relative to 5:2% gelapin ($p < 0.05$). [b] Statistically significant difference relative to 5:5% gelapin ($p < 0.05$). [c] Statistically significant difference relative to 5:10% gelapin ($p < 0.05$). [d] Statistically significant difference relative to 10:2% gelapin ($p < 0.05$). [e] Statistically significant difference relative to 10:5% gelapin ($p < 0.05$). [f] Statistically significant difference relative to 10:10% gelapin ($p < 0.05$). [g] Statistically significant difference relative to 9.6 mg/mL fibrin ($p < 0.05$). [h] Statistically significant difference relative to collagen ($p < 0.05$). [i] Statistically significant difference relative to collagen–genipin ($p < 0.05$).

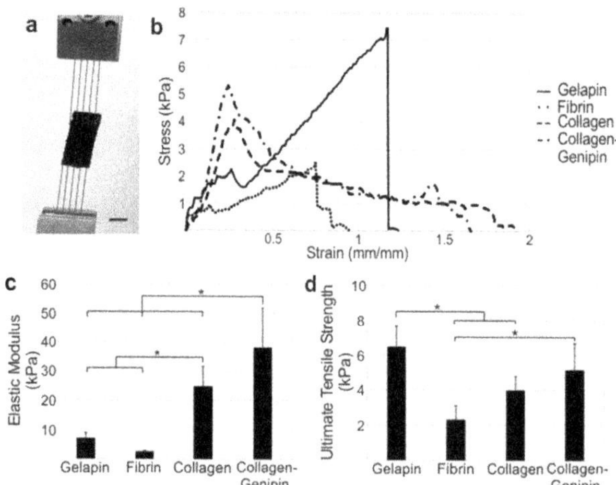


**Figure 2.** Mechanical properties of extracellular matrix hydrogels alone. (**a**) Tensile setup for a gelapin hydrogel sample. (**b**) Average stress–strain graphs of 10:2% gelapin, 9.6 mg/mL fibrin, 4 mg/mL collagen,

and 4 mg/mL—2% collagen–genipin demonstrate the (**c**) elastic moduli and (**d**) ultimate tensile strength varied significantly between optimized hydrogels. * Denotes significance between groups ($p < 0.05$). Scale bar = 5 mm.

The ideal fibrinogen concentration for the fibrin gel was determined. Increasing the concentration of fibrinogen by 2-fold did not result in any significant difference in the mechanical properties of fibrin hydrogels. The addition of genipin to the fibrinogen solutions prevented gel formation, likely due to the inhibition of fibrinogen cleavage by attachment of the genipin molecules (as further discussed in the Discussion). The average elastic modulus, ultimate tensile strength, failure strength, and elongation at failure of fibrin gels composed of 4.8 mg/mL fibrinogen were $2.68 \pm 0.693$ kPa, $1.98 \pm 0.413$ kPa, $1.79 \pm 0.229$ kPa, and $70.5 \pm 15.9\%$, respectively; whereas, in fibrinogen gels these properties were $2.54 \pm 0.247$ kPa, $2.31 \pm 0.781$ kPa, $2.22 \pm 0.680$ kPa, and $81.1 \pm 6.95\%$ for 9.6 mg/mL, respectively. Given the mechanics and the consistency of gel formation and handleability, the 9.6 mg/mL fibrinogen concentration was identified as the ideal concentration to utilize in creating the optimized fibrin coating.

Mechanical properties of collagen gels at 4 mg/mL with and without the addition of 2% *w*/*w* genipin were examined (Table 1). Collagen gels had an average elastic modulus, ultimate tensile strength, failure strength, and elongation of $24.8 \pm 6.77$ kPa, $3.99 \pm 0.808$ kPa, $0.78 \pm 0.386$ kPa, and $118 \pm 40.2\%$, respectively. The average elastic modulus, ultimate tensile strength, failure strength, and elongation of collagen–genipin gels were $38.2 \pm 13.6$ kPa, $5.15 \pm 1.51$ kPa, $0.808 \pm 0.566$ kPa, and $102 \pm 42.3\%$, respectively. No significant differences were found between either group, though the introduction of genipin crosslinking yielded a higher average ultimate tensile strength ($p = 0.149$) and elasticity modulus ($p = 0.068$). Both groups were further tested for cellular toxicity and vessel longitudinal mechanics.

The tensile mechanics between optimized hydrogels of each group varied significantly as shown by the average stress–strain curves (Figure 2). Collagen-based gels were significantly stiffer than fibrin and gelapin ($p < 0.05$) (Figure 2b). The addition of genipin to collagen gels resulted in a significantly higher elastic modulus compared to collagen alone ($p < 0.05$). The ultimate tensile strength of gelapin was significantly stronger than fibrin and collagen alone ($p < 0.05$).

***Genipin Crosslinking Inhibits Fibroblast Viability and Proliferation.*** The cellular proliferation and morphology of fibroblast-embedded hydrogels were assessed through fluorescent live cell imaging over a 7 d period (Figure 3). After 1 d, fibrin gels were noticeably more cell populated demonstrating the cellular compatibility of the fibrinogen and thrombin precursors. Gelapin and collagen–genipin gels showed limited cell survival. Fibroblast morphology exhibited limited cell spreading and live cells were present in all gel types. Following 3 d of culture, PtFibs readily proliferated in fibrin and collagen matrices. However, fibrin encouraged morphological changes from condensed rounded cell bodies towards an elongated spindle shape. The differences in cell proliferation and shape between fibrin and collagen gels became more apparent on day 7 where fibrin yielded highly populated gels consisting of interconnected networks of elongated PtFibs. Conversely, collagen gel cellularity was similar to day 3 and cell morphology remained rounded with limited cell-to-cell connections.

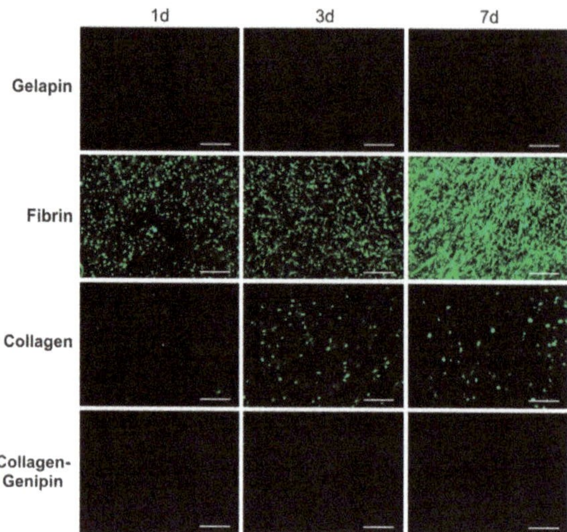



**Figure 3.** Fibrin stimulates fibroblast proliferation and healthy elongated morphology. Live cell fluorescence images of fibroblasts (green) embedded into the hydrogel groups reveal greater cellular viability in fibrin followed by collagen hydrogels over a 7 d culture. Scale bar = 500 μm.

***Fibrin Strengthens Longitudinal Vessel Mechanics With Time.*** Longitudinal tensile mechanics of engineered adventitia vessels altered based on coating type and culture duration as shown by average stress–strain plots and mechanical properties (Figure 4, Table 2). One day after vessel formation, elastic moduli and failure strengths were similar between all groups. However, initially, collagen-coated vessels exhibited the highest ultimate tensile strength of 6.08 ± 2.99 kPa, which was significantly greater in comparison to vessels coated in gelapin ($p < 0.05$).

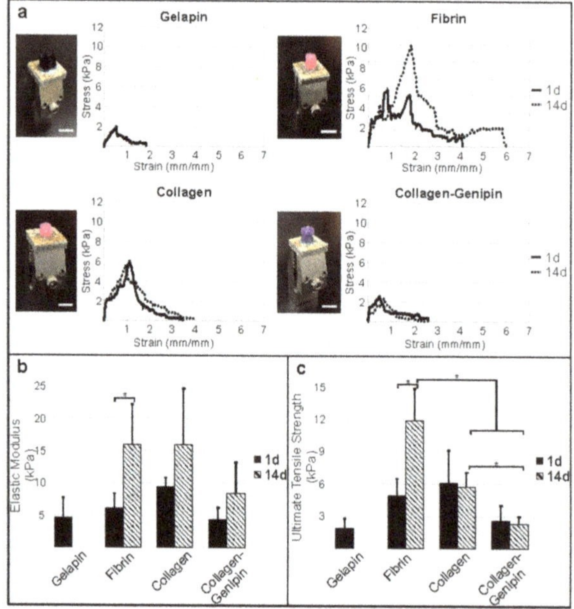



**Figure 4.** Fibrin hydrogel coating with cells strengthens vessel longitudinal mechanics over time. (**a**) Tensile setup and average stress–strain graphs of adventitia vessels coated with each hydrogel type

cultured for 1 d and 14 d. Fibrin gel significantly increased vessel longitudinal (**b**) elastic modulus and (**c**) ultimate tensile strength over the culture duration. Additionally, fibrin was significantly stronger than collagen and collagen–genipin at 14 d. * Denotes significance between groups ($p < 0.005$). Scale bar = 5 mm.

**Table 2.** Average longitudinal tensile properties of hydrogel-coated adventitia vessels.

| Material Coating | Culture Duration | E (kPa) | UTS (kPa) | Max Force (N) | FS (kPa) | Elongation (%) |
|---|---|---|---|---|---|---|
| Gelapin (10:2%) | 1 d | 4.55 ± 3.12 [d] | 1.89 ± 0.827 [d] | 0.103 ± 0.594 | 0.493 ± 0.127 [b] | 99.2 ± 15.6 [b] |
| | 14 d | -- | -- | -- | -- | -- |
| Fibrin (9.6 mg/mL) | 1 d | 6.10 ± 2.22 [c] | 4.91 ± 1.56 [c] | 0.141 ± 0.045 [c] | 1.30 ± 0.56 [a,c] | 248 ± 95.7 [a] |
| | 14 d | 15.9 ± 6.20 [b] | 11.9 ± 2.91 [b,e,g] | 0.203 ± 0.056 [b,e,g] | 2.51 ± 1.29 [b] | 214 ± 119 |
| Collagen (4 mg/mL) | 1 d | 9.34 ± 1.42 [a,f] | 6.08 ± 2.99 [a,f] | 0.185 ± 0.084 [e] | 0.848 ± 0.331 | 173 ± 30.6 [e] |
| | 14 d | 15.9 ± 8.67 | 5.67 ± 1.37 [c,g] | 0.087 ± 0.035 [c,d] | 0.922 ± 0.236 | 242 ± 20.4 [d] |
| Collagen–Genipin (4 mg/mL—2%) | 1 d | 4.36 ± 1.80 [d] | 2.62 ± 1.38 [d] | 0.069 ± 0.039 | 0.645 ± 0.278 | 134 ± 88.4 |
| | 14 d | 8.40 ± 4.76 | 2.31 ± 0.626 [c,e] | 0.063 ± 0.022 [c] | 0.550 ± 0.118 | 121 ± 43.1 |

[a] Statistically significant difference relative to gelapin 1 d ($p < 0.05$). [b] Statistically significant difference relative to fibrin 1 d ($p < 0.05$). [c] Statistically significant difference relative to fibrin 14 d ($p < 0.05$). [d] Statistically significant difference relative to collagen 1 d ($p < 0.05$). [e] Statistically significant difference relative to collagen 14 d ($p < 0.05$). [f] Statistically significant difference relative to collagen–genipin 1 d ($p < 0.05$). [g] Statistically significant difference relative to collagen–genipin 14 d ($p < 0.05$).

Significant differences in mechanical properties were observed following 14 d in culture. Particularly, the extended culture period resulted in a 2-fold increase in the ultimate tensile strength of fibrin-coated vessels from 4.91 ± 1.56 kPa to 11.9 ± 2.91 kPa ($p < 0.005$; Figure 4). The elastic modulus of fibrin vessels significantly increased from 6.10 ± 2.22 kPa on day 1 to 15.9 ± 6.20 kPa on day 14 ($p < 0.05$). However, the mechanics of gelapin, collagen, and collagen–genipin gels did not increase with time in culture. The elastic modulus and ultimate tensile strength of gelapin vessels following 1 d of culture were 4.55 ± 3.12 kPa and 1.89 ± 0.827 kPa, respectively. Gelapin coated vessels were not able to be tested as the exterior coating was fully degraded after 14 d. Interestingly, collagen-coated vessels had a significant decrease in maximum force from 0.185 ± 0.084 N to 0.087 ± 0.035 N, although no significant difference in ultimate tensile strength from 6.08 ± 2.99 kPa on day 1 to 5.67 ± 1.37 kPa was observed.

No significant differences in mechanical properties were observed between collagen–genipin vessel time points. The elastic moduli and ultimate tensile strength of collagen–genipin vessels were 4.36 ± 1.80 kPa and 2.62 ± 1.38 kPa on day 1 and 8.40 ± 4.76 kPa and 2.31 ± 0.626 kPa. Overall, fibrin-coated vessels cultured for longer periods of time had significantly higher ultimate tensile, maximum force, and failure strength relative to those coated in collagen or collagen–genipin ($p < 0.001$).

*Fibrin-Coated Vessels Exhibited a Significant Increase in Circumferential Mechanics and Hemodynamics.* Adventitia vessel circumferential mechanics varied significantly with culture duration (Figure 5). Fibrin-coated vessels cultured for 1 d had an elastic modulus of 21.1 ± 3.52 kPa and ultimate tensile strength of 30.5 ± 8.51 kPa. Following 14 d of culture, the vessel circumferential elastic modulus and ultimate tensile strength significantly increased to 28.6 ± 5.73 kPa ($p = 0.027$) and 42.5 ± 9.69 kPa ($p = 0.05$), respectively (Figure 5c,d).

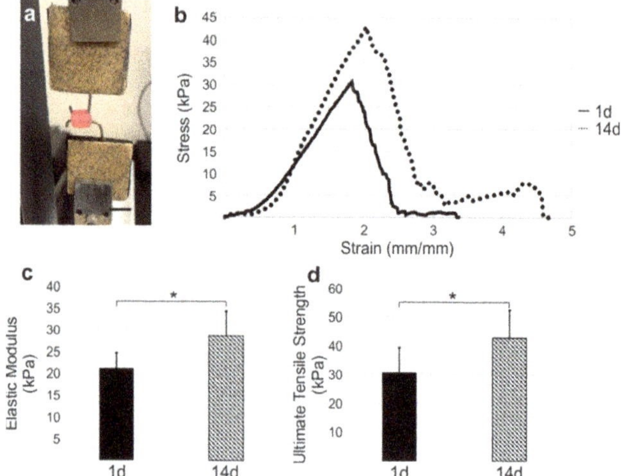

**Figure 5.** Circumferential strength of fibrin-coated vessels increased over time. (**a**) Circumferential tensile setup and (**b**) average stress–strain graphs of fibrin coated adventitia vessels after 1 d and 14 d of culture. Prolonged culture resulted in significantly higher (**c**) elastic moduli and (**d**) ultimate tensile strength. Scale bar = 10 mm. * Denotes significance between groups ($p < 0.05$).

Finally, hemodynamic analysis was performed on tunica media vessels assembled with the optimized fibroblast-embedded fibrin hydrogel coating. Vessels cultured for 16 weeks exhibited a maximum burst pressure of $229 \pm 23.8$ mmHg, representing a physiological blood pressure level.

***ECM Structural Proteins Evident in Vessels.*** Hematoxylin and eosin, Masson's trichrome, and Picrosirius Red stains were performed to visualize the cellular and extracellular matrix content and organization within the adventitia vessels coated with fibrin and collagen without genipin as the optimized groups (Figure 6). Histology showed that the base fibrin hydrogel that is a part of the original ring structure was evident along the borders of the rings. The cells inside the ring exhibited circumferential alignment. The cell-loaded fibrin and collagen coatings formed distinct layers along the abluminal surface of the tissue rings. Collagen deposition was observed within the cellular layer of the engineered vessels of both fibrin and collagen-coated groups (Figure 6b,c,e,f). Collagen fiber and cell orientation concurrently aligned circumferentially along the vessel.

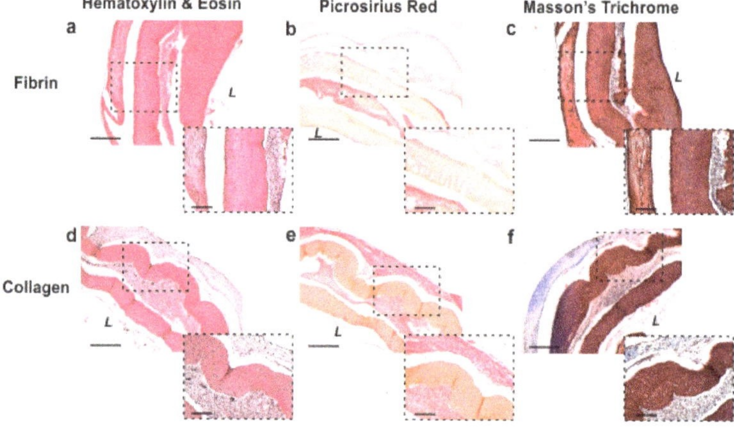


**Figure 6.** Adventitia vessel histology reveals cellular organization and collagen production. Hematoxylin and eosin staining of (**a**) fibrin- and (**d**) collagen-coated vessels show a robust cellular layer

bordered by fibrin hydrogel and localization of the exterior coating on the abluminal surface. (**b,e**) Picrosirius Red and (**c,f**) Masson's Trichrome showed collagen presence within the tissue. Collagen was observed in both fibrin and collagen hydrogel groups with evidence of more collagen in the collagen group. *L* indicates lumen area. Scale bar of 4× images = 500 μm. Scale bar of magnified 10× images (dashed border) = 200 μm.

## 4. Discussion

This work evaluated the use of a cell-loaded hydrogel as a coating for soft tissue engineering applications, demonstrated here for vascular tissue engineering. The main purpose was to determine how to strengthen and improve the cohesiveness, and hence, more importantly, the strength, of engineered soft tissue. This is especially pertinent to our particular engineered vessel methodology wherein independent rings are used to form the final structure. Genipin was explored as a crosslinking agent to further strengthen the tissues; however, in this study, we found that its presence hindered cell viability and hence optimized coatings did not incorporate genipin. Cells were added to the hydrogel coatings to test their contribution to the stiffness of the coating due to the cells' own mechanical properties of their cell membrane. In addition, cells, especially fibroblasts, deposit ECM components such as collagen which further strengthen the coating. Hydrogels with material properties closer to soft tissue mechanics were chosen here for that purpose, specifically different combinations of gelatin, collagen, and fibrin gel. Proprietary hydrogel mixtures from the dental clinic were also investigated called "Silicone Soft Relining System" and "Take1 Advanced" both by Kerr (provided by co-author BF); however, these hydrogels were too stiff to apply as a coating on the soft tissues. The advantage of the gelatin, collagen, and fibrin hydrogels is their tunability of properties with cross-linkers. Standard cross-linker glutaraldehyde was excluded from this study due to its toxicity. Instead, genipin was explored as a viable hydrogel cross-linker. Genipin is a naturally occurring cross-linker derived from the gardenia plant [16,17].

Our results correlate with others' findings, showing increased gelapin stiffness with increased genipin concentration (Table 1) [18,19]. Interestingly, the ultimate tensile strength was inversely related to the genipin concentration of gelapin gels most likely through genipin aggregate formation [20,21]. Higher genipin concentrations had a negative impact on the formation of fibrin hydrogels, although recent studies have demonstrated fibrin–genipin gel formation at lower genipin concentrations [22,23]. The use of a higher level 2% *w/w* genipin may have impeded thrombin enzymatic cleavage of fibrinogen during fibrin gel formation as previously reported [20]. Gelapin and collagen–genipin gels showed decreased cell viability and proliferation (Figure 3). Genipin may reduce the availability of integrin-binding sites which may correspond to the limited viability and proliferation observed in our study [23,24]. While collagen and fibrin were studied extensively as scaffolding materials for soft tissue engineering [2,25,26], direct comparisons of fibroblast compatibility and subsequent changes in tissue mechanical properties over time have remained limited. Despite bearing the lowest mechanical properties as a material alone, fibrin-coated adventitia vessels had the highest longitudinal mechanics after extended culture. Interestingly, minimal collagen production was observed in the cell-loaded fibrin coating group, which may indicate that the fibrin fibers inhibited collagen fibril formation in the coating. Although collagen is typically the primary source of strength in the ECM, this finding suggests that in this case, another mechanism led to increased mechanics, perhaps through increased cell proliferation. TGF-β1 was added to the cultures, which enhances cellular proliferation [27,28].

Vessels were cultured for 1 d and 14 d to assess the effect of time on ECM protein deposition over time by the fibroblasts, primarily collagen. Time also allowed for matrix remodeling, creating a more organized collagen network indicative of a normal, more functional ECM. The effects of longer vessel culture were seen in the increase in collagen

(Figure 5). In addition, vessel mechanics were increased with longer culture, evidenced by an increase in elastic modulus in the fibrin, collagen, and collagen–genipin groups. The ultimate tensile strength was increased with longer culture in the fibrin coating group (Figures 4 and 6). Lastly, prolonging the culture duration to 16 weeks demonstrated the engineered vessels' ability to withstand physiologic pressures with an average burst pressure of 229 ± 23.8 mmHg. As reported in our previous work, our non-coated engineered vessels exhibited a burst pressure of 51.3 ± 2.19 mmHg [1], thus showing that with the newly developed hydrogel coating, a significant increase in burst pressure to physiological blood pressure levels can be achieved while maintaining soft tissue mechanical compliance. Hydrogel degradation was not evident at the later time culture time point as demonstrated by consistency in average wall thickness and tensile mechanics over time. The optimal coating was used in the hemodynamic testing, which did not include genipin.

The use of fibrin in the base ring structure and in the coating may raise questions about the potential for coagulation activation as fibrin is a key component of the coagulation cascade [29]. The advantages of fibrin in biomedical applications are its tunability by modulating its base concentrations of fibrinogen and thrombin, its rapid degradability, and its demonstrated ability to promote collagen production [30]. Fibrin gel has historically been used for many tissue engineering applications, including for vascular tissue engineering [3,9,31–33]. Issues of coagulation have not been reported. In our lab, a previous test of platelet adhesion to our fibrin-based vessel yielded non-detectable platelet adhesion [2].

Collagen, as a structural protein, is able to add strength to tissues. In this study, collagen gel was inferior to fibrin gel due to fibrin gel's ability to better support cell viability and proliferation. Hence, collagen gel was not deemed the ideal coating material.

Creating vascular tissues using tubular cell-loaded hydrogels in cylindrical molds has become common practice in vascular tissue engineering [9,31]. The difference in this work is that our main tissues are formed with stable vascular tissue rings that were then coated with a cell-loaded hydrogel coating. We fabricated a custom silicone half-cylinder mold to ensure even coating circumferentially around the outer surface of the vessels. Localization of the collagen coating to the abluminal surface demonstrated the effectiveness and precision of the developed casting method (Figure 5e,f).

## 5. Conclusions

Here, we demonstrated the efficacy of a cell-loaded hydrogel as a coating to improve the mechanical strength of tissue-engineered soft tissues. Two mechanisms were assessed to strengthen the hydrogel coatings: genipin crosslinking and the inclusion of fibroblasts. Incorporation of genipin increased the elastic modulus of gelatin- and collagen-based hydrogels; however, cell viability was adversely affected. Fibrin hydrogels without genipin facilitated fibroblast attachment and proliferation resulting in the highest tensile strength over prolonged culture. Future work aims to assess the utility of our strengthened engineered vascular tissues as potential patient grafts.

**Author Contributions:** Conceptualization, B.T.W. and M.T.L.; Methodology, B.T.W. and M.T.L.; Validation, B.T.W. and M.T.L.; Formal Analysis, B.T.W. and M.T.L.; Investigation, B.T.W.; Resources, B.F. and M.T.L.; Data Curation, B.T.W.; Writing—Original Draft Preparation, B.T.W. and M.T.L.; Writing—Review and Editing, B.T.W., B.F. and M.T.L.; Supervision, M.T.L.; Project Administration, M.T.L.; Funding Acquisition, M.T.L. All authors have read and agreed to the published version of the manuscript.

**Funding:** B.T.W. was supported by the 5T32HL120822-10 NIH T32 Detroit Cardiovascular Training Program. This work was in part supported by the Fisher Family Foundation grant administered through the Vascular Surgery Department at the Henry Ford Hospital in Detroit, MI, USA.

**Institutional Review Board Statement:** The study was conducted according to the guidelines of the Declaration of Helsinki and approved by the Institutional Review Board in accordance with both Wayne State University (IRB # 054514M1E, approval date 11 June 2019) and Henry Ford Health System (IRB # SiddiquiLam_8958, approval date 27 July 2017) Institutional Review Board (IRB) guidelines.

**Informed Consent Statement:** Written informed consent was obtained from the patient(s) to publish this paper.

**Data Availability Statement:** No new data were created or analyzed in this study. Data sharing is not applicable to this article.

**Acknowledgments:** We thank the staff at Plymouth Family Dentistry for their help in preparing the dental hydrogels that were tested, with special thanks to Kristen Perras for her assistance with the materials.

**Conflicts of Interest:** The authors declare no conflict of interest.

# References

1. Patel, B.; Wonski, B.T.; Saliganan, D.M.; Rteil, A.; Kabbani, L.S.; Lam, M.T. Decellularized dermis extracellular matrix alloderm mechanically strengthens biological engineered tunica adventitia-based blood vessels. *Sci. Rep.* **2021**, *11*, 11384. [CrossRef] [PubMed]
2. Patel, B.; Xu, Z.; Pinnock, C.B.; Kabbani, L.S.; Lam, M.T. Self-assembled Collagen-Fibrin Hydrogel Reinforces Tissue Engineered Adventitia Vessels Seeded with Human Fibroblasts. *Sci. Rep.* **2018**, *8*, 3294. [CrossRef] [PubMed]
3. Pinnock, C.B.; Meier, E.M.; Joshi, N.N.; Wu, B.; Lam, M.T. Customizable engineered blood vessels using 3D printed inserts. *Methods* **2016**, *99*, 20–27. [CrossRef] [PubMed]
4. Strobel, H.A.; Dikina, A.D.; Levi, K.; Solorio, L.D.; Alsberg, E.; Rolle, M.W. Cellular Self-Assembly with Microsphere Incorporation for Growth Factor Delivery within Engineered Vascular Tissue Rings. *Tissue Eng. Part A* **2017**, *23*, 143–155. [CrossRef] [PubMed]
5. Syedain, Z.H.; Tranquillo, R.T. TGF-beta1 diminishes collagen production during long-term cyclic stretching of engineered connective tissue: Implication of decreased ERK signaling. *J. Biomech.* **2011**, *44*, 848–855. [CrossRef]
6. Jeong, Y.; Yao, Y.; Yim, E.K.F. Current understanding of intimal hyperplasia and effect of compliance in synthetic small diameter vascular grafts. *Biomater. Sci.* **2020**, *8*, 4383–4395. [CrossRef]
7. Post, A.; Diaz-Rodriguez, P.; Balouch, B.; Paulsen, S.; Wu, S.; Miller, J.; Hahn, M.; Cosgriff-Hernandez, E. Elucidating the role of graft compliance mismatch on intimal hyperplasia using an ex vivo organ culture model. *Acta Biomater.* **2019**, *89*, 84–94. [CrossRef]
8. L'Heureux, N.; Paquet, S.; Labbe, R.; Germain, L.; Auger, F.A. A completely biological tissue-engineered human blood vessel. *FASEB J.* **1998**, *12*, 47–56. [CrossRef]
9. Syedain, Z.H.; Meier, L.A.; Bjork, J.W.; Lee, A.; Tranquillo, R.T. Implantable arterial grafts from human fibroblasts and fibrin using a multi-graft pulsed flow-stretch bioreactor with noninvasive strength monitoring. *Biomaterials* **2011**, *32*, 714–722. [CrossRef]
10. Noori, A.; Ashrafi, S.J.; Vaez-Ghaemi, R.; Hatamian-Zaremi, A.; Webster, T.J. A review of fibrin and fibrin composites for bone tissue engineering. *Int. J. Nanomed.* **2017**, *12*, 4937–4961. [CrossRef]
11. Chow, M.J.; Turcotte, R.; Lin, C.P.; Zhang, Y. Arterial extracellular matrix: A mechanobiological study of the contributions and interactions of elastin and collagen. *Biophys. J.* **2014**, *106*, 2684–2692. [CrossRef] [PubMed]
12. Bello, A.B.; Kim, D.; Kim, D.; Park, H.; Lee, S.H. Engineering and Functionalization of Gelatin Biomaterials: From Cell Culture to Medical Applications. *Tissue Eng. Part B Rev.* **2020**, *26*, 164–180. [CrossRef] [PubMed]
13. Echave, M.C.; Saenz del Burgo, L.; Pedraz, J.L.; Orive, G. Gelatin as Biomaterial for Tissue Engineering. *Curr. Pharm. Des.* **2017**, *23*, 3567–3584. [CrossRef] [PubMed]
14. Vyborny, K.; Vallova, J.; Koci, Z.; Kekulova, K.; Jirakova, K.; Jendelova, P.; Hodan, J.; Kubinova, S. Genipin and EDC crosslinking of extracellular matrix hydrogel derived from human umbilical cord for neural tissue repair. *Sci. Rep.* **2019**, *9*, 10674. [CrossRef] [PubMed]
15. Sung, H.W.; Huang, R.N.; Huang, L.L.; Tsai, C.C. In vitro evaluation of cytotoxicity of a naturally occurring cross-linking reagent for biological tissue fixation. *J. Biomater. Sci. Polym. Ed.* **1999**, *10*, 63–78. [CrossRef] [PubMed]
16. Wang, C.; Lau, T.T.; Loh, W.L.; Su, K.; Wang, D.A. Cytocompatibility study of a natural biomaterial crosslinker–Genipin with therapeutic model cells. *J. Biomed. Mater. Res. B Appl. Biomater.* **2011**, *97*, 58–65. [CrossRef]
17. Yu, Y.; Xu, S.; Li, S.; Pan, H. Genipin-cross-linked hydrogels based on biomaterials for drug delivery: A review. *Biomater. Sci.* **2021**, *9*, 1583–1597. [CrossRef]
18. Gattazzo, F.; De Maria, C.; Rimessi, A.; Dona, S.; Braghetta, P.; Pinton, P.; Vozzi, G.; Bonaldo, P. Gelatin-genipin-based biomaterials for skeletal muscle tissue engineering. *J. Biomed. Mater. Res. B Appl. Biomater.* **2018**, *106*, 2763–2777. [CrossRef]
19. Wang, S.; Li, K.; Zhou, Q. High strength and low swelling composite hydrogels from gelatin and delignified wood. *Sci. Rep.* **2020**, *10*, 17842. [CrossRef]

20. Gamboa-Martinez, T.C.; Luque-Guillen, V.; Gonzalez-Garcia, C.; Gomez Ribelles, J.L.; Gallego-Ferrer, G. Crosslinked fibrin gels for tissue engineering: Two approaches to improve their properties. *J. Biomed. Mater. Res. A* **2015**, *103*, 614–621. [CrossRef]

21. Ge, L.; Xu, Y.; Liang, W.; Li, X.; Li, D.; Mu, C. Short-range and long-range cross-linking effects of polygenipin on gelatin-based composite materials. *J. Biomed. Mater. Res. A* **2016**, *104*, 2712–2722. [CrossRef] [PubMed]

22. Gupta, N.; Cruz, M.A.; Nasser, P.; Rosenberg, J.D.; Iatridis, J.C. Fibrin-Genipin Hydrogel for Cartilage Tissue Engineering in Nasal Reconstruction. *Ann. Otol. Rhinol. Laryngol.* **2019**, *128*, 640–646. [CrossRef] [PubMed]

23. Panebianco, C.J.; Rao, S.; Hom, W.W.; Meyers, J.H.; Lim, T.Y.; Laudier, D.M.; Hecht, A.C.; Weir, M.D.; Weiser, J.R.; Iatridis, J.C. Genipin-crosslinked fibrin seeded with oxidized alginate microbeads as a novel composite biomaterial strategy for intervertebral disc cell therapy. *Biomaterials* **2022**, *287*, 121641. [CrossRef]

24. Panebianco, C.J.; DiStefano, T.J.; Mui, B.; Hom, W.W.; Iatridis, J.C. Crosslinker concentration controls TGFbeta-3 release and annulus fibrosus cell apoptosis in genipin-crosslinked fibrin hydrogels. *Eur. Cell Mater.* **2020**, *39*, 211–226. [CrossRef]

25. Cummings, C.L.; Gawlitta, D.; Nerem, R.M.; Stegemann, J.P. Properties of engineered vascular constructs made from collagen, fibrin, and collagen-fibrin mixtures. *Biomaterials* **2004**, *25*, 3699–3706. [CrossRef] [PubMed]

26. Lai, V.K.; Frey, C.R.; Kerandi, A.M.; Lake, S.P.; Tranquillo, R.T.; Barocas, V.H. Microstructural and mechanical differences between digested collagen-fibrin co-gels and pure collagen and fibrin gels. *Acta Biomater.* **2012**, *8*, 4031–4042. [CrossRef] [PubMed]

27. Reed, M.J.; Vernon, R.B.; Abrass, I.B.; Sage, E.H. TGF-beta 1 induces the expression of type I collagen and SPARC, and enhances contraction of collagen gels, by fibroblasts from young and aged donors. *J. Cell. Physiol.* **1994**, *158*, 169–179. [CrossRef]

28. Xiao, L.; Du, Y.; Shen, Y.; He, Y.; Zhao, H.; Li, Z. TGF-beta 1 induced fibroblast proliferation is mediated by the FGF-2/ERK pathway. *Front. Biosci.* **2012**, *17*, 2667–2674. [CrossRef]

29. Chapin, J.C.; Hajjar, K.A. Fibrinolysis and the control of blood coagulation. *Blood Rev.* **2015**, *29*, 17–24. [CrossRef]

30. Huang, A.H.; Niklason, L.E. Engineering of arteries in vitro. *Cell. Mol. Life Sci.* **2014**, *71*, 2103–2118. [CrossRef]

31. Helms, F.; Lau, S.; Klingenberg, M.; Aper, T.; Haverich, A.; Wilhelmi, M.; Boer, U. Complete Myogenic Differentiation of Adipogenic Stem Cells Requires Both Biochemical and Mechanical Stimulation. *Ann. Biomed. Eng.* **2020**, *48*, 913–926. [CrossRef] [PubMed]

32. Liang, M.S.; Andreadis, S.T. Engineering fibrin-binding TGF-beta1 for sustained signaling and contractile function of MSC based vascular constructs. *Biomaterials* **2011**, *32*, 8684–8693. [CrossRef] [PubMed]

33. Syedain, Z.H.; Graham, M.L.; Dunn, T.B.; O'Brien, T.; Johnson, S.L.; Schumacher, R.J.; Tranquillo, R.T. A completely biological "off-the-shelf" arteriovenous graft that recellularizes in baboons. *Sci. Transl. Med.* **2017**, *9*, eaan4209. [CrossRef] [PubMed]

 *bioengineering*

Article

# Culturing of Cardiac Fibroblasts in Engineered Heart Matrix Reduces Myofibroblast Differentiation but Maintains Their Response to Cyclic Stretch and Factor β1

Meike C. Ploeg [1], Chantal Munts [1], Tayeba Seddiqi [1], Tim J. L. ten Brink [2], Jonathan Breemhaar [3], Lorenzo Moroni [2], Frits. W. Prinzen [1] and Frans. A. van Nieuwenhoven [1,*]

[1] Department of Physiology, Cardiovascular Research Institute Maastricht (CARIM), Maastricht University, 6200 MD Maastricht, The Netherlands
[2] Institute for Technology-Inspired Regenerative Medicine (MERLN), Maastricht University, 6200 MD Maastricht, The Netherlands
[3] MosaMeat, 6229 PM Maastricht, The Netherlands
[*] Correspondence: f.vannieuwenhoven@maastrichtuniversity.nl

**Abstract:** Isolation and culturing of cardiac fibroblasts (CF) induces rapid differentiation toward a myofibroblast phenotype, which is partly mediated by the high substrate stiffness of the culture plates. In the present study, a 3D model of Engineered Heart Matrix (EHM) of physiological stiffness (Youngs modulus ~15 kPa) was developed using primary adult rat CF and a natural hydrogel collagen type 1 matrix. CF were equally distributed, viable and quiescent for at least 13 days in EHM and the baseline gene expression of myofibroblast-markers alfa-smooth muscle actin (Acta2), and connective tissue growth factor (Ctgf) was significantly lower, compared to CF cultured in 2D monolayers. CF baseline gene expression of transforming growth factor-beta1 (Tgfβ1) and brain natriuretic peptide (Nppb) was higher in EHM-fibers compared to the monolayers. EHM stimulation by 10% cyclic stretch (1 Hz) increased the gene expression of Nppb (3.0-fold), Ctgf (2.1-fold) and Tgfβ1 (2.3-fold) after 24 h. Stimulation of EHM with TGFβ1 (1 ng/mL, 24 h) induced Tgfβ1 (1.6-fold) and Ctgf (1.6-fold). In conclusion, culturing CF in EHM of physiological stiffness reduced myofibroblast marker gene expression, while the CF response to stretch or TGFβ1 was maintained, indicating that our novel EHM structure provides a good physiological model to study CF function and myofibroblast differentiation.

**Keywords:** stiffness; stretch; cardiac fibroblast; three dimensional

Citation: Ploeg, M.C.; Munts, C.; Seddiqi, T.; ten Brink, T.J.L.; Breemhaar, J.; Moroni, L.; Prinzen, F.W.; van Nieuwenhoven, F.A. Culturing of Cardiac Fibroblasts in Engineered Heart Matrix Reduces Myofibroblast Differentiation but Maintains Their Response to Cyclic Stretch and Transforming Growth Factor β1. *Bioengineering* **2022**, *9*, 551. https://doi.org/10.3390/bioengineering9100551

Academic Editors: Ngan F. Huang, Brandon J. Tefft and Ning Sun

Received: 8 September 2022
Accepted: 10 October 2022
Published: 14 October 2022

## 1. Introduction

The cardiac extracellular matrix (ECM) is a network of structural proteins, mostly collagen fibers, which provides structural stability, tensile strength but also alignment cues, biochemical signals and mechanical support to surrounding cells [1–3]. Cardiac fibroblasts (CF) are the cells producing the structural and regulating components of the ECM [4,5] and are therefore important for maintaining the integrity of the ECM [6,7]. In response to injury CF become activated, then differentiate to so called myofibroblasts [8,9] showing special morphological and functional characteristics, such as the expression of alpha smooth muscle actin (αSMA, encoded by the ACTA2 gene) [9,10]. Myofibroblasts are key players in cardiac structural remodeling [11–14], producing excessive collagen, resulting in cardiac fibrosis and increased myocardial stiffness [15,16].

CF can be isolated from the heart and cultured to study their function in vitro. For many years, these studies have been performed in monolayers (defined as two dimensional (2D)-cultures). In this model, cells attach to a flat surface and cell-matrix attachments are restricted to one plane, while in vivo these attachments are present all around the cells [17–19]. Two-dimensional culture conditions also limit cell–cell interactions, as cells grow in monolayers [20]. In addition, culturing CF on hard plastic cell culture plates promotes CF activation and differentiation into myofibroblasts [21,22]. Stiffness is an important stimulus in this process [23], shown in cardiac fibroblasts [24] but also other types of fibroblasts [25–27]. Therefore, several groups started generating 3D culture systems to allow in vitro investigation of the cell–matrix interactions in a more physiologically relevant environment [17–19].

The aim of the present study was the development of a 3D cell culture model of engineered heart matrix (EHM) of physiological stiffness and to compare CF function in EHM structures vs. 2D monolayers and their response to TGFβ1 and mechanical stimulation. To this purpose, primary adult rat CF were cultured in a natural collagen type 1 hydrogel and stiffness was determined. To gain insight into the CF activation state, we measured gene expression of genes related to CF activation and myofibroblast differentiation: alfa-smooth muscle actin (Acta2) [28], connective tissue growth factor (Ctgf) [29] transforming growth factor beta 1 (TGFβ1) [30,31] and brain natriuretic peptide (Nppb) [32]. Baseline gene expression levels of these genes in EHM cultures were compared with CF cultured in 2D monolayers both on cell culture plastic and on Bioflex silicone bottom plates. Finally, EHM responses to cyclic stretch (10%, 1 Hz) and TGFβ1, both established stimuli for CF-activation, were determined.

## 2. Materials and Methods

### 2.1. Isolation and Culturing of CF

CF were isolated from cardiac ventricles (combined left and right) of adult surplus rats from any age, weight, sex or breed ($n$ = 48). Most of the rats used were either from the Lewis or Wistar strain and aged between 5 and 52 weeks. Rat cardiac ventricular fibroblasts were isolated and cultures as previously described [33–35] in Dulbecco's modified eagles medium (DMEM; no. 22320, Gibco, Thermo Fisher Scientific, Waltham, MA, USA) supplemented with 10% ($v/v$) fetal bovine serum (FBS, Gibco), gentamicin (50 µg/mL, Gibco), 1% ($v/v$) Insulin-Transferrin-Selenium-Sodium Pyruvate (ITS-A, Gibco), basic fibroblast growth factor (1 ng/mL, Gibco) and vitamin C (500 uM, Sigma Aldrich, Saint Louis, MO, USA) (CF growth medium, CFGM) on standard cell culture flasks (Cellstar, Greiner Bio-One, Frickenhausen, Germany). The vast majority of these cells are fibroblast-like cells and these primary fibroblasts were used between passage 1 and 3. Experiments were performed with approval of the Animal Ethical Committee of Maastricht University (DEC-2007-116, 31 July 2007) and conformed to the national legislation for the protection of animals used for scientific purposes.

### 2.2. Assembly of the Ring Formation Molds

Sylgard-184 silicone elastomer base and curing agent (Dow Chemical, Midland, MI, USA) were mixed together and poured into a well of a 12-well culture plate. Custom made 3D-printed casts (Mosa Meat, Maastricht, The Netherlands) were used to provide the shape with an outer diameter of 22 mm. An area with a 12 mm diameter was created to load 250 uL of gel. A 2 mm central pole created the ring shape (Figure 1). The mold was allowed to cure at room temperature for 3 days after which the 3D-printed casts were removed. The custom-made molds were cleaned and sterilized by autoclavation.

### 2.3. Collagen Hydrogel

Collagen type 1 from rat tail (5 mg/mL, Ibidi, Gräfelfing, Germany) was diluted with sterile water, 10× DMEM, 20× NaHCO$_3$ and CF until the desired final collagen concentration (between 1 and 3 mg collagen/mL) and cell density, with the final gel containing 1× DMEM.

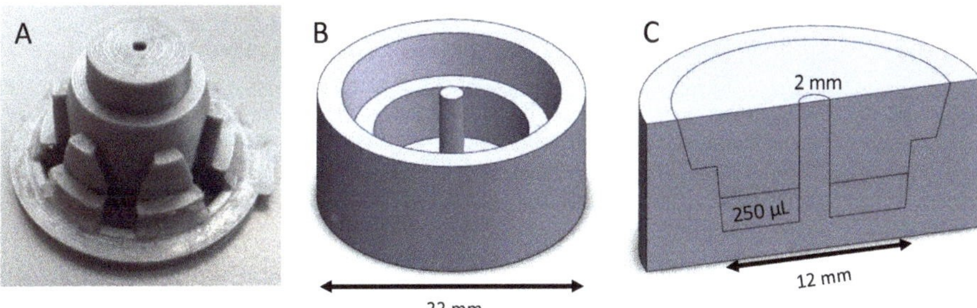

**Figure 1.** Silicone mold design. Custom made 3D-printed casts to provide the shape for the silicone mold (**A**); The silicone mold had a diameter of 22 mm (**B**); and a centrally placed post 2 mm (**C**). The initial gel volume was 250 µL and after gelation 500 µL of the medium was added on top.

### 2.4. Engineered Heart Matrix Ring (EHM-Ring) Formation

CF were grown in CFGM on Cellstar cell culture flasks to about 80–90% confluency, detached using trypsin-EDTA (Gibco) and taken up in CFGM. Cells were centrifuged at 1500 rpm for 5 min and then resuspended in CFGM to reach a concentration of 10 million cells/mL. The cells were diluted to 1 million cells/mL into the hydrogel mixture. The hydrogel –cell mixture (250 µL) was reverse pipetted into the molds and put into a 5% $CO_2$ incubator at 37 °C. After approximately 1 h, to polymerize the gel, 500 µL of CF maintenance medium (CFMM, consisting of CFGM with 1% FBS) was added on top of the gels.

### 2.5. Engineered Heart Matrix Fiber (EHM Fiber; Flexcell Tissue Train)

Using the optimized EHM ring protocol longitudinal EHM fibers were formed in the Flexcell Tissue Train system. This was achieved by pipetting 200 µL of the hydrogel–cell mixture between the two collagen-1-coated anchors of the Tissue Train culture plates (6-well plates, Flexcell Dunn Labortechnik, Asbach, Germany) atop of the Trough Loaders (Figure 2). After approximately 1 hour of polymerization of the gel, 4 mL of CFGM was added on top of the gels. The next day the CFGM was replaced by CFMM. Subsequently, gels were subjected to cyclic stretch (10%, 1 Hz), (Flexcell FX-5000 strain unit, Dunn Labortechnik) for 4 h or 24 h. Control, non-stretched gels were subjected to identical conditions but without stretch being applied.

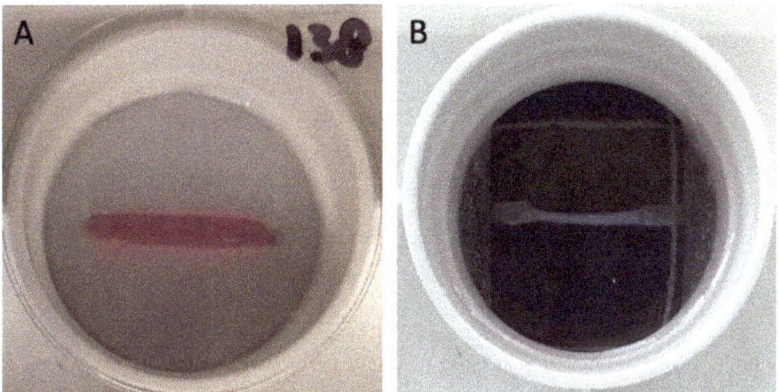

**Figure 2.** Images of EHM fibers in Flexcell Tissue Train culture plates. Image (**A**) shows the (pink) gel after pipetting in the 6 well Flexcell Tissue Train culture plate, still in the Trough Loader (white bottom) providing the mold for the gel; Image (**B**) shows the EHM fiber between the two anchors after polymerization of the gel. The black and white grid underneath represents 3 cm by 3 cm. The diameter of the well was 3.5 cm.

## 2.6. Measurement of EHM Stiffness

To determine the stiffness of the EHM ring, mechanical analysis was performed using an Electroforce-3200 Series III multiaxial tensile tester (TA Instruments, Asse, Belgium) combined with a 1000 gf (10 N) load cell (1 kg/cm$^2$), as previously described [36,37]. Test setups and data acquisition were directed through the WinTest 7 operational software (TA Instruments). The displacement mode of loading materials was controlled through vertical, axial movement with a motorized extension arm (DispE, −40/40 mm). The EHM ring was locked into place using a custom-made Radial Tensile Strength tool (MERLN, Institute for Technology-Inspired Regenerative Medicine, Maastricht University, Maastricht, The Netherlands), based on previous research [38,39].

Uniaxial displacement was applied at a rate of 1% strain per second (0.04 mm/sec.) until EHM ring failure. Load (N) and displacement (mm) were recorded over time at a rate of 20 points/second. The obtained raw datasets were processed in Microsoft Excel and converted to a dataset representing exerted stress ($\sigma$, kPa) over strain ($\varepsilon$, %). Noise within the stress–strain curve was reduced using a moving average analysis, where the interval average was set at 20 points to equalize all measured data points per second. Young's moduli were calculated from the slope of 15% strain residing within the elastic region of the stress/strain curve.

## 2.7. Histology and Immunohistochemistry (IHC)

EHM rings or fibers were washed in PBS (Thermo Fisher), fixed in 4% paraformaldehyde (Klinipath) for 20 min, stained in eosin (J.T.Baker) for 1 h and stored overnight in 70% ethanol (Sigma Aldrich). Finally, the EHM ring or fiber was embedded into paraffin wax. Sections of 5 μm were cut using a rotary microtome.

Hematoxylin and eosin (H&E) staining was performed to gain insight in the cellular distribution and the extracellular matrix (ECM) structure. After rehydration, the slides were placed in Hematoxylin (5 min), washed with running tap water (10 min), placed in eosin (1 min) and washed with demi water. Dehydration steps were performed and the slides were closed with Entallan. Images were obtained using a Leica Microscope (5×, 10× or 20× magnification).

Vimentin and CNA35 IHC were performed to visualize vimentin-positive cells, implicated on being CF and/or collagen matrix. Slides were stained with vimentin antibody (1/150 dilution, ab92547, Abcam, Cambridge, UK) followed by appropriate secondary antibody (1/500 dilution) or adding CNA35 (1/100 dilution) [40,41]. Sections were further incubated in 4′,6′-diamidino-2-phenylindole (DAPI)-containing mounting medium (Vector Laboratories, Burlingame, USA) to stain nuclei. Images were obtained using a Leica fluorescent microscope (40× magnification) or a Leica SPE confocal microscope (63× magnification).

## 2.8. Gene Expression Analysis

Total RNA was isolated from cells using an RNA isolation kit (Omega Biotek, Norcross, GA, USA) and reversed transcribed into cDNA using the iScript cDNA synthesis kit (Biorad, Hercules, CA, USA) according to the manufacturer's instructions and previously described [32]. Real-time PCR was performed on an CFX96 Touch Real-Time PCR Detection System using iQ SYBR-Green Supermix (Biorad). Gene expression levels of Alpha-smooth muscle actin (Acta2), Connective tissue growth factor (Ctgf), Transforming growth factor, beta 1 (Tgfβ1) and Brain Natriuretic Peptide (Nppb) were normalized using the housekeeping gene Cyclophilin-A (Cyclo), and their relative expression was calculated using the comparative threshold cycle (Ct) method by calculating $2^{\Delta Ct}$ (e.g., $2^{(Cyclophilin\ Ct–Nppb\ Ct)}$). The gene expression values were multiplied by 1000 (formula $1000 \times 2^{\Delta Ct}$), to enhance readability. The sequences of the specific primers used are provided below (Table 1).

**Table 1.** Gene specific primer sequences used for real-time qPCR.

| Gene | Forward Primer | Reverse Primer |
| --- | --- | --- |
| Alpha-Smooth muscle actin (Acta2) | AAGGCCAACCGGGAGAAAAT | AGTCCAGCACAATACCAGTTGT |
| Connective tissue growth factor (Ctgf) | CACAGAGTGGAGCGCCTGTTC | GATGCACTTTTTGCCCTTCTTAATG |
| Transforming growth factor, beta 1 (Tgfβ1) | GCACCATCCATGACATGAAC | GCTGAAGCAGTAGTTGGTATC |
| Brain Natriuretic Peptide (Nppb) | AGACAGCTCTCAAAGGACCA | CTATCTTCTGCCCAAAGCAG |
| Cyclophilin-A (Cyclo) | CAAATGCTGGACCAAACACAA | TTCACCTTCCCAAAGACCACAT |

*2.9. Statistics*

Data are presented as average, average ± standard deviation or individual data points (indicating separate CF isolations) and were analyzed with the Wilcoxon matched pairs test, the Kruskal–Wallis test, or the Dunn posthoc test where appropriate (Graphpad PRISM V9). Differences were considered statistically significant when $p < 0.05$.

## 3. Results
### 3.1. Optimizing CF Cell Density in EHM Rings

Different CF cell densities were cultured in EHM rings using a hydrogel collagen concentration of 1 mg/mL in CFMM. EHM ring formation (compaction) was clearly less in the two lower cell densities (Figure 3). The two higher cell densities showed considerably more compaction leading to an EHM ring with a wall thickness of approximately 1.3 mm when using 2000 cells/µL, while using 400 cells/µL led to a wall thickness of approximately 2.1 mm.

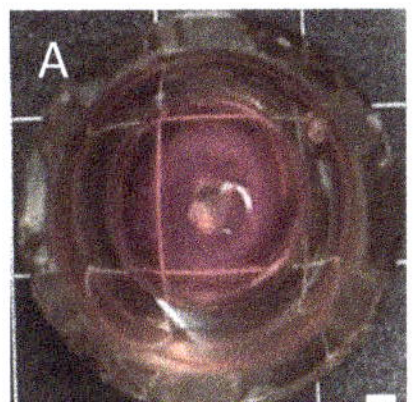
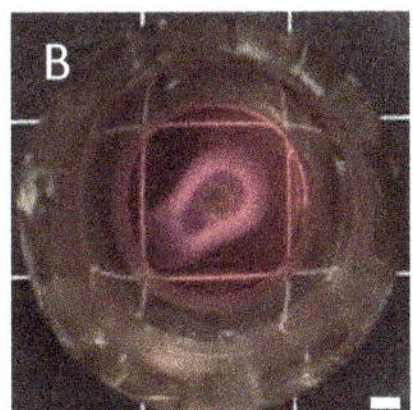
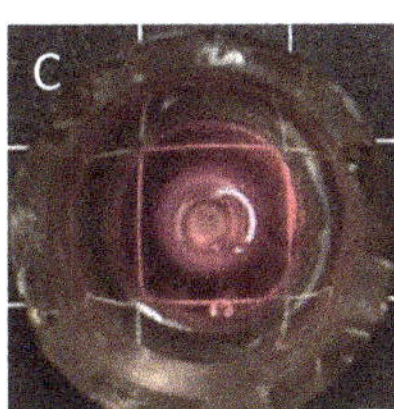
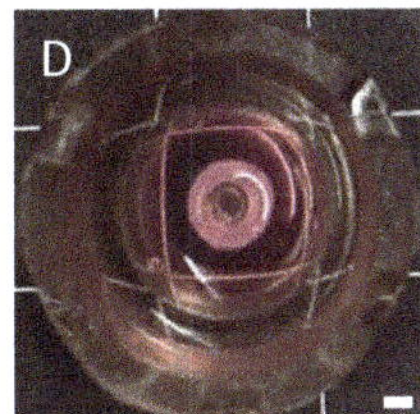
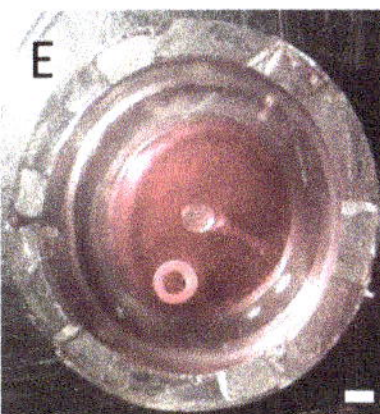


**Figure 3.** Images of EHM rings taken 24 h after casting the gel using different cell densities (**A–D**): 100; 200; 400 and 2000 CF cells/µL using a 1 mg/mL collagen concentration. The images show the top view of the EHM-rings within the clear silicone mold on a black background with white grid (1 cm by 1 cm). Surrounding the central pole is a light pink circle, most left vaguely showing, becoming clearer with increasing the cell density; The final image (**E**) shows the EHM-ring (8000 cells/µL, 1 mg/mL collagen concentration) after 7 days of culturing, next to the central pole. The EHM ring has been removed from the central pole manually to enhance visibility. Bars represent 2 mm.

### 3.2. Optimizing the EHM Initial Hydrogel Collagen Concentration

Two different initial hydrogel collagen concentrations (1 and 3 mg/mL) were used to vary the stiffness of the EHM ring [42]. Gel compaction of the 3 mg/mL collagen EHM-ring was reduced resulting in an increased wall thickness compared to EHM ring of 1 mg/mL collagen ($4.9 \pm 0.9$ mm vs $1.8 \pm 0.6$ mm, $n = 3$) (Supplementary Figure S1). This is reflected by the HE staining (Supplementary Figure S1) where the cell density was lower in the 3 mg/mL collagen compared to the 1 mg/mL collagen EHM ring ($65 \pm 5$ vs $205 \pm 25$ CF/mm$^2$, $n = 2$). RNA isolation from the 3 mg/mL collagen EHM ring resulted in a lower RNA yield when compared to the 1 mg/mL collagen EHM ring (Supplementary Figure S2). Moreover, qPCR analyses revealed increased mRNA expression of Acta2 and Ctgf in the 3 mg/mL collagen EHM-ring, indicating processes towards myofibroblast differentiation, while there was no effect on Tgfβ1 mRNA expression (Supplementary Figure S2). Based on these results, an optimal collagen concentration of 1.5 mg/mL was chosen for further experiments. Using these conditions, we cultured EHM rings for up to 13 days with an average RNA yield of 9.6 pg/cell, measured at day 1, 6, 10 and 13, indicating consistency in viable and quiescent cells over the culturing period. Both histological and immunohistological analysis showed even CF distribution and structural alignment within the EHM structure (Figure 4).

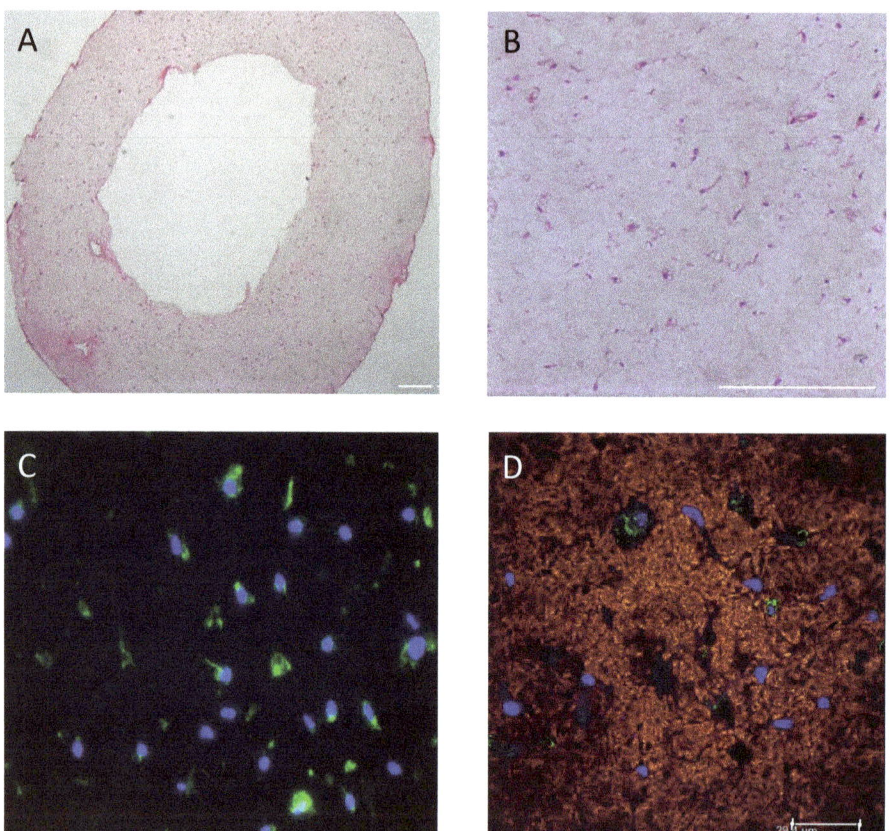

**Figure 4.** Histological and immunohistological analysis of EHM-ring sections. HE staining of an EHM ring section showing even distribution of the cells throughout the EHM ring structure. EHM ring was formed using 1.5 mg/mL collagen concentration and 1000 CF cells/µL gel and was fixed and stained 24 h after casting. Darker dots implicate CF. Magnification 5× (**A**); or 20× (**B**). Bars represent 200 µm. Vimentin (green) and DAPI (blue) staining (**C**); and Z-stack image of CNA35 (red) and Vimentin (green) staining nuclei are blue (DAPI) (**D**) (bar represents 29.4 µm).

### 3.3. EHM Stiffness

EHM stiffness was measured at two different timepoints: 24 h and 48 h after casting the ring structures. The EHM ring was locked into place using a custom-made Radial Tensile Strength tool at 0% elongation (Figure 5A) and elongation at break (Figure 5B). After 24 h the average Young's modulus was 15 kPa and increased to 25 kPa after 48 h (Figure 5C). The stress–strain ($\sigma/\varepsilon$) curve of the moving average showed the elastic region, followed by the plastic region and breaking point (Figure 5D). These results indicate that the stiffness of our EHM rings was in the same order of magnitude as the passive stiffness of myocardial tissue in vivo, which has been described to be around 10 kPa [43–45]. For comparison, the Bioflex plates have been estimated to have a stiffness of ~1000 kPa and plastic culture plates have a stiffness in the gigapascal range [45].

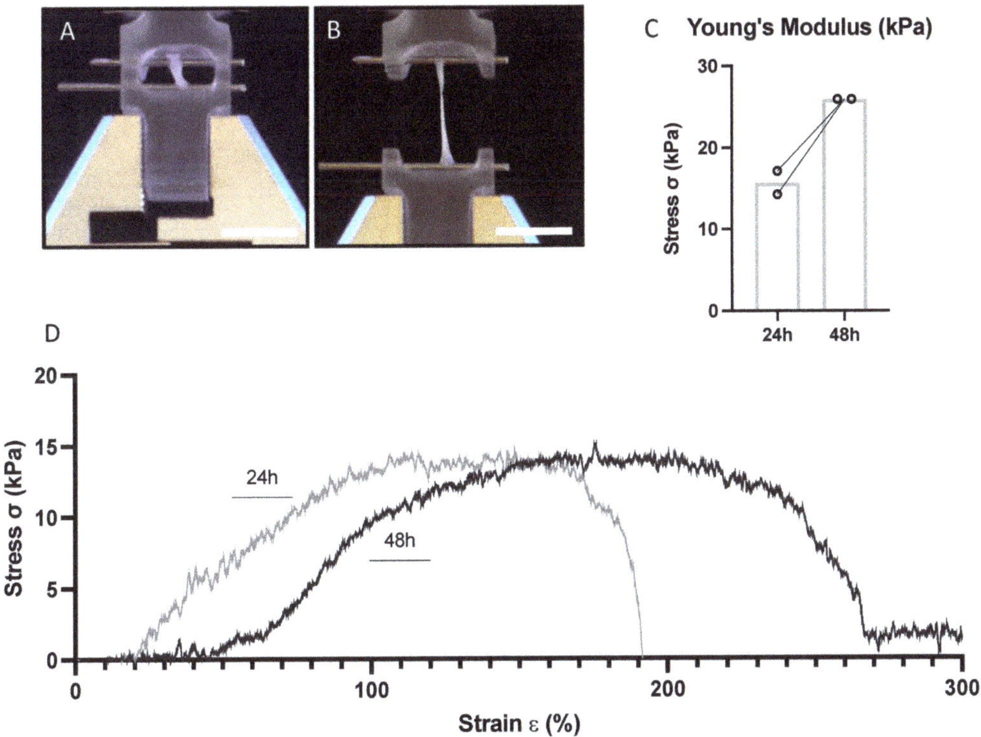



**Figure 5.** Stiffness measurements of the EHM ring (1.5 mg/mL collagen concentration; 1.000 cells/μL gel) at two different timepoints (24 h and 48 h). Images of tensile analysis at 0% elongation (**A**) and at elongation at break; (**B**). Bar represents 10 mm. Tensile stiffness displayed as Young's Modulus ($n = 2$); (**C**); Stress–strain ($\sigma/\varepsilon$) curve of the moving average (20 points/strain value) ($n = 1$) after 24 h and 48 h (**D**).

### 3.4. Baseline CF Gene Expression in 3D (EHM) and 2D (Monolayer) Culturse

Gene expression of Nppb and Tgfβ1, was higher in EHM compared to both regular hard plastic cell culture plates and Bioflex cell culture plates (Figure 6). The opposite was true for the gene expression of Ctgf and Acta2, where the expression was much lower in EHM fibers compared to both hard plastic and Bioflex cell culture plates (Figure 6). The mRNA expression of Ctgf and Acta2 decreased with decreasing the stiffness of the culture substrate.

### 3.5. Stretch of EHM Fibers in the Flexcell Tissue Train System

Rat CF in EHM fibers exposed to 4 h cyclic stretch (10%, 1 Hz) showed a significantly higher gene expression (2.2-fold) of Nppb compared to non-stretched controls (Figure 7A, EHM). Since similar experiments have previously been performed in 2D CF monolayers on Bioflex plates [32], the effects of cyclic stretch were also compared between Bioflex plates and EHM fibers. No effect of 4 h cyclic stretch (10%, 1 Hz) was found on the gene expression of Tgfβ1, Ctgf and Acta2 in the EHM fibers. In EHM fibers exposed to 24 h of cyclic stretch (10%, 1 Hz) the increase in Nppb remained (3.0-fold). In addition, there was an increased mRNA expression of Ctgf (2.1-fold) and Tgfβ1 (2.3-fold) compared to non-stretched controls (Figure 7B, EHM). Acta2 gene expression showed a 1.6-fold increase after 24h, but this difference did not reach statistical significance. Baseline gene expression differences as indicated in Figure 6 were also seen when comparing the non-stretch conditions of flex plates and EHM culture conditions.

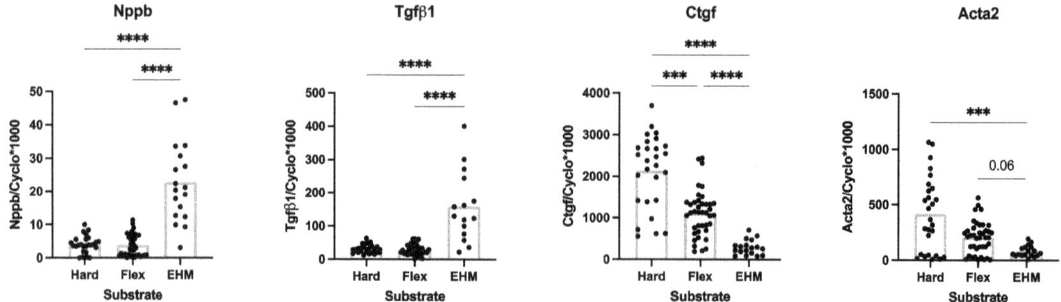

**Figure 6.** Baseline gene expression levels of Nppb, Tgfβ1, Ctgf and Acta2 in CF cultured on the different substrates: 2D monolayer on hard plastic cell culture plates (Hard), 2D monolayer on Bioflex cell culture plates (Flex) and 3D EHM fibers (EHM). Data are presented as relative mRNA levels normalized to house-keeping gene cyclophylin ($n$ = 15–40). *** $p < 0.001$; **** $p < 0.0001$. Bar indicates mean.

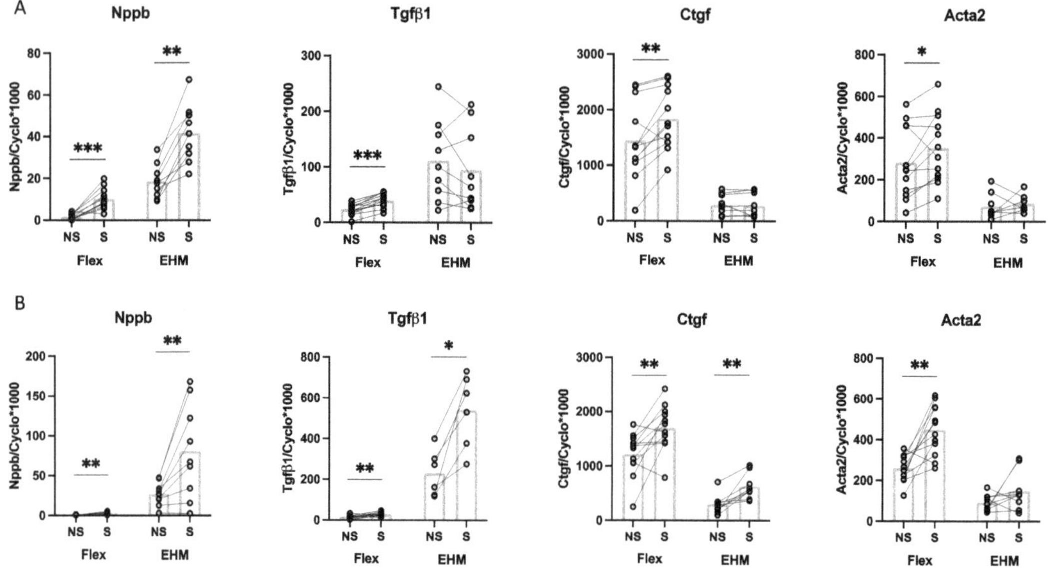

**Figure 7.** The effect of 4 h (**A**); and 24 h (**B**) cyclic stretch (10%, 1 Hz) on gene expression in CF cultured on 2D monolayer Bioflex plates (Flex) (data from [32]) or 3D EHM fibers (EHM) conditions. Presented are the relative gene expression levels of Nppb, Tgfβ1, Ctgf and Acta2 in 2D ($n$ = 5–12) and 3D ($n$ = 6–9) in stretch (S) and non-stretch (NS) conditions. * $p < 0.05$; ** $p < 0.01$; *** $p < 0.001$. Bar indicates mean.

### 3.6. Effect of TGFβ1 Stimulation on CF Gene Expression in 2D (Monolayer) and 3D (EHM) Cultures

TGFβ1-stimulation (1 ng/mL) of CF for 24 h in 2D monolayers on hard plates showed a significant reduction in Nppb gene expression, no effect on Tgfβ1 gene expression and a significant induction of Acta2 and Ctgf gene expression (Figure 8A, Hard). Similar effects of TGFβ1 stimulation on Nppb, Tgfβ1, Ctgf and Acta2 gene expression were observed in 2D monolayers on Bioflex plates (Figure 8A, Flex). Stimulation with TGFβ1 in 3D EHM rings showed no effect on Nppb expression, although as described above Nppb baseline expression levels were higher in the EHM than in the 2D cultures. TGFβ1 stimulation in EHM significantly increased Tgfβ1 gene expression (1.6-fold) (Figure 8A, EHM). In addition, Ctgf showed 1.6-fold increased expression upon TGFβ1 stimulation, although this did not reach statistical significance (Figure 8B). No clear effect of TGFβ1 stimulation on Acta2 was observed in EHM (Figure 8). These results indicate that although the baseline gene expression levels are different in 3D EHM cultures, CF are still capable of responding to TGFβ1 in a similar way as they would in 2D monolayers.

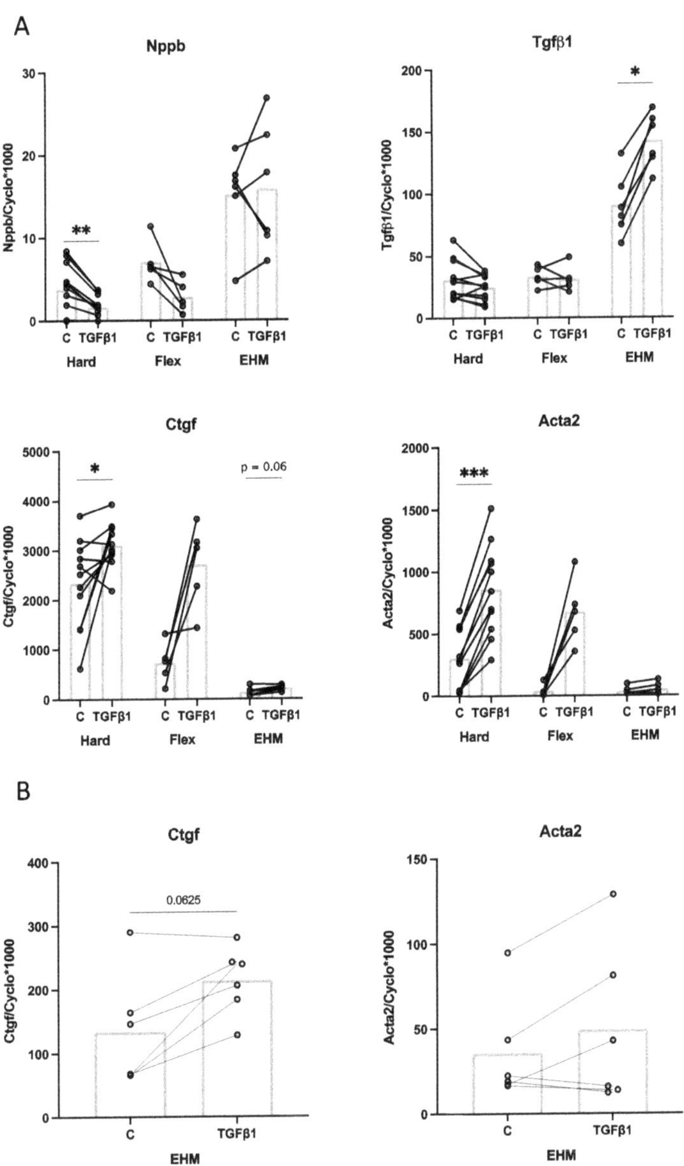

**Figure 8.** The effect of TGFβ1 stimulation of rat CF for 24 h cultured on different substrates. Presented are the relative gene expression levels of Nppb, Tgfβ1, Ctgf and Acta2 after stimulation with TGFβ1 (TGFβ1) or control (C) conditions for 24 h in 2D hard plastic cell culture plates (Hard, *n* = 11), 2D Bioflex plates (Flex, *n* = 5) and 3D EHM (EHM, *n* = 6) (**A**); Graphs of the relative gene expression levels of Ctgf and Acta2 in EHM were enhanced to improve readability (**B**). * *p* < 0.05; ** *p* < 0.01; *** *p* < 0.001. Bar indicates mean.

## 4. Discussion

In this study we designed self-assembling EHM rings and EHM fibers composed of rat tail collagen 1 and adult rat ventricular CF, as models for 3D culturing. The stiffness of our EHM rings was ~15 kPa. Comparison of CF gene expression between 2D monolayers and 3D EHM revealed reduced gene expression of Acta2 and Ctgf in EHM, indicating a more quiescent CF state. Cyclic stretch and TGFβ1 stimulation of EHM structures showed CF activation, comparable to 2D cultures of CF. These data indicate that the EHMs provide a more physiological model to study CF function.

### 4.1. Influence of Substrate Stiffness on Baseline CF Gene Expression

Our finding of lower baseline gene expression of Acta2 and Ctgf in 3D EHM cultures compared to 2D monolayers is in line with results from other studies showing that culturing CF on hard plastic cell culture plates promotes myofibroblast differentiation [45–47]. Moreover, human CF cultured in low stiffness GelMA gels also showed a reduced expression of ACTA2 [48]. A more quiescent state of fibroblasts cultured in 3D was also found in human fetal lung fibroblasts [49] and tendon fibroblasts [50]. Therefore, both cardiac and non-cardiac CF respond to 3D culturing in a similar way. Taken together, fibroblasts appear to remain in a more quiescent state, i.e., less prone to differentiate into myofibroblasts, when cultured in EHM compared to 2D monolayers. Given the large difference in stiffness between the EHM (~15 kPa) and culture plates (>1000 kPa), it is likely that the quiescent state of CF in EHM, results from the more physiological stiffness of the 3D culture substrates [46,51].

### 4.2. Possible Influence of Culture Medium Differences on Baseline CF Gene Expression

Aside from differences in stiffness between EHM and 2D monolayers, differences due to the use of different culture media cannot be excluded and may affect baseline CF gene expression. The EHM fibers were cultured in CFMM, containing 1% fetal bovine serum (FBS), while the experiments in 2D monolayers were performed in serum-free conditions. FBS is known to have an effect on the expression of many genes [35]. However, for Acta2 it is known that culturing in the presence of serum increases the expression [24,35], while we showed a lower Acta2 expression in our EHM-fibers (CFMM, 1% serum), compared to 2D monolayers, indicating that this lower Acta2 mRNA expression is unlikely to be caused by serum. Galie et al. investigated the effect of serum conditions (5% or 10% FBS) on Tgfβ1 expression in rat cardiac fibroblasts cultured in 3D collagen matrix. Culturing the gels for 24 h resulted in a lower Tgfβ1 mRNA expression in 10% serum compared to 5% serum, measurements during other timepoints (6, 48 and 120 h) showed no difference between 5% or 10% serum on Tgfβ1 mRNA expression [24]. These results implicate that the higher Tgfβ1 mRNA expression we showed in the EHM fibers was most likely not caused by the 1% serum used in those culture conditions. It is important to note here that our FBS percentage of 1% was much lower than those used by Galie and colleagues [24].

### 4.3. Effect of Cyclic Stretch and TGFβ1 on CF Gene Expression

Cyclic stretch for 24 h increased the gene expression of Nppb, Tgfβ1 and Ctgf in both the 2D monolayer and 3D EHM. Acta2 expression was significantly increased in stretched 2D monolayers, but the 1.6-fold increase in EHM was not statistically significant ($p = 0.20$). This lack of significance was most likely caused by low statistical power and large individual variation. Taken together, these data indicate that stretch initiates fibroblast activation, ultimately leading to myofibroblast differentiation. Our results showing an increased mRNA expression of Acta2 in EHM after exposure to cyclic stretch are supported by previous research in NIH 3T3 fibroblasts [52] and marrow-derived progenitor cells [53], using a collagen and fibrin 3D construct, respectively. Primary murine dermal fibroblasts within a collagen 3D construct exposed to 24 h cyclic stretch support our results in showing an increased expression of Tgfβ1 and Ctgf [54]. However, rat cardiac fibroblast-seeded collagen gels exposed to 5% cyclic strain showed a decrease in Tgfβ1 mRNA expression compared to controls [55]. Differences could be attributed to the type of mechanical stimulation applied, compression vs. cyclic stretch. We have previously shown Nppb as being a sensitive marker for stretch in 2D monolayers [32]; this statement was reinforced when a similar strong increase in Nppb gene expression in EHM fibers, both after 4 h and 24 h of cyclic stretch, was shown.

Gene expression of Nppb and Tgfβ1 in EHM was higher in both stretch and non-stretch conditions compared to the 2D monolayers, in both 4 and 24 h conditions. The opposite was true for the expression of Acta2 and Ctgf, suggesting a quiescent state, which has previously been shown [48]. Although the expression of Nppb and Tgfβ1 in non-stretch conditions was higher, CF in EHM were still able to respond to the stimulus of cyclic stretch by even further increasing the expression of Nppb and Tgfβ1, an interesting finding which merits further investigation.

Stimulation of EHM rings with TGFβ1 showed increased gene expression of Tgfβ1 and Ctgf similar as seen in the 2D cultures. No effect of TGFβ1 stimulation on Acta2 and Nppb expression was found. Our results of increased Ctgf expression after TGFβ1 stimulation are supported by previous research in human cardiac fibroblasts [56], adult rabbit cardiac fibroblasts [29] and human lung fibroblasts [57]. TGFβ1-induced Tgfβ1 expression in CF is something we [29] and others [58] have shown previously. TGFβ1 is a well-known stimulus for myofibroblast differentiation [59–61], and our finding of TGFβ1–induced increase in Acta2 mRNA expression in 2D cultures therefore fits into the literature [35,62]. This is further supported by Bracco Gartner et al. [48] in 3D cultures. By contrast, our EHM cultures did not show a clear TGFβ1-mediated induction of Acta2. Possibly this lack of effect indicates that the cells are less prone to TGFβ1-induced myofibroblast differentiation when cultured in EHM. Another possible explanation could be the high baseline expression of Nppb in EHM cultures. We have previously shown that BNP inhibits the TGFβ1-induced Acta2 expression [32]. It could be that the high baseline expression of Nppb translates to high levels of BNP within the culture media, inhibiting the TGFβ1-induced Acta2 expression in EHM, hence the lack of effect we see here. Future research is necessary to investigate the role of Nppb within EHM culture. Taken together, CF remain quiescent in EHM, but exhibit a clear response after stimulation with stretch or TGFβ1.

In the present study, we described the development of 3D engineered heart matrix (EHM) ring and fiber structures, with viable and quiescent CF embedded in a type1 collagen gel of physiological stiffness. Markers of CF differentiation toward myofibroblasts were lower in EHM, compared to 2D cultures, while CF activation by stretch or TGFβ1 was maintained, indicating that these EHM structures are a good model to study the process of CF activation and differentiation toward myofibroblasts.

**Supplementary Materials:** The following supporting information can be downloaded at: https://www.mdpi.com/article/10.3390/bioengineering9100551/s1, Figure S1: Effect of initial collagen concentration on EHM compaction. Figure S2: Effect of initial collagen concentration on RNA yield per cell and gene expression of Acta2, Ctgf and Tgfβ1.

**Author Contributions:** Conceptualization, M.C.P. and F.A.v.N.; methodology, J.B. and T.J.L.t.B.; performed experiments and produced data, M.C.P., C.M., T.S. and T.J.L.t.B.; data analysis, M.C.P., C.M., T.S., F.A.v.N. and T.J.L.t.B.; writing—original draft preparation, M.C.P. and F.A.v.N.; writing—review and editing, M.C.P., F.A.v.N., C.M., L.M. and F.W.P. All authors have read and agreed to the published version of the manuscript.

**Funding:** This research received no external funding.

**Institutional Review Board Statement:** Not applicable.

**Informed Consent Statement:** Not applicable.

**Data Availability Statement:** The data presented in this study are available from the corresponding author upon request.

**Acknowledgments:** The authors would like to thank Ger J. van der Vusse for support and valuable suggestions during study conceptualization and manuscript writing.

**Conflicts of Interest:** The authors declare no conflict of interest.

## References


1. Ariyasinghe, N.R.; Lyra-Leite, D.M.; McCain, M.L. Engineering cardiac microphysiological systems to model pathological extracellular matrix remodeling. *Am. J. Physiol. Heart Circ. Physiol.* **2018**, *315*, H771–H789. [CrossRef] [PubMed]
2. Walker, C.A.; Spinale, F.G. The structure and function of the cardiac myocyte: A review of fundamental concepts. *J. Thorac. Cardiovasc. Surg.* **1999**, *118*, 375–382. [CrossRef]
3. Ross, R.S.; Borg, T.K. Integrins and the myocardium. *Circ. Res.* **2001**, *88*, 1112–1119. [CrossRef] [PubMed]
4. Eghbali, M. Cardiac Fibroblasts: Function, Regulation of Gene Expression, and Phenotypic Modulation. *Basic Res. Cardiol.* **1992**, *87*, 183–189. [CrossRef] [PubMed]
5. Porter, K.E.; Turner, N.A. Cardiac fibroblasts: At the heart of myocardial remodeling. *Pharmacol. Ther.* **2009**, *123*, 255–278. [CrossRef]
6. Fan, D.; Takawale, A.; Lee, J.; Kassiri, Z. Cardiac fibroblasts, fibrosis and extracellular matrix remodeling in heart disease. *Fibrogenes. Tissue Repair* **2012**, *5*, 15. [CrossRef]
7. van Nieuwenhoven, F.A.; Turner, N.A. The role of cardiac fibroblasts in the transition from inflammation to fibrosis following myocardial infarction. *Vascul. Pharmacol.* **2013**, *58*, 182–188. [CrossRef]
8. Powell, D.W.; Mifflin, R.C.; Valentich, J.D.; Crowe, S.E.; Saada, J.I.; West, A.B.; Myofibroblasts, I. Paracrine cells important in health and disease. *Am. J. Physiol.* **1999**, *277*, C1–C19. [CrossRef]
9. Tsuruda, T.; Boerrigter, G.; Huntley, B.K.; Noser, J.A.; Cataliotti, A.; Costello-Boerrigter, L.C.; Chen, H.H.; Burnett, J.C., Jr. Brain natriuretic Peptide is produced in cardiac fibroblasts and induces matrix metalloproteinases. *Circ. Res.* **2002**, *91*, 1127–1134. [CrossRef]
10. van den Borne, S.W.; Diez, J.; Blankesteijn, W.M.; Verjans, J.; Hofstra, L.; Narula, J. Myocardial remodeling after infarction: The role of myofibroblasts. *Nat. Rev. Cardiol.* **2010**, *7*, 30–37. [CrossRef]
11. Brown, R.D.; Ambler, S.K.; Mitchell, M.D.; Long, C.S. The cardiac fibroblast: Therapeutic target in myocardial remodeling and failure. *Annu. Rev. Pharmacol. Toxicol.* **2005**, *45*, 657–687. [CrossRef] [PubMed]
12. Camelliti, P.; Borg, T.K.; Kohl, P. Structural and functional characterisation of cardiac fibroblasts. *Cardiovasc. Res.* **2005**, *65*, 40–51. [CrossRef] [PubMed]
13. Banerjee, I.; Yekkala, K.; Borg, T.K.; Baudino, T.A. Dynamic interactions between myocytes, fibroblasts, and extracellular matrix. *Ann. N. Y. Acad. Sci.* **2006**, *1080*, 76–84. [CrossRef] [PubMed]
14. Pedrotty, D.M.; Klinger, R.Y.; Kirkton, R.D.; Bursac, N. Cardiac fibroblast paracrine factors alter impulse conduction and ion channel expression of neonatal rat cardiomyocytes. *Cardiovasc. Res.* **2009**, *83*, 688–697. [CrossRef] [PubMed]
15. Creemers, E.E.; Pinto, Y.M. Molecular mechanisms that control interstitial fibrosis in the pressure-overloaded heart. *Cardiovasc. Res.* **2011**, *89*, 265–272. [CrossRef]
16. Jalil, J. Structural vs. contractile protein remodeling and myocardial stiffness in hypertrophied rat left ventricle. *J. Mol. Cell. Cardiol.* **1988**, *20*, 1179–1187. [CrossRef]
17. Li, Y.; Asfour, H.; Bursac, N. Age-dependent functional crosstalk between cardiac fibroblasts and cardiomyocytes in a 3D engineered cardiac tissue. *Acta Biomater* **2017**, *55*, 120–130. [CrossRef]
18. Abbott, A. Cell culture: Biology's new dimension. *Nature* **2003**, *424*, 870–872. [CrossRef]
19. Baker, B.M.; Chen, C.S. Deconstructing the third dimension: How 3D culture microenvironments alter cellular cues. *J. Cell Sci.* **2012**, *125*, 3015–3024. [CrossRef]
20. Hinz, B.; Gabbiani, G. Cell-matrix and cell-cell contacts of myofibroblasts: Role in connective tissue remodeling. *Thromb. Haemost.* **2003**, *90*, 993–1002. [CrossRef]
21. Landry, N.M.; Rattan, S.G.; Dixon, I.M.C. An Improved Method of Maintaining Primary Murine Cardiac Fibroblasts in Two-Dimensional Cell Culture. *Sci. Rep.* **2019**, *9*, 12889. [CrossRef] [PubMed]
22. Santiago, J.J.; Dangerfield, A.L.; Rattan, S.G.; Bathe, K.L.; Cunnington, R.H.; Raizman, J.E.; Bedosky, K.M.; Freed, D.H.; Kardami, E.; Dixon, I.M. Cardiac fibroblast to myofibroblast differentiation in vivo and in vitro: Expression of focal adhesion components in neonatal and adult rat ventricular myofibroblasts. *Dev. Dyn.* **2010**, *239*, 1573–1584. [CrossRef] [PubMed]
23. Hinz, B. Tissue stiffness, latent TGF-beta1 activation, and mechanical signal transduction: Implications for the pathogenesis and treatment of fibrosis. *Curr. Rheumatol. Rep.* **2009**, *11*, 120–126. [CrossRef] [PubMed]
24. Galie, P.A.; Westfall, M.V.; Stegemann, J.P. Reduced serum content and increased matrix stiffness promote the cardiac myofibroblast transition in 3D collagen matrices. *Cardiovasc. Pathol.* **2011**, *20*, 325–333. [CrossRef] [PubMed]
25. Chen, J.H.; Chen, W.L.; Sider, K.L.; Yip, C.Y.; Simmons, C.A. beta-catenin mediates mechanically regulated, transforming growth factor-beta1-induced myofibroblast differentiation of aortic valve interstitial cells. *Arterioscler. Thromb. Vasc. Biol.* **2011**, *31*, 590–597. [CrossRef]
26. Liu, F.; Mih, J.D.; Shea, B.S.; Kho, A.T.; Sharif, A.S.; Tager, A.M.; Tschumperlin, D.J. Feedback amplification of fibrosis through matrix stiffening and COX-2 suppression. *J. Cell Biol.* **2010**, *190*, 693–706. [CrossRef] [PubMed]

27. Olsen, A.L.; Bloomer, S.A.; Chan, E.P.; Gaca, M.D.; Georges, P.C.; Sackey, B.; Uemura, M.; Janmey, P.A.; Wells, R.G. Hepatic stellate cells require a stiff environment for myofibroblastic differentiation. *Am. J. Physiol. Gastrointest. Liver Physiol.* **2011**, *301*, G110–G118. [CrossRef]

28. Mayer, D.C.; Leinwand, L.A. Sarcomeric gene expression and contractility in myofibroblasts. *J. Cell Biol.* **1997**, *139*, 1477–1484. [CrossRef]

29. Blaauw, E.; Lorenzen-Schmidt, I.; Babiker, F.A.; Munts, C.; Prinzen, F.W.; Snoeckx, L.H.; van Bilsen, M.; van der Vusse, G.J.; van Nieuwenhoven, F.A. Stretch-induced upregulation of connective tissue growth factor in rabbit cardiomyocytes. *J. Cardiovasc Transl Res.* **2013**, *6*, 861–869. [CrossRef]

30. Tarbit, E.; Singh, I.; Peart, J.N.; Rose'Meyer, R.B. Biomarkers for the identification of cardiac fibroblast and myofibroblast cells. *Heart Fail. Rev.* **2019**, *24*, 1–15. [CrossRef]

31. Swaney, J.S.; Roth, D.M.; Olson, E.R.; Naugle, J.E.; Meszaros, J.G.; Insel, P.A. Inhibition of cardiac myofibroblast formation and collagen synthesis by activation and overexpression of adenylyl cyclase. *Proc. Natl. Acad. Sci. USA* **2005**, *102*, 437–442. [CrossRef] [PubMed]

32. Ploeg, M.C.; Munts, C.; Prinzen, F.W.; Turner, N.A.; van Bilsen, M.; van Nieuwenhoven, F.A. Piezo1 Mechanosensitive Ion Channel Mediates Stretch-Induced Nppb Expression in Adult Rat Cardiac Fibroblasts. *Cells* **2021**, *10*, 1745. [CrossRef] [PubMed]

33. Turner, N.A.; Porter, K.E.; Smith, W.H.T.; White, H.L.; Ball, S.G.; Balmforth, A.J. Chronic β2-adrenergic receptor stimulation increases proliferation of human cardiac fibroblasts via an autocrine mechanism. *Cardiovasc. Res.* **2003**, *57*, 784–792. [CrossRef]

34. van Nieuwenhoven, F.A.; Hemmings, K.E.; Porter, K.E.; Turner, N.A. Combined effects of interleukin-1α and transforming growth factor-β1 on modulation of human cardiac fibroblast function. *Matrix Biol.* **2013**, *32*, 399–406. [CrossRef]

35. van Nieuwenhoven, F.A.; Munts, C.; Op't Veld, R.C.; Gonzalez, A.; Diez, J.; Heymans, S.; Schroen, B.; van Bilsen, M. Cartilage intermediate layer protein 1 (CILP1): A novel mediator of cardiac extracellular matrix remodelling. *Sci. Rep.* **2017**, *7*, 16042. [CrossRef]

36. van Kampen, K.A.; Fernández-Pérez, J.; Baker, M.; Mota, C.; Moroni, L. Fabrication of a mimetic vascular graft using melt spinning with tailorable fiber parameters. *Biomater. Adv.* **2022**, *139*, 212972. [CrossRef]

37. van Kampen, K.A.; Olaret, E.; Stancu, I.C.; Moroni, L.; Mota, C. Controllable four axis extrusion-based additive manufacturing system for the fabrication of tubular scaffolds with tailorable mechanical properties. *Mater. Sci. Eng. C Mater. Biol. Appl.* **2021**, *119*, 111472. [CrossRef]

38. Geelhoed, W.J.; Lalai, R.A.; Sinnige, J.H.; Jongeleen, P.J.; Storm, C.; Rotmans, J.I. Indirect Burst Pressure Measurements for the Mechanical Assessment of Biological Vessels. *Tissue Eng. Part. C Methods* **2019**, *25*, 472–478. [CrossRef]

39. Shazly, T.; Rachev, A.; Lessner, S.; Argraves, W.S.; Ferdous, J.; Zhou, B.; Moreira, A.M.; Sutton, M. On the Uniaxial Ring Test of Tissue Engineered Constructs. *Exp. Mech.* **2014**, *55*, 41–51. [CrossRef]

40. de Jong, S.; van Middendorp, L.B.; Hermans, R.H.; de Bakker, J.M.; Bierhuizen, M.F.; Prinzen, F.W.; van Rijen, H.V.; Losen, M.; Vos, M.A.; van Zandvoort, M.A. Ex vivo and in vivo administration of fluorescent CNA35 specifically marks cardiac fibrosis. *Mol. Imaging* **2014**, *13*, 7290. [CrossRef]

41. Baues, M.; Klinkhammer, B.M.; Ehling, J.; Gremse, F.; van Zandvoort, M.; Reutelingsperger, C.P.M.; Daniel, C.; Amann, K.; Babickova, J.; Kiessling, F.; et al. A collagen-binding protein enables molecular imaging of kidney fibrosis in vivo. *Kidney Int.* **2020**, *97*, 609–614. [CrossRef] [PubMed]

42. Helary, C.; Bataille, I.; Abed, A.; Illoul, C.; Anglo, A.; Louedec, L.; Letourneur, D.; Meddahi-Pelle, A.; Giraud-Guille, M.M. Concentrated collagen hydrogels as dermal substitutes. *Biomaterials* **2010**, *31*, 481–490. [CrossRef] [PubMed]

43. Berry, M.F.; Engler, A.J.; Woo, Y.J.; Pirolli, T.J.; Bish, L.T.; Jayasankar, V.; Morine, K.J.; Gardner, T.J.; Discher, D.E.; Sweeney, H.L. Mesenchymal stem cell injection after myocardial infarction improves myocardial compliance. *Am. J. Physiol. Heart Circ. Physiol.* **2006**, *290*, H2196–H2203. [CrossRef] [PubMed]

44. Engler, A.J.; Carag-Krieger, C.; Johnson, C.P.; Raab, M.; Tang, H.Y.; Speicher, D.W.; Sanger, J.W.; Sanger, J.M.; Discher, D.E. Embryonic cardiomyocytes beat best on a matrix with heart-like elasticity: Scar-like rigidity inhibits beating. *J. Cell Sci.* **2008**, *121*, 3794–3802. [CrossRef] [PubMed]

45. Herum, K.M.; Lunde, I.G.; McCulloch, A.D.; Christensen, G. The Soft- and Hard-Heartedness of Cardiac Fibroblasts: Mechanotransduction Signaling Pathways in Fibrosis of the Heart. *J. Clin. Med.* **2017**, *6*, 53. [CrossRef]

46. Herum, K.M.; Choppe, J.; Kumar, A.; Engler, A.J.; McCulloch, A.D. Mechanical regulation of cardiac fibroblast profibrotic phenotypes. *Mol. Biol. Cell* **2017**, *28*, 1871–1882. [CrossRef]

47. Kong, M.; Lee, J.; Yazdi, I.K.; Miri, A.K.; Lin, Y.D.; Seo, J.; Zhang, Y.S.; Khademhosseini, A.; Shin, S.R. Cardiac Fibrotic Remodeling on a Chip with Dynamic Mechanical Stimulation. *Adv. Healthc Mater.* **2019**, *8*, e1801146. [CrossRef]

48. Bracco Gartner, T.C.L.; Deddens, J.C.; Mol, E.A.; Magin Ferrer, M.; van Laake, L.W.; Bouten, C.V.C.; Khademhosseini, A.; Doevendans, P.A.; Suyker, W.J.L.; Sluijter, J.P.G.; et al. Anti-fibrotic Effects of Cardiac Progenitor Cells in a 3D-Model of Human Cardiac Fibrosis. *Front. Cardiovasc. Med.* **2019**, *6*, 52. [CrossRef]

49. Hackett, T.L.; Vriesde, N.; Al-Fouadi, M.; Mostaco-Guidolin, L.; Maftoun, D.; Hsieh, A.; Coxson, N.; Usman, K.; Sin, D.D.; Booth, S.; et al. The Role of the Dynamic Lung Extracellular Matrix Environment on Fibroblast Morphology and Inflammation. *Cells* **2022**, *11*, 185. [CrossRef]

50. Taylor, S.E.; Vaughan-Thomas, A.; Clements, D.N.; Pinchbeck, G.; Macrory, L.C.; Smith, R.K.; Clegg, P.D. Gene expression markers of tendon fibroblasts in normal and diseased tissue compared to monolayer and three dimensional culture systems. *BMC Musculoskelet Disord* **2009**, *10*, 27. [CrossRef]

51. Pelham, R.J., Jr.; Wang, Y. Cell locomotion and focal adhesions are regulated by substrate flexibility. *Proc. Natl. Acad. Sci. USA* **1997**, *94*, 13661–13665. [CrossRef] [PubMed]

52. Lee, P.Y.; Liu, Y.C.; Wang, M.X.; Hu, J.J. Fibroblast-seeded collagen gels in response to dynamic equibiaxial mechanical stimuli: A biomechanical study. *J. Biomech.* **2018**, *78*, 134–142. [CrossRef] [PubMed]

53. Nieponice, A.; Maul, T.M.; Cumer, J.M.; Soletti, L.; Vorp, D.A. Mechanical stimulation induces morphological and phenotypic changes in bone marrow-derived progenitor cells within a three-dimensional fibrin matrix. *J. Biomed. Mater. Res. A* **2007**, *81*, 523–530. [CrossRef] [PubMed]

54. Peters, A.S.; Brunner, G.; Krieg, T.; Eckes, B. Cyclic mechanical strain induces TGFbeta1-signalling in dermal fibroblasts embedded in a 3D collagen lattice. *Arch. Dermatol. Res.* **2015**, *307*, 191–197. [CrossRef] [PubMed]

55. Galie, P.A.; Russell, M.W.; Westfall, M.V.; Stegemann, J.P. Interstitial fluid flow and cyclic strain differentially regulate cardiac fibroblast activation via AT1R and TGF-beta1. *Exp. Cell Res.* **2012**, *318*, 75–84. [CrossRef] [PubMed]

56. Chen, M.M.; Lam, A.; Abraham, J.A.; Schreiner, G.F.; Joly, A.H. CTGF expression is induced by TGF- beta in cardiac fibroblasts and cardiac myocytes: A potential role in heart fibrosis. *J. Mol. Cell Cardiol.* **2000**, *32*, 1805–1819. [CrossRef]

57. Watts, K.L.; Spiteri, M.A. Connective tissue growth factor expression and induction by transforming growth factor-beta is abrogated by simvastatin via a Rho signaling mechanism. *Am. J. Physiol. Lung Cell Mol. Physiol.* **2004**, *287*, L1323–L1332. [CrossRef]

58. Flanders, K.; Holder, M.G.; Winokur, T.S. Autoinduction of mRNA and protein expression for transforming growth factor-?s in cultured cardiac cells. *J. Mol. Cell. Cardiol.* **1995**, *27*, 805–812. [CrossRef]

59. Hinz, B. Formation and function of the myofibroblast during tissue repair. *J. Invesigt. Dermatol.* **2007**, *127*, 526–537. [CrossRef]

60. Tomasek, J.J.; Gabbiani, G.; Hinz, B.; Chaponnier, C.; Brown, R.A. Myofibroblasts and mechano-regulation of connective tissue remodelling. *Nat. Rev. Mol. Cell Biol.* **2002**, *3*, 349–363. [CrossRef]

61. Biernacka, A.; Dobaczewski, M.; Frangogiannis, N.G. TGF-beta signaling in fibrosis. *Growth Factors* **2011**, *29*, 196–202. [CrossRef] [PubMed]

62. Watson, C.J.; Phelan, D.; Xu, M.; Collier, P.; Neary, R.; Smolenski, A.; Ledwidge, M.; McDonald, K.; Baugh, J. Mechanical stretch up-regulates the B-type natriuretic peptide system in human cardiac fibroblasts: A possible defense against transforming growth factor-beta mediated fibrosis. *Fibrogenes. Tissue Repair* **2012**, *5*, 9. [CrossRef] [PubMed]

 *bioengineering*

*Article*

# Fibroblast-Generated Extracellular Matrix Guides Anastomosis during Wound Healing in an Engineered Lymphatic Skin Flap

Alvis Chiu [1], Wenkai Jia [1], Yumeng Sun [1], Jeremy Goldman [2] and Feng Zhao [1,*]

[1] Stem Cell and Tissue Engineering Lab, Department of Biomedical Engineering, College of Engineering, Texas A&M University, College Station, TX 77843, USA

[2] Vascular Materials Lab, Department of Biomedical Engineering, College of Engineering, Michigan Technological University, Houghton, MI 49931, USA

* Correspondence: fengzhao@tamu.edu

**Abstract:** A healthy lymphatic system is required to return excess interstitial fluid back to the venous circulation. However, up to 49% of breast cancer survivors eventually develop breast cancer-related lymphedema due to lymphatic injuries from lymph node dissections or biopsies performed to treat cancer. While early-stage lymphedema can be ameliorated by manual lymph drainage, no cure exists for late-stage lymphedema when lymph vessels become completely dysfunctional. A viable late-stage treatment is the autotransplantation of functional lymphatic vessels. Here we report on a novel engineered lymphatic flap that may eventually replace the skin flaps used in vascularized lymph vessel transfers. The engineered flap mimics the lymphatic and dermal compartments of the skin by guiding multi-layered tissue organization of mesenchymal stem cells and lymphatic endothelial cells with an aligned decellularized fibroblast matrix. The construct was tested in a novel bilayered wound healing model and implanted into athymic nude rats. The in vitro model demonstrated capillary invasion into the wound gaps and deposition of extracellular matrix fibers, which may guide anastomosis and vascular integration of the graft during wound healing. The construct successfully anastomosed in vivo, forming chimeric vessels of human and rat cells. Overall, our flap replacement has high potential for treating lymphedema.

**Keywords:** lymphatic; extracellular matrix; decellularized matrix; tissue engineering; in vitro model; self-assembled vessels

**Citation:** Chiu, A.; Jia, W.; Sun, Y.; Goldman, J.; Zhao, F. Fibroblast-Generated Extracellular Matrix Guides Anastomosis during Wound Healing in an Engineered Lymphatic Skin Flap. *Bioengineering* **2023**, *10*, 149. https://doi.org/10.3390/bioengineering10020149

Academic Editor: Brandon J. Tefft

Received: 16 December 2022
Revised: 4 January 2023
Accepted: 17 January 2023
Published: 22 January 2023



## 1. Introduction

It is estimated that 287,850 women developed breast cancer in 2022 in the US, with 10% not surviving [1]. Among the survivors, 20% [2] to 49% [3] will develop breast cancer-related lymphedema (BCRL), leading to heaviness, numbness, and tightness in the affected limb. This disease not only drastically lowers the patient's quality of life [1,4], but also creates an alarming healthcare burden for mental health services, disease monitoring, and disease treatment [5]. BCRL is a frequent consequence of axillary lymph node dissection or sentinel lymph node biopsy in cancer patients, which can interrupt the resorption of excess interstitial fluid of the arm and its transport back to the venous circulation. This permanently decreases lymph transport capacity and often causes fluid buildup, painful arm swelling and susceptibility to infections. Currently, there is no cure for BCRL, only preventative or palliative treatments. During the early stages of lymphedema, before the onset of fibrosis due to the accumulation of protein and lipids, swelling can be improved by manually compressing the limb using garments, bandages, or intermittent pneumatic compression therapy [6–8]. Unfortunately, these modalities require intense lifelong effort and are ineffective for advanced fibrotic limbs where the tissue has already irreversibly remodeled from the excess interstitial fluid [6,9].

Vascularized lymph node transfer is the most commonly used procedure for reducing lymphedema in advanced lymphedema patients [10]. It is accomplished by transplanting skin or adipose flaps containing functional lymph nodes harvested from a healthy donor site [11,12]. The implanted lymph nodes spontaneously anastomose with recipient site lymphatic and venous vessels and drain lymph fluid from those respective systems [11–13]. However, this procedure has an 18% chance of causing seroma [14,15], and harvesting lymph nodes from the preferred groin donor site can create chronic lymphedema [14,16]. Moreover, recent evidence has shown that the lymph nodes were actually irrelevant to the lymphedema reduction, as the therapeutic agents of lymph node transfers were in fact the surrounding lymphatic vessels [17]. In 2016, Koshima et al. completed the first pilot study of vascularized lymph vessel transfer (VLVT), which aimed to minimize donor site morbidity by not sacrificing lymph nodes [13,18]. The effectiveness of this treatment has been proven by multiple clinical trials that showed relief of symptoms, weaning of compression garments, and improvement of quality of life [17–21]. Although donor site lymphedema has not been reported for VLVT [18], donor site availability still significantly limits its application. Therefore, there is a crucial need for skin and adipose flap substitutes.

Various novel therapeutics have been proposed to induce the growth of new lymphatic vessels across the dissected areas to improve lymph drainage. Vascular endothelial growth factor-C therapies have shown effectiveness in reducing lymphedema by directly signaling lymphatic endothelial cells (LECs) to sprout more capillaries from existing capillaries [22,23]. Nanofibrous collagen scaffolds have been utilized to create temporary bridges that direct interstitial fluid flow over the damaged areas to guide lymphangiogenesis [24]. Mesenchymal stem cell (MSC) therapy reduced lymphedema by secreting lymphangiogenic and immunomodulatory factors that augment the wound microenvironment [25]. Although these prolymphangiogenic therapies have achieved success in reducing acute lymphedema in animal models, long-term functional lymphangiogenesis is inhibited by the chronic inflammation and fibrosis present in mild to severe lymphedema [9,25,26]. Therefore, these approaches may be inadequate for treating chronic or severe cases.

Since stimulating peri-wound lymphangiogenesis to restore lymphatic continuity may be challenging for advanced lymphedema patients, the implantation of tissue engineered lymphatic vessels may serve as a more effective treatment modality. LECs can self-assemble into lymphatic microvascular networks in the presence of various supporting cells. Co-culturing LECs with adipose-derived MSCs formed stable networks for up to 4 weeks [27]. Similarly, 3D lymphatic networks with native ultrastructure have been generated by seeding LECs on fibroblast sheets [28]. More recently, LECs co-cocultured with fibroblasts on a collagen sheet were shown to anastomose with host lymphatics in a mouse model [29]. However, the aforementioned approaches lacked guidance cues, so the lymphatic vessels formed were randomly oriented, which made them deficient in the natural lymphatic vessel alignment (axiality) that facilitates unidirectional lymph flow in native tissues [20]. Decellularized adipose tissue has also been used as a scaffold to generate lymphatic networks with anastomosis capacity [30]. While decellularization may preserve the lymphatic axiality of the tissue as newly seeded LECs colonize existing vessel channels, decellularized scaffolds suffer from problems of donor scarcity, host responses, and pathogen transfers when allogeneic or xenogeneic tissues are used [31]. Compared with reconstituted or decellularized scaffolds, cell-derived extracellular matrix (ECM) has several advantages: (1) Sterile culture conditions eliminate pathogen contamination risks. (2) Cell-derived ECM can be engineered with controlled topography and porosity to guide lymph axiality. (3) Cell-derived ECM modulates host immune responses, reducing fibrosis [31]. Therefore, cell-derived ECM offers a promising alternative to scaffolds derived from natural tissues.

The objective of this study was to utilize cell-derived ECM to develop an engineered lymphatic flap that mimics the dermal and lymphatic components of a native ultra-thin skin flap used in VLVT. The construct consists of three layers: (1) decellularized fibroblast ECM, (2) human MSCs, and (3) self-assembled lymphatic capillaries. During VLVT, due to the bulk of the skin flap, a pedicled artery and vein that spans the graft needs to be surgically anastomosed to perfuse the graft [18]. We expect that, by eliminating the unnecessary epidermal, blood, and subcutaneous components and only keeping the relevant cells for lymphedema reduction, our construct may function without a dedicated blood supply. The construct is created on a nanograted polydimethylsiloxane (PDMS) substrate that guides fibroblast and fibroblast-secreted ECM alignment, which further directs MSC alignment, and in turn, LEC alignment. Aligned LECs form unidirectional capillaries, mimicking the native lymphatic axiality of skin flaps [20,32,33]. They also have the cellular and ultrastructural features of native human dermal lymphatic capillaries [28]. Upon implantation, MSCs reduce inflammation [34], secrete lymphangiogenic factors [35], and serve as supporting cells for the lymphatic capillaries [36]. The fibroblast-derived ECM itself also promotes wound healing, reduces fibrosis, and provides mechanical support for the flap [36,37]. This approach can incorporate autologous cells, thereby eliminating the use of allogeneic or xenogeneic tissues while conserving lymph axiality. This skin flap replacement has the potential to increase graft availability and reduce lymphatic damage-related surgical procedures and complications.

## 2. Materials & Methods

### 2.1. Cell Culture

All cells used were obtained from commercial sources. Clonetics™ Dermal Lymphatic Microvascular Endothelial Cells (LECs) and Poietics™ Normal Human Bone Marrow Derived Mesenchymal Stem Cells (MSCs) were obtained from Lonza, Basel, Switzerland. Normal Human Adult Primary Dermal Fibroblasts (HDFs) were obtained from American Type Culture Collection, Manassas, VA, USA. LECs at passage 5, were cultured in endothelial growth medium 2 (EGM-2, Lonza). Passage 7 HDFs were cultured in fibroblast growth media consisting of 60% Dulbecco's Modified Eagle Medium, 20% F-12, 20% fetal bovine serum (FBS) and 1% penicillin/streptomycin. Passage 4 MSCs were cultured in Minimum Essential Medium $\alpha$ with 20% FBS, 1% penicillin/streptomycin, and 1% L-glutamine (Thermo Fisher Scientific, Waltham, MA, USA). All cells were cultured in a humidified, 37 °C, 5% $CO_2$ incubator.

### 2.2. Wound Healing Model

The wound healing model used 4 well culture-inserts (Ibidi, Fitchburg, WI, USA) to create bilayered lymphvascularized tissues separated by wound gaps. Each insert is cylindrical with an outer diameter of 17 mm and inner diameter of 13 mm and has 4 hollow quadrants dividing the center into 4 wells each with a growth area of 0.35 $cm^2$ per well. The quadrants are separated by walls that create a $500 \pm 100$ μm cell-free gap between each quadrant. The inserts were placed into 12-well plates. MSCs were seeded at 5000 cells/$cm^2$ and HDFs at 3000 cells/$cm^2$ in their respective media in either the MSC-HDF or HDF-HDF configuration (Figure 1). These seeding densities were optimized to make both supporting cell types reach confluency on the same day while minimizing aggregate formation. In MSC-HDF, each wound gap is flanked by one MSC side and one HDF side. For HDF-HDF, both sides are HDF. Four wound gaps were created using one insert for a total of 12 gaps per timepoint for each configuration. Cell culture media was changed every 3 days until both MSCs and HDFs reached confluency after 9 days of culture. LECs at 20,000 cells/$cm^2$ were seeded on top of the basal cells after basal cell confluency. The culture-insert was removed 10 h after LEC seeding to allow for 10 h of lymphangiogenesis. The culture media was then switched to a mixture of 50% fibroblast growth media and 50% EGM-2 and was changed every 2 days. Wound gaps were examined 4, 6, and 8 days after insert removal.

### 2.3. Immunofluorescence Staining

Skin explants of implanted engineered flaps and cell cultures of the wound healing model were fixed in 4% paraformaldehyde in PBS containing calcium and magnesium for 20 min and blocked with a solution of 2% bovine serum albumin (Sigma-Aldrich, St. Louis, MO, USA). All antibodies were diluted in blocking solution and applied overnight on a rocker at 4 °C. For the wound healing assay, LECs were stained using anti-CD31 antibodies, and ECM was stained using anti-collagen I antibodies. For the animal tissue sectioning, LECs were stained for podoplanin (PDPN). To distinguish between rat and human LECs, human nuclear antigen (HNA) was targeted. All IgG secondary antibodies (Alexa Fluor 488, 568, and 647 goat anti-rabbit and goat anti-mouse) were purchased from Thermo Fisher Scientific diluted 1:500 in blocking solution. All imaging was done on a Zeiss Observer 3 microscope using either a Zeiss 20× air or 40× water objective and an Axiocam 503 mono camera (Zeiss, Oberkochen, Germany).

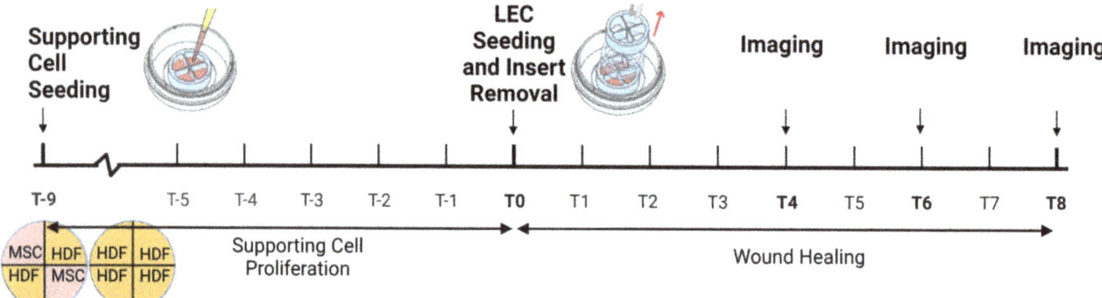


**Figure 1.** Timeline of in vitro wound healing model. The culture-insert is first attached to a well plate. The supporting cells, human dermal fibroblasts (HDFs) and mesenchymal stem cells (MSCs), are cultured inside for 9 days in either the MSC-HDF or HDF-HDF configuration. In MSC-HDF, the supporting cells alternate between MSCs and HDFs, whereas HDF-HDF only contain HDFs. Subsequently, lymphatic endothelial cells (LECs) are seeded on top for 10 h of vessel assembly, or lymphangiogenesis, before culture-insert removal. Three inserts were used for each configuration at each imaging timepoint.

### 2.4. Quantitating Endothelial Invasion

Images of the wound healing model were taken according to the aforementioned method and stitched using the MosaicJ plugin for the open-source image processing program ImageJ. A 600 × 6500 µm rectangle containing the original wound gap was selected for analysis for vertical wound gaps. The dimensions of the analysis area were flipped for horizontal wound gaps. The exact position of the analysis rectangle was first determined by macroscopically locating the wound gap and then placing the box around the middle of the two sides of invading vessels. The angle of the box was adjusted accordingly as the culture-inserts were placed by hand. Vessel area was calculated by thresholding CD31 signal and using the analyze particles command. The area was measured in triplicates by two researchers independently.

### 2.5. Lymphatic Skin Flap Replacement Fabrication

The engineered flap was fabricated following our published protocol [32,38]. Briefly, PDMS substrates were cast from molds with 350 nm grating width and 130 nm grating depth and coated with collagen I. HDFs were seeded on top and cultured for 3 weeks in fibroblast growth media with a media change every 3 days. This produces uniform 70 µm-thick cell sheets [38], which were gently peeled off and decellularized with 0.5% sodium dodecyl sulfate, 10 mM Tris (Bio-rad Laboratories, Hercules, CA, USA), and 25 mM Ethylenediaminetetraacetic acid (Sigma-Aldrich) solution for 15 min. Afterwards, the ECM sheets were thoroughly washed with PBS and incubated in culture media for 48 h. The ECM sheets produced by our decellularization protocol retain collagen I, elastin, and fibronectin fibers and have an elastic modulus of around 250 Pa and a viscous modulus of 300 Pa [38]. The fibers are uniformly aligned with a dimeter of $78 \pm 9$ nm [38]. MSCs were seeded at 10,000 cells/cm$^2$ on the decellularized ECM sheet and cultured under hypoxia (2% O$_2$) for 7 days. Subsequently, LECs were seeded at 20,000 cells/cm$^2$ on top of the MSC/ECM constructs. The co-cultures were maintained at 20% $O_2$ for up to 7 days in EGM-2 with a media change every 2 days. This method has shown to generate vessel networks with a mean vessel diameter of 11 μm, a mean vessel length of 220 μm, an intercapillary distance of 19 μm, and a 14% area vessel coverage [32,33].

*2.6. Animal Model*

All animal experiments were performed following protocols approved by the institutional committee for animal use and care regulations at Michigan Technological University. Athymic Rowett nude (RNU) rats (male, 6–8 weeks old, 200–250 g weight) were purchased from Charles River Laboratories (Wilmington, MA, USA). Rats were anesthetized with isoflurane and their backs were shaved with clippers. A small incision was made on the rat dorsum to expose the subcutaneous space, in which a stack of three 2 $cm^2$ disks of the engineered lymphatic flap was implanted. Incisions were closed with suture clips. The rats were euthanized at day 7 post implantation and the skin was harvested from the back that contained the engineered lymphatics. A region of skin from another random area from the back was also harvested to serve as a control. A total of 2 samples were harvested from each of 3 rats. The skin explants were stained as mentioned above.

*2.7. Statistical Analysis*

CD31 coverage data from each timepoint (day 4, 6, 8) for MSC-HDF and HDF-MDF gaps were compared using one-way ANOVA and Tukey's post hoc test on GraphPad Prism. The results were reported as mean $\pm$ standard deviation. Results were considered statistically significant for $p \leq 0.05$. Datapoints from detaching or aggregating tissues were excluded to prevent the detachment or aggregation from confounding the results.

**3. Results**

*3.1. Flap Anastomoses with Self-Assembled Lymphatic Capillaries*

To investigate the anastomosis capacity of the engineered flap with host lymphatics, an in vitro model was designed to replicate the integration of the flap with the host. Basal cells (MSCs in the present case) and then LECs were seeded on HDF-derived ECM to simulate flap transplantation (Figure 2A). The basal cells and LECs were also seeded directly on the culture plates (Figure 2B–left side), where LECs underwent lymphangiogenesis and formed capillaries on top of the basal cells on both sides of a 500 μm wide wound gap (Figure 3A). LECs failed to form capillaries when seeded alone on the culture plate (Figure 4C), indicating the necessity of basal supporting cells and their deposited ECM for lymphangiogenesis. Thus, the bilayered co-culture setup created a network of self-assembled capillaries on top of a traditional scratch assay (Figure 2B), where capillary sprouting was both enabled and confined by basal cell expansion. The basal cell layer beneath the LECs modeled two different tissues. LEC/MSC represented the lymphatic flap, as the engineered flap is built upon MSCs (Figure 2C). LEC/HDF modeled the host recipient site as the fibroblasts recapitulated a dermal organoid. In either LEC/MSC or LEC/HDF, the wound-mimicking gap fully closed after 4 days, as shown by complete collagen coverage of the gap (Figure 2D). Capillaries on the wound edge then sprouted over the basal cells into the wound gap (Figure 3A). Once inside the gap, the capillaries exhibited minimal branching and a straight orientation. HDF-HDF interfaces had significantly more capillary coverage at day 6 and 8 compared to MSC-HDF (Figure 3B). The invasion of capillaries in HDF-HDF into the wound gap generally followed collagen I fibers generated by the migration and proliferation patterns of the basal cells (Figures 2D and 5), and capillaries from opposing sides migrated towards each other (Figure 3A). In contrast to the results of HDF-HDF, capillaries in MSC-HDF underwent pruning and regression from day 4 to 8, resulting in a decrease of capillary coverage across the gap (Figures 2D and 3A).

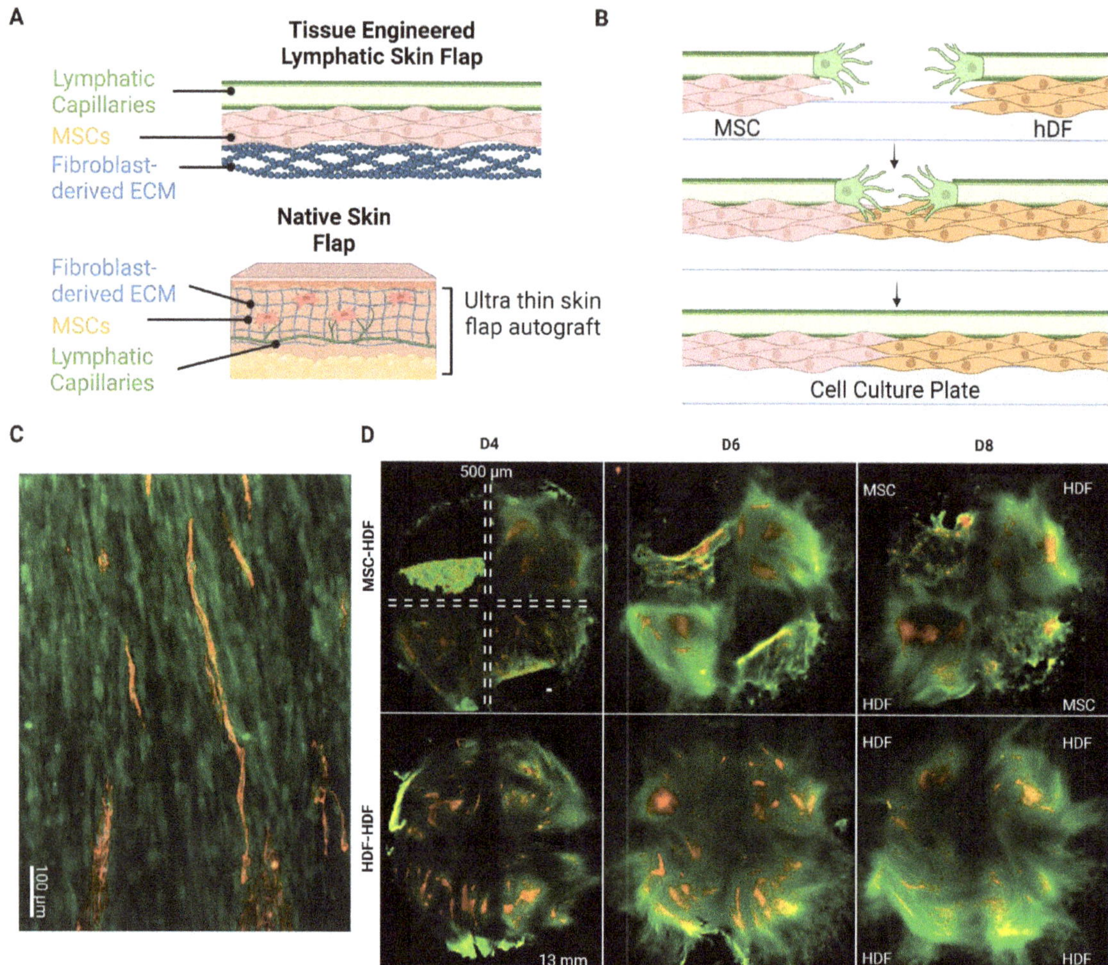

**Figure 2.** Engineered lymphatic flap and wound healing model. (**A**) Composition of engineered lymphatic flap. The three layers of the engineered flap (lymphatic capillaries, MSCs, and HDF extracellular matrix (ECM)) mimics the ultra-thin skin flaps used in vascularized lymph vessel transfer (VLVT). (**B**) Mechanism of bilayered wound healing assay. Capillary formation and sprouting is limited by the basal ECM. Supporting cell proliferation and ECM secretion allows lymphangiogenesis over the inhospitable culture plate substrate. (**C**) Immunostaining of engineered flap. LECs (Red). Collagen I (Green). Cell Nuclei (Blue). Scale bar: 100 μm. (**D**) Natural collagen swirls forming in the process of wound healing. Lymphatic capillaries colocalized with these ECM patterns. Wound gaps with detached tissues were excluded for analysis. LECs (red) Collagen I (Green). Wound gap: 500 μm. Scale bar: 13 mm.

The unexpected lack of capillary coverage in MSC-HDF and the guiding effect of collagen fibers in HDF-HDF led us to investigate the role of basal cell-deposited ECM in lymphangiogenesis. The original protocol was modified by increasing the lymphangiogenic period between LEC seeding and culture insert removal from 10 h to 48 h to allow for more buildup of the MSC layer. While this increased capillary invasion and produced capillaries that crossed over the entire wound gap in both MSC-HDF and HDF-HDF (Figure 4), this also exacerbated the aggregation of MSCs, causing capillaries to be pulled into aggregates (Figure S1), so a statistical analysis could not be performed.

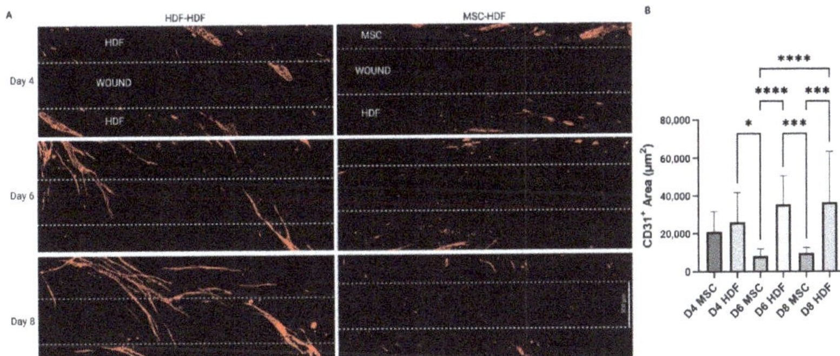

**Figure 3.** Time lapse images of capillary invasion into the wound gap after a 10-h lymphangiogenic period. (**A**) After basal cell coverage of the wound gap at day 4, self-assembled capillaries started protruding into the filled gap over the basal cells in HDF/HDF interfaces. Capillaries from opposing sides pointed towards each other. Capillary invasion was not observed in MSC-HDF interfaces. (**B**) Capillary coverage is present in wound gaps between HDF and either MSC or HDF. HDF-HDF had significantly more vessel invasion than MSC-HDF on day 6 and 8. Sample sizes: D4 MSC ($n = 8$) D4 HDF ($n = 10$), D6 MSC ($n = 8$), D6 HDF ($n = 12$), D8 MSC ($n = 8$), D8 HDF ($n = 11$). * $p \leq 0.05$, *** $p \leq 0.001$, **** $p \leq 0.0001$. Scale bar: 500 μm.

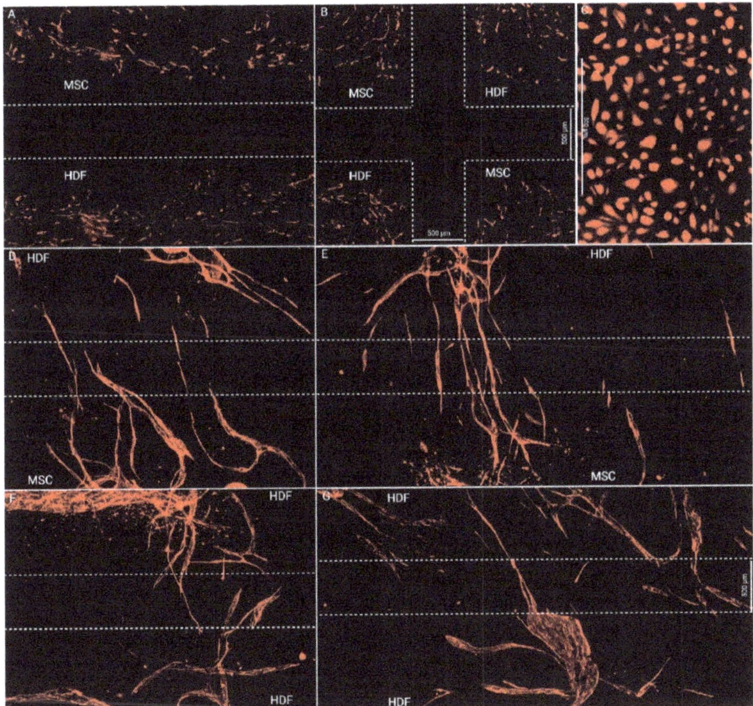

**Figure 4.** Images of capillary invasion into the wound gap after a 48-h lymphangiogenic period. (**A,B**) Day 4. Supporting cells have closed the wound gap. Some capillaries have formed, but not yet invaded the wound gap. Capillaries were still short and immature. (**C**) LECs fail to form capillaries when seeded directly on the culture dish. Capillary invasion over wound gaps must therefore be over basal cells. (**D–G**) Day 8. Capillaries have invaded the wound gap. MSC-HDF gaps had more connecting capillaries than HDF-HDF. Scale bar: 500 μm.

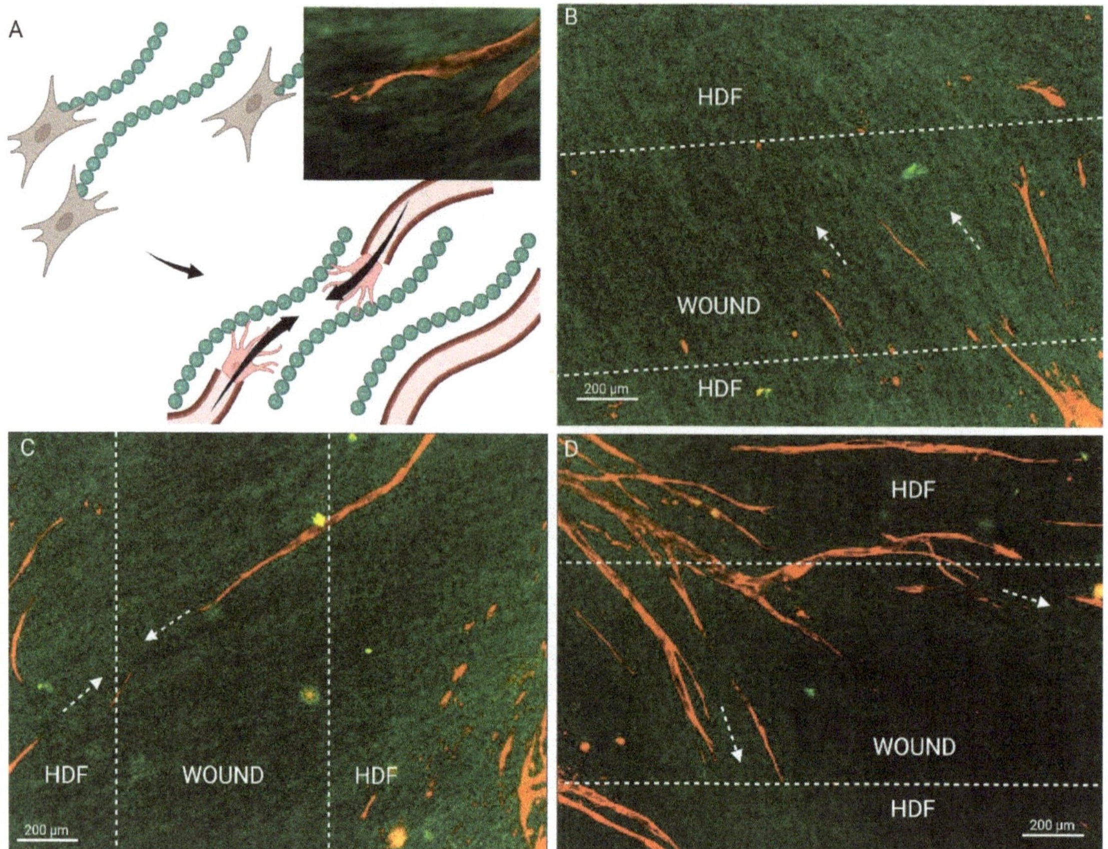


**Figure 5.** Fibroblasts guide capillary integration through natural collagen I alignment during in vitro graft integration. (**A**) ECM-mediated anastomosis model, where end-to-end anastomosis of capillaries from opposing wound edges are guided by collagen tracks. (**B–D**) Fibroblast proliferation and migration patterns in the wound gap leave behind collagen tracks that the capillaries follow. Arrows mark the direction of the collagen fibers as well as capillary invasion. LECs (red) Collagen I (Green). Scale bar: 200 µm.

## 3.2. Engineered Flap Spontaneously Anastomoses with Rat Model

To determine the anastomosis capacity of the engineered flap after implantation, the flap was subcutaneously implanted in athymic nude rats, which can prevent xenograft rejection. Seven days post implantation, chimeric capillaries made of rat and human LECs were observed (Figure 6). Human nuclear antigen (marker of human cells) negative rat LECs were found incorporated into human capillaries, indicating the fusion of implanted human capillaries with host rat capillaries. The sectioned lumens were also open.

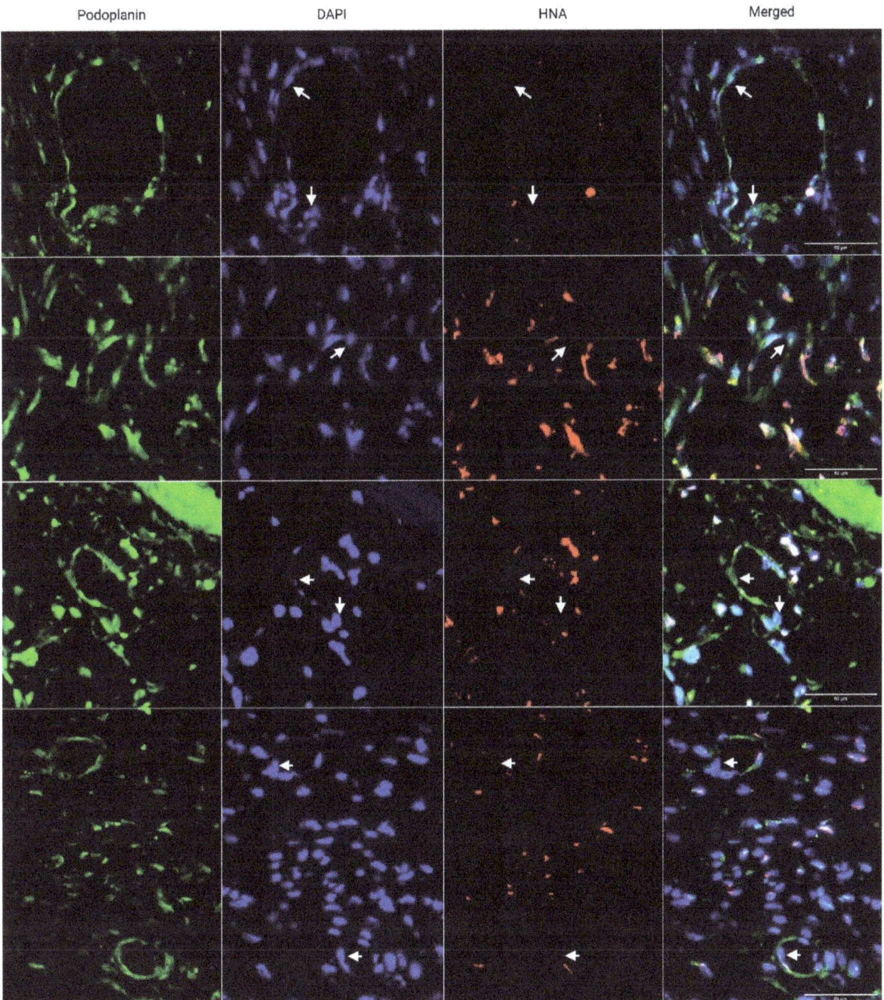

**Figure 6.** Immunostaining of implanted engineered lymphatic flap. Rat-human chimeric capillaries were found 7 days after subcutaneous implantation in athymic nude rats ($n = 3$). Rat, Human PDPN (green), HNA (red). Each row represents a different site of anastomosis. Arrows point to HNA⁻ LECs within HNA⁺ vessels. Scale bar: 50 μm.

## 4. Discussion

The supply of skin autografts for VLVT is limited. We have fabricated an engineered lymphatic flap replacement that is both scalable and personalizable to the patient. In vitro results showed that the proliferation of supporting cells in the engineered lymphatic flap paved the way for lymphangiogenesis, which leads to the connection of graft and host capillaries. Subcutaneous implantation also indicated that the engineered flap is capable of spontaneous anastomosis in vivo. Therefore, our construct has high potential for ameliorating advanced fibrotic lymphedema.

ECM deposition by graft and host cells plays an underappreciated role in graft vascular integration. MSC-HDF was expected to have more vessel invasion due to the chemotactic and lymphangiogenic effects of MSC-secreted paracrine factors [25]. The unexpectedly low levels of capillary invasion in MSC-HDF may therefore have been caused by insufficient matrix deposition over the wound gap by MSCs as HDFs proliferated and deposited matrix faster than MSCs (Figure 2D), which correlated with higher capillary invasion in HDF-HDF (Figure 3B). Extension of the lymphangiogenic period may have rescued the lack of capillary coverage of the MSC-HDF wound gaps (Figure 4) by allowing more accumulation of MSC-deposited ECM, which may have improved the quantity and quality of capillaries on the wound edges that participated in the invasion. The MSC layer may have also increased in thickness and cell density, which can contribute to the overall ECM deposition rate into the wound gap. In addition to providing a substrate for lymphangiogenesis, ECM can signal surface topographical cues known to guide cell alignment by localizing focal adhesions along grooves [39]. We previously found that aligned ECM nanofibrous scaffolds guided capillary alignment in a multi-layered vascular construct [32]. An aligned electrospun gelatin scaffold has also shown a similar ability to direct angiogenesis [40]. Coincidentally, thin oriented collagen fibers are also deposited during skin wound healing [41]. Therefore, the ECM fibers secreted by HDFs during wound healing may direct and promote capillary invasion much like an aligned scaffold would in addition to providing mechanical support to the wound. We posit that continuous ECM fibers from two sides of a wound facilitate capillary invasion and end-to-end anastomosis as capillaries from the two opposing sides follow the same fibers and come into contact (Figure 5). This may be a novel mechanism of wound healing and graft revascularization and integration. Further research into understanding the mechanisms of graft integration could improve the success rate of tissued engineered grafts.

Our model is in agreement with past wound healing models. The morphology of the invading capillaries from our system is comparable to that of animal wound healing and ex vivo models. Our 2D assay reflects cross sections of the healing fibrin clot of a full-thickness porcine wound model [42]. Similar to our results, the porcine capillaries extended from the granulation tissue into the provincial matrix by day 4 of wound repair. An explant model using artery and vein sections cultured 500~1000 μm apart reported similar anastomosis patterns [43]. A micropatterned substrate covered in thymosin β4-hydrogel was used to induce directed capillary sprouting between two explants, and the outgrowths connected the two parent explants by day 21. The guided vessels resembled the capillaries in Figure 4D–G, further supporting our ECM-guidance hypothesis. The initial progressive increase in capillary area coverage of HDF-HDF is also consistent with murine [44] and human [45] skin autograft revascularization, where the graft integration layer increases in vascular density after implantation. However, Tefft et al. [46] reported no capillary invasion in their in vitro wound healing model of granulation tissue formation. Their model was created by seeding fibroblasts and human umbilical vein endothelial cells in a 3D collagen and fibrin gel in a microfluidic chamber. An incision was made after 3 days of capillary assembly by stabbing the gel with a dissection knife. This model also achieved a wound closure time of 4 days. However, unlike our results, their blood capillaries remained at the periphery of the wound as fibroblasts filled in the wound with fibronectin and collagen III. This could be because the fibroblasts initially migrated circumferentially around the wound during wound contraction before migrating to the center of the void. This would have left collagen tracks tangential to the wound, leading capillaries to sprout around it.

Upon implantation, the engineered flap spontaneously connected with host lymphatics in a rat model (Figure 6). Spontaneous anastomosis has been demonstrated in clinical lymph-interpositional-flap transfers, where the lymph vessel stump of a flap is placed closely to those of a recipient site without microsurgical anastomosis [10,11]. This is believed to be the primary mechanism by which implanted flaps connect to and restore recipient site lymph flow. Compatibility of lymph axiality of the graft and the recipient site is critical for lymph flow restoration and reducing lymphedema development risk. Our engineered flap is designed with highly aligned capillaries, giving it high lymph axiality and therefore high potential to restore lymph flow after anastomosis. The ECM fibers in the flap may have also provided interstitial flow guidance for lymphangiogenesis similar to nanofibrillar collagen scaffolds [15].

As this study was a proof-of-concept study, there were several limitations. First, although capillaries spanning across the wound gap were observed, end-to-end anastomosis of capillaries originating from opposite side was not confirmed. Capillaries may have primarily sprouted from one side to the other, so cell origin will be clarified by labeling the LECs. Second, the identity of the supporting cells was not verified. The cells covering the wound gap may have been mostly HDFs, as they proliferated faster. Aggregation was also prevalent when culture time increased in the MSC tissue because of the difficulties in using a tri-culture. We previously found that the optimal culture time for lymphangiogenesis was 7 days of co-culture with MSCs [32], but this caused aggregation and detachment in the model. While we minimized aggregation, viable wound gap samples still decreased from 12 to 8 in MSC-HDF and from 12 to 11 in HDF-HDF. However, this should not have affected the results as wound interfaces with detaching sides were excluded. The peeling and aggregation may have been caused by cellular detachment from the culture plate caused by the removal of the culture-inserts. In addition, live cell tracking can investigate the formation process of oriented collagen I fibers, and disruption of aligned collagen formation is needed to prove its role in capillary invasion guidance. This may be achieved by seeding basal cells in the wound gap to induce contact inhibition on peri-wound basal cells while still maintaining a basal cell layer for capillaries to invade over.

## 5. Conclusions

We engineered a lymphatic flap that is designed to replace skin flaps in VLVT. Using the fabrication techniques for the engineered flap, we developed a novel wound healing assay that reproduces the process of wound closure and graft integration. This model has revealed that collagen fibers deposited during wound healing may guide angiogenesis, facilitate anastomosis, and graft perfusion. Understanding the mechanism of graft integration will improve the success rate of transplants and engineered tissue constructs. The engineered lymphatic flap itself is capable of spontaneous connection with host lymphatics. Our findings propose a potential novel therapeutic for treating lymphedema patients with dysfunctional lymphatic vessels by increasing lymph drainage to the venous and lymphatic systems.

**Supplementary Materials:** The following supporting information can be downloaded at: https://www.mdpi.com/article/10.3390/bioengineering10020149/s1, Figure S1: Aggregation in wound healing model at day 10. The 10-hour lymphangiogenic period was extended to 48 hours, resulting in thickened basal cell layer and more complete tube formation. However, basal cell contraction pulled the surface capillaries inward, sometimes over the wound gaps.

**Author Contributions:** Conceptualization, A.C., F.Z. and W.J.; methodology, A.C., J.G. and W.J.; software, A.C.; validation, A.C. and Y.S.; formal analysis, A.C. and Y.S.; investigation, A.C. and W.J.; resources, A.C. and W.J.; data curation, A.C.; writing—original draft preparation, A.C.; writing—review and editing, F.Z.; visualization, A.C.; supervision, F.Z.; project administration, F.Z.; funding acquisition, F.Z. All authors have read and agreed to the published version of the manuscript.

**Funding:** This study was supported by the National Institutes of Health (R01HL146652 and R15CA202656) and the National Science Foundation (1703570, 2106048) to F.Z.

**Institutional Review Board Statement:** The study was conducted according to the guidelines of the Declaration of Helsinki and approved by the Institutional Review Board of Michigan Technological University (protocol code L0294, approved April 2020).

**Data Availability Statement:** The data presented in this study are available in the article and Supplementary Materials.

**Conflicts of Interest:** The authors declare no conflict of interest.

## References


1. Siegel, R.L.; Miller, K.D.; Fuchs, H.E.; Jemal, A. Cancer statistics, 2022. *CA A Cancer J. Clin.* **2022**, *72*, 7–33. [CrossRef] [PubMed]
2. Hara, Y.; Otsubo, R.; Shinohara, S.; Morita, M.; Kuba, S.; Matsumoto, M.; Yamanouchi, K.; Yano, H.; Eguchi, S.; Nagayasu, T. Lymphedema After Axillary Lymph Node Dissection in Breast Cancer: Prevalence and Risk Factors—A Single-Center Retrospective Study. *Lymphat. Res. Biol.* **2022**, *20*, 600–606. [CrossRef] [PubMed]
3. Liu, Y.-F.; Liu, J.-E.; Mak, Y.W.; Zhu, Y.; Qiu, H.; Liu, L.-H.; Yang, S.-S.; Chen, S.-H. Prevalence and predictors of breast cancer-related arm lymphedema over a 10-year period in postoperative breast cancer patients: A cross-sectional study. *Eur. J. Oncol. Nurs.* **2021**, *51*, 101909. [CrossRef] [PubMed]
4. Togawa, K.; Ma, H.; Smith, A.W.; Neuhouser, M.L.; George, S.M.; Baumgartner, K.B.; McTiernan, A.; Baumgartner, R.; Ballard, R.M.; Bernstein, L. Self-reported symptoms of arm lymphedema and health-related quality of life among female breast cancer survivors. *Sci. Rep.* **2021**, *11*, 10701. [CrossRef] [PubMed]
5. Shih, Y.-C.T.; Xu, Y.; Cormier, J.N.; Giordano, S.H.; Ridner, S.H.; Buchholz, T.A.; Perkins, G.H.; Elting, L.S. Incidence, treatment costs, and complications of lymphedema after breast cancer among women of working age: A 2-year follow-up study. *J. Clin. Oncol.* **2009**, *27*, 2007–2014. [CrossRef] [PubMed]
6. Sleigh, B.C.; Manna, B. Lymphedema. In *StatPearls*; StatPearls Publishing LLC.: Treasure Island, FL, USA, 2022.
7. Zaleska, M.T.; Olszewski, W.L. The Effectiveness of Intermittent Pneumatic Compression in Therapy of Lymphedema of Lower Limbs: Methods of Evaluation and Results. *Lymphat. Res. Biol.* **2019**, *17*, 60–69. [CrossRef]
8. Paramanandam, V.S.; Dylke, E.; Clark, G.M.; Daptardar, A.A.; Kulkarni, A.M.; Nair, N.S.; Badwe, R.A.; Kilbreath, S.L. Prophylactic Use of Compression Sleeves Reduces the Incidence of Arm Swelling in Women at High Risk of Breast Cancer–Related Lymphedema: A Randomized Controlled Trial. *J. Clin. Oncol.* **2022**, *40*, 2004–2012. [CrossRef] [PubMed]
9. Kataru, R.P.; Wiser, I.; Baik, J.E.; Park, H.J.; Rehal, S.; Shin, J.Y.; Mehrara, B.J. Fibrosis and secondary lymphedema: Chicken or egg? *Transl. Res.* **2019**, *209*, 68–76. [CrossRef]
10. Yoshimatsu, H.; Visconti, G.; Karakawa, R.; Hayashi, A. Lymphatic System Transfer for Lymphedema Treatment: Transferring the Lymph Nodes with Their Lymphatic Vessels. *Plast. Reconstr. Surg. Glob. Open* **2020**, *8*, e2721. [CrossRef]
11. Ito, R.; Zelken, J.; Yang, C.-Y.; Lin, C.-Y.; Cheng, M.-H. Proposed pathway and mechanism of vascularized lymph node flaps. *Gynecol. Oncol.* **2016**, *141*, 182–188. [CrossRef]
12. Moon, K.-C.; Kim, H.-K.; Lee, T.-Y.; You, H.-J.; Kim, D.-W. Vascularized lymph node transfer for surgical treatments of upper versus lower extremity lymphedema. *J. Vasc. Surg. Venous Lymphat. Disord.* **2021**, *10*, 170–178. [CrossRef] [PubMed]
13. Narushima, M.; Mihara, M.; Yamamoto, T.; Hara, H.; Ohshima, A.; Kikuchi, K.; Todokoro, K.; Seki, Y.; Iida, T.; Nakagawa, M.; et al. Lymphadiposal Flaps and Lymphaticovenular Anastomoses for Severe Leg Edema: Functional Reconstruction for Lymph Drainage System. *J. Reconstr. Microsurg.* **2015**, *32*, 050–055. [CrossRef] [PubMed]
14. Hamdi, M.; Ramaut, L.; De Baerdemaeker, R.; Zeltzer, A. Decreasing donor site morbidity after groin vascularized lymph node transfer with lessons learned from a 12-year experience and review of the literature. *J. Plast. Reconstr. Aesthetic Surg.* **2020**, *74*, 540–548. [CrossRef]
15. Ciudad, P.; Manrique, O.J.; Date, S.; Sacak, B.; Bs, W.-L.C.; Kiranantawat, K.; Lim, S.Y.; Chen, H.-C. A head-to-head comparison among donor site morbidity after vascularized lymph node transfer: Pearls and pitfalls of a 6-year single center experience. *J. Surg. Oncol.* **2016**, *115*, 37–42. [CrossRef] [PubMed]
16. Pons, G.; Masia, J.; Loschi, P.; Nardulli, M.L.; Duch, J. A case of donor-site lymphoedema after lymph node–superficial circumflex iliac artery perforator flap transfer. *J. Plast. Reconstr. Aesthetic Surg.* **2014**, *67*, 119–123. [CrossRef] [PubMed]
17. Zeng, W.; Babchenko, O.; Chen, W.F. Microsurgery: Vascularized Lymph Vessel Transfer. In *Peripheral Lymphedema: Pathophysiology, Modern Diagnosis and Management*; Liu, N., Ed.; Springer Singapore: Singapore, 2021; pp. 211–222.
18. Orfahli, L.M.; Fahradyan, V.; Chen, W.F. Vascularized lymph vessel transplant (VLVT): Our experience and lymphedema treatment algorithm. *Ann. Breast Surg.* **2022**, *6*, 8. [CrossRef]
19. Yamamoto, T.; Iida, T.; Yoshimatsu, H.; Fuse, Y.; Hayashi, A.; Yamamoto, N. Lymph Flow Restoration after Tissue Replantation and Transfer: Importance of Lymph Axiality and Possibility of Lymph Flow Reconstruction without Lymph Node Transfer or Lymphatic Anastomosis. *Plast. Reconstr. Surg.* **2018**, *142*, 796–804. [CrossRef]
20. Yamamoto, T.; Yamamoto, N.; Kageyama, T.; Sakai, H.; Fuse, Y.; Tsukuura, R. Lymph-interpositional-flap transfer (LIFT) based on lymph-axiality concept: Simultaneous soft tissue and lymphatic reconstruction without lymph node transfer or lymphatic anastomosis. *J. Plast. Reconstr. Aesthetic Surg.* **2021**, *74*, 2604–2612. [CrossRef]
21. Pandey, S.K.; Fahradyan, V.; Orfahli, L.M.; Chen, W.F. Supermicrosurgical lymphaticovenular anastomosis vs. vascularized lymph vessel transplant—Technical optimization and when to perform which. *Plast. Aesthetic Res.* **2021**, *8*, 47. [CrossRef]
22. Hartiala, P.; Suominen, S.; Suominen, E.; Kaartinen, I.; Kiiski, J.; Viitanen, T.; Alitalo, K.; Saarikko, A.M. Phase 1 Ⓡ Study: Short-term Safety of Combined Adenoviral VEGF-C and Lymph Node Transfer Treatment for Upper Extremity Lymphedema. *J. Plast. Reconstr. Aesthetic Surg.* **2020**, *73*, 1612–1621. [CrossRef]
23. Visuri, M.T.; Honkonen, K.M.; Hartiala, P.; Tervala, T.V.; Halonen, P.J.; Junkkari, H.; Knuutinen, N.; Ylä-Herttuala, S.; Alitalo, K.; Saarikko, A.M. VEGF-C and VEGF-C156S in the pro-lymphangiogenic growth factor therapy of lymphedema: A large animal study. *Angiogenesis* **2015**, *18*, 313–326. [CrossRef] [PubMed]

24. Nguyen, D.; Zaitseva, T.S.; Zhou, A.; Rochlin, D.; Sue, G.; Deptula, P.; Tabada, P.; Wan, D.; Loening, A.; Paukshto, M.; et al. Lymphatic regeneration after implantation of aligned nanofibrillar collagen scaffolds: Preliminary preclinical and clinical results. *J. Surg. Oncol.* **2021**, *125*, 113–122. [CrossRef] [PubMed]

25. Ogino, R.; Hayashida, K.; Yamakawa, S.; Morita, E. Adipose-Derived Stem Cells Promote Intussusceptive Lymphangiogenesis by Restricting Dermal Fibrosis in Irradiated Tissue of Mice. *Int. J. Mol. Sci.* **2020**, *21*, 3885. [CrossRef] [PubMed]

26. Avraham, T.; Yan, A.; Zampell, J.C.; Daluvoy, S.V.; Haimovitz-Friedman, A.; Cordeiro, A.P.; Mehrara, B.J. Radiation therapy causes loss of dermal lymphatic vessels and interferes with lymphatic function by TGF-beta1-mediated tissue fibrosis. *Am. J. Physiol.-Cell Physiol.* **2010**, *299*, C589–C605. [CrossRef] [PubMed]

27. Knezevic, L.; Schaupper, M.; Mühleder, S.; Schimek, K.; Hasenberg, T.; Marx, U.; Priglinger, E.; Redl, H.; Holnthoner, W. Engineering Blood and Lymphatic Microvascular Networks in Fibrin Matrices. *Front. Bioeng. Biotechnol.* **2017**, *5*, 25. [CrossRef] [PubMed]

28. Gibot, L.; Galbraith, T.; Bourland, J.; Rogic, A.; Skobe, M.; Auger, F.A. Tissue-engineered 3D human lymphatic microvascular network for in vitro studies of lymphangiogenesis. *Nat. Protoc.* **2017**, *12*, 1077–1088. [CrossRef]

29. Landau, S.; Newman, A.; Edri, S.; Michael, I.; Ben-Shaul, S.; Shandalov, Y.; Ben-Arye, T.; Kaur, P.; Zheng, M.H.; Levenberg, S. Investigating lymphangiogenesis in vitro and in vivo using engineered human lymphatic vessel networks. *Proc. Natl. Acad. Sci. USA* **2021**, *118*, e2101931118. [CrossRef]

30. Zhang, Q.; Wu, Y.; Schaverien, M.V.; Hanson, S.E.; Chang, E.I.; Butler, C.E. Abstract 128: Engineering Lymphatic Vessels For Secondary Lymphedema Treatment. *Plast. Reconstr. Surg.–Glob. Open* **2020**, *8*, 85–86. [CrossRef]

31. Lu, H.; Hoshiba, T.; Kawazoe, N.; Koda, I.; Song, M.; Chen, G. Cultured cell-derived extracellular matrix scaffolds for tissue engineering. *Biomaterials* **2011**, *32*, 9658–9666. [CrossRef] [PubMed]

32. Qian, Z.; Sharma, D.; Jia, W.; Radke, D.; Kamp, T.; Zhao, F. Engineering stem cell cardiac patch with microvascular features representative of native myocardium. *Theranostics* **2019**, *9*, 2143–2157. [CrossRef] [PubMed]

33. Zhang, L.; Qian, Z.; Tahtinen, M.; Qi, S.; Zhao, F. Prevascularization of natural nanofibrous extracellular matrix for engineering completely biological three-dimensional prevascularized tissues for diverse applications. *J. Tissue Eng. Regen. Med.* **2017**, *12*, e1325–e1336. [CrossRef]

34. Kouroupis, D.; Bowles, A.C.; Willman, M.A.; Orfei, C.P.; Colombini, A.; Best, T.M.; Kaplan, L.D.; Correa, D. Infrapatellar fat pad-derived MSC response to inflammation and fibrosis induces an immunomodulatory phenotype involving CD10-mediated Substance P degradation. *Sci. Rep.* **2019**, *9*, 10864. [CrossRef]

35. Robering, J.W.; Weigand, A.; Pfuhlmann, R.; Horch, R.E.; Beier, J.P.; Boos, A.M. Mesenchymal stem cells promote lymphangiogenic properties of lymphatic endothelial cells. *J. Cell. Mol. Med.* **2018**, *22*, 3740–3750. [CrossRef]

36. Chen, L.; Xing, Q.; Zhai, Q.; Tahtinen, M.; Zhou, F.; Chen, L.; Xu, Y.; Qi, S.; Zhao, F. Pre-vascularization Enhances Therapeutic Effects of Human Mesenchymal Stem Cell Sheets in Full Thickness Skin Wound Repair. *Theranostics* **2017**, *7*, 117–131. [CrossRef]

37. Kim, H.S.; Hwang, H.J.; Kim, H.J.; Choi, Y.; Lee, D.; Jung, H.H.; Do, S.H. Effect of Decellularized Extracellular Matrix Bioscaffolds Derived from Fibroblasts on Skin Wound Healing and Remodeling. *Front. Bioeng. Biotechnol.* **2022**, *10*, 865545. [CrossRef]

38. Xing, Q.; Vogt, C.; Leong, K.W.; Zhao, F. Highly Aligned Nanofibrous Scaffold Derived from Decellularized Human Fibroblasts. *Adv. Funct. Mater.* **2014**, *24*, 3027–3035. [CrossRef] [PubMed]

39. Uttayarat, P.; Toworfe, G.K.; Dietrich, F.; Lelkes, P.I.; Composto, R.J. Topographic guidance of endothelial cells on silicone surfaces with micro- to nanogrooves: Orientation of actin filaments and focal adhesions. *J. Biomed. Mater. Res. Part A* **2005**, *75*, 668–680. [CrossRef] [PubMed]

40. Montero, R.B.; Vial, X.; Nguyen, D.T.; Farhand, S.; Reardon, M.; Pham, S.M.; Tsechpenakis, G.; Andreopoulos, F.M. bFGF-containing electrospun gelatin scaffolds with controlled nano-architectural features for directed angiogenesis. *Acta Biomater.* **2012**, *8*, 1778–1791. [CrossRef]

41. Doillon, C.J.; Dunn, M.G.; Bender, E.; Silver, F.H. Collagen Fiber Formation in Repair Tissue: Development of Strength and Toughness. *Collagen Relat. Res.* **1985**, *5*, 481–492. [CrossRef] [PubMed]

42. Tonnesen, M.G.; Feng, X.; Clark, R.A. Angiogenesis in Wound Healing. *J. Investig. Dermatol. Symp. Proc.* **2000**, *5*, 40–46. [CrossRef]

43. Chiu, L.L.Y.; Montgomery, M.; Liang, Y.; Liu, H.; Radisic, M. Perfusable branching microvessel bed for vascularization of engineered tissues. *Proc. Natl. Acad. Sci. USA* **2012**, *109*, E3414–E3423. [CrossRef] [PubMed]

44. O'Ceallaigh, S.; Herrick, S.E.; Bluff, J.E.; McGrouther, D.A.; Ferguson, M.W.J. Quantification of Total and Perfused Blood Vessels in Murine Skin Autografts Using a Fluorescent Double-Labeling Technique. *Plast. Reconstr. Surg.* **2006**, *117*, 140–151. [CrossRef] [PubMed]

45. Deegan, A.J.; Lu, J.; Sharma, R.; Mandell, S.P.; Wang, R.K. Imaging human skin autograft integration with optical coherence tomography. *Quant. Imaging Med. Surg.* **2021**, *11*, 784–796. [CrossRef]

46. Tefft, J.B.; Chen, C.S.; Eyckmans, J. Reconstituting the dynamics of endothelial cells and fibroblasts in wound closure. *APL Bioeng.* **2021**, *5*, 016102. [CrossRef] [PubMed]

 *bioengineering*

*Article*

# Keratin Promotes Differentiation of Keratinocytes Seeded on Collagen/Keratin Hydrogels

Kameel Zuniga [1,*], Neda Ghousifam [2], John Sansalone [2], Kris Senecal [3], Mark Van Dyke [4] and Marissa Nichole Rylander [2]

1 Department of Biomedical Engineering, The University of Texas at Austin, Austin, TX 78712, USA
2 Department of Mechanical Engineering, The University of Texas at Austin, Austin, TX 78712, USA
3 Natick Soldier Center, U.S. Army Soldier & Biological Chemical Command, Natick, MA 01760, USA
4 College of Biomedical Engineering, The University of Arizona, Tucson, AZ 85721, USA
* Correspondence: kameelzuniga@gmail.com

**Abstract:** Keratinocytes undergo a complex process of differentiation to form the stratified stratum corneum layer of the skin. In most biomimetic skin models, a 3D hydrogel fabricated out of collagen type I is used to mimic human skin. However, native skin also contains keratin, which makes up 90% of the epidermis and is produced by the keratinocytes present. We hypothesized that the addition of keratin (KTN) in our collagen hydrogel may aid in the process of keratinocyte differentiation compared to a pure collagen hydrogel. Keratinocytes were seeded on top of a 100% collagen or 50/50 C/KTN hydrogel cultured in either calcium-free (Ca-free) or calcium+ (Ca+) media. Our study demonstrates that the addition of keratin and calcium in the media increased lysosomal activity by measuring the glucocerebrosidase (GBA) activity and lysosomal distribution length, an indication of greater keratinocyte differentiation. We also found that the presence of KTN in the hydrogel also increased the expression of involucrin, a differentiation marker, compared to a pure collagen hydrogel. We demonstrate that a combination (i.e., containing both collagen and kerateine or "C/KTN") hydrogel was able to increase keratinocyte differentiation compared to a pure collagen hydrogel, and the addition of calcium further increased the differentiation of keratinocytes. This multi-protein hydrogel shows promise in future models or treatments to increase keratinocyte differentiation into the stratum corneum.

**Keywords:** keratinocytes; differentiation; collagen; keratin; hydrogel; lysosome; glucocerebrosidase; involucrin

**Citation:** Zuniga, K.; Ghousifam, N.; Sansalone, J.; Senecal, K.; Van Dyke, M.; Rylander, M.N. Keratin Promotes Differentiation of Keratinocytes Seeded on Collagen/Keratin Hydrogels. *Bioengineering* **2022**, *9*, 559. https://doi.org/10.3390/bioengineering9100559

Academic Editors: Ngan F. Huang and Brandon J. Tefft

Received: 5 September 2022
Accepted: 9 October 2022
Published: 15 October 2022



## 1. Introduction

Keratinocytes undergo a complex process of differentiation to develop the cornified layer that protects the body from foreign invasion. Because the epidermis is continually renewing every two weeks, keratinization requires a balance between proliferation, differentiation, and apoptosis [1]. This involves major changes in the intracellular organization of the keratinocytes, and the subsequent development of the stratum corneum. In the basal layer, the keratinocytes are in a proliferative state and express cell-specific markers, such as cytokeratin 14 (K14) [1–4]. These epidermal stem cells give rise to daughter cells that migrate up into the spinous layer [3], and as they detach from the basal layer, they become stratified as the calcium concentration increases. Seo et al. discovered that 290 genes are up-regulated, most of which are related to differentiation, in response to increased calcium [5]. These genes include involucrin and the keratin 1 gene, which includes the activator 1 (AP1) protein, a calcium-responsive promoter [6–8]. In addition, the G-protein calcium-sensing receptor (CaSR) allows keratinocytes to sense the levels of extracellular calcium, allowing the cells to differentiate based on the existing extracellular calcium gradient found in the epidermal layer, and leading to increased intracellular calcium levels [9]. Although the mechanism of why calcium regulates keratinocyte differentiation is not fully understood, the calcium gradient throughout the epidermis allows sequential differentiation through outside–in and inside–out signaling of calcium, allowing the formation of desomosomes, adherens junctions, and tight junctions for stratification [4]. These membrane complexes trigger changes in actin distribution and further increase intracellular calcium, increasing the expression of differentiation markers such as involucrin, a component of the cornified envelope of the epidermis [4,10–12].

Literature shows that lysosomal activity is a requirement in keratinocyte differentiation and cornification in 3D in vitro models by triggering mitochondrial metabolism with reactive oxygen species production, which causes autophagy lysosomal degradation, creating a feedback loop [13]. As the nucleus disappears and cells flatten, keratinocytes become dead corneocytes. The corneocytes are mechanically strong due to the crosslinking of the keratin networks; moreover, because they have lost their organelles, it is hypothesized that the flattening and volume decrease exhibited by these cells lead to further mechanical resilience by extracellular lipid organization [3], creating a hardened structure of the stratum corneum.

Two-dimensional cell monolayers do not recapitulate the complex 3D architecture of tissue or the inherent cell–cell and cell–matrix interactions critical for understanding responses to injury and wound healing. Most 3D in vitro models consist of an extracellular matrix seeded with fibroblasts, in which keratinocytes are cultured on top. They also consist of an air–liquid interface (ALI), another stimulant for keratinocyte differentiation, which is attained by exposing seeded keratinocytes to the air without being submerged in culture media. This can be accomplished using an organotypic transwell skin model [14–16]. The extracellular matrix of 3D in vitro skin models is often in the form of collagen type I hydrogels, and currently, there are no existing skin models that employ keratin. Although keratin does not exist in the dermal layer, it does comprise almost 90% of the epidermis [17]. The addition of keratin in a skin model may promote the differentiation of keratinocytes by forming a cytoskeleton similar to that of native keratinocytes [18]. In native skin, when the keratinocytes flatten, the keratin filaments start to align into disulfide crosslinks and undergo proteolytic degradation [17,19]. Therefore, the presence of keratin in the initial microenvironment may aid in the process of keratinocyte differentiation.

In the current study, we sought to determine whether the inclusion of keratin extracted from human hair in frequently used collagen hydrogels induces the differentiation of keratinocytes, specifically for the telomerase-immortalized keratinocyte cell line (Ker-CT cells) compared to the influence of a pure collagen environment. Although primary keratinocytes are frequently used for modeling skin, telomerase-immortalized keratinocytes engineered by the Rheinwald group by the transduction of primary keratinocytes with the human telomerase reverse transcriptase gene present normal differentiation in monolayer and organotypic skin models [20–26]. We also chose to employ hair keratin, which is more commercially available, unlike skin. Although both epithelial and hair keratins have a structural subunit containing keratins of differing molecular weight and composition named type I (acidic) and type II (basic–neutral), which interact to form heterodimers and polymerize to form intermediate filaments (IFs), hard keratin is more often used in biomedical research [27,28]. Soft keratins from epithelia exist in the form of loosely packed and disorganized IFs in an amorphous matrix, whereas hair keratins contain a much higher cysteine residue content with IFs organized in an ordered array in an $\alpha$-keratin matrix, allowing tougher and more stable structures by intermolecular disulfide formation [27,29]. This natural fiber system is similar to synthetic polymers in that it contains structural macromolecules and crosslinkers, making hair keratins more practical and stable for biomedical research.

Our study aims were achieved by creating acellular collagen and collagen/keratin multiprotein hydrogels and growing keratinocytes as a monolayer on top of the gel in calcium-free and calcium+ media. To assess differentiation, lysosomal activity was investigated through a glucocerebrosidase (GBA) assay for both primary keratinocytes and Ker-CT cells, and LysoTracker staining was employed to visualize lysosomes (Ker-CT only).

For our initial study of GBA activity only, we also tested the influence of fetal bovine serum (FBS) within the media in addition to calcium (Ca+/FBS), since skin models include FBS in their culture medium to induce differentiation. We also assessed the differentiation of Ker-CT cells by staining specifically for CK14 (proliferation marker) and involucrin (differentiation marker). We were able to demonstrate that the inclusion of keratin in the hydrogel increased lysosomal activity and involucrin expression. Calcium inclusion in the media also further increased the activities of both key markers, indicating higher levels of differentiation. Although we were unable to observe larger cell sizes with the inclusion of keratin, compared to pure collagen, in the hydrogels, the presence of a 3D matrix significantly increased the size of the Ker-CT cells compared to seeding on a 2D plastic surface, indicating the importance of a 3D environment when differentiating keratinocytes.

## 2. Materials and Methods

### 2.1. Collagen Extraction

Collagen type I was isolated from rat tail tendons donated at the conclusion of other studies at the University of Texas Austin (IACUC Protocol ID: AUP-2022-00040). Rat tail tendons were placed in a 1 N HCl solution (Fisher Scientific, Hampton, NH, USA) at approximately at pH 2.0 and stirred for 16 h at room temperature. The solution was then transferred to 50 mL centrifuge tubes (Fisher Scientific, Hampton, NH, USA) and centrifuged at $30,000 \times g$ (16,000 rpm) for 60 min at 4 °C to pellet the insoluble components. The supernatant was decanted from each tube and transferred into separate 50 mL centrifuge tubes, stored at $-20$ °C overnight, and freeze dried for 48 h. Samples were dissolved in an appropriate amount of 0.01% glacial acetic acid (Fisher Scientific, Hampton, NH, USA) for a final concentration of 8 mg/mL of stock solution [30].

### 2.2. Kerateine Extraction

Kerateine (KTN) was kindly donated by Dr. Mark Van Dyke, and was extracted from human hair obtained from a commercial source. Briefly, human hair was washed, chopped into small pieces, and soaked in a solution of 0.5 M thioglycolic acid (Sigma-Aldrich, St. Louis, MO, USA) in deionized (DI) water for 12 h at 36 °C with gentle stirring [31]. Hair was then filtered from the liquid with a 500 μm sieve (W. S. Tyler, Mentor, OH, USA), and the reducing solution retained. Free proteins were further extracted in excess 100 mM Tris base for 1 h, followed by DI water for 1 h with gentle shaking at 37 °C. Extracts were collected with the 500 μm sieve and combined with the reductant solution. This entire process was repeated one additional time, and all extracts were combined, centrifuged, and filtered. The combined extracts were then purified and concentrated using tangential flow filtration, frozen, and lyophilized on a Labconco benchtop freeze drying system under ambient conditions.

### 2.3. Cell Culture

Both keratinocyte cell lines were seeded in T-75 Eppendorf HEPA-filtered flasks (Eppendorf, Hamburg, Germany). Primary normal human epidermal keratinocytes (NHEKs, passage 3–6, PromoCell, Heidelberg, Germany) and telomerase-immortalized epidermal keratinocytes (Ker-CT cells, passage 30–40, ATCC, Manassas, VA, USA) were cultured in complete proliferation media composed of keratinocyte growth medium 2 (PromoCell, Heidelberg, Germany) supplemented with 0.004 mL/mL bovine pituitary extract (BPE), 0.125 ng/mL recombinant human epidermal growth factor (EGF), 5 μg/mL recombinant human insulin, 0.33 μg/mL hydrocortisone, 0.39 μg/mL epinephrine, 10 μg/mL recombinant human transferrin, and 0.06 mM $CaCl_2$. This media formulation will hereafter be described as Ca-free media. When assessed for differentiation, cells were cultured in proliferation media formulation, without BPE and EGF, with 2.5 mM $CaCl_2$ and 0.05 μg/mL ascorbic acid, and referred to as Ca+ media. For the GBA assay, we also included 10% fetal bovine serum (FBS) to the Ca+ media for testing, referred to as Ca+/FBS media. Cells were maintained in 5% $CO_2$ atmosphere at 37 °C in a sterile cell culture incubator, and corresponding complete media was changed every 2 days. Cells were detached from the flask by washing with 1× phosphate buffered saline (PBS) and replaced with 4 mL of accutase (PromoCell, Heidelberg, Germany), followed by incubation for 15 min. Detached cell solution was neutralized with complete media, transferred to a 15 mL conical tube, and centrifuged at 200× *g* for 3–5 min. The pellet was isolated and resuspended in 1 mL of complete medium for cell counting.

### 2.4. Collagen/Kerateine Hydrogel Fabrication and Cell Seeding

To mimic healthy skin tissue, 4 mg/mL collagen solution was used [32–35]. The 100% collagen hydrogels were prepared from 8 mg/mL collagen stock solution and mixed and neutralized with 10× DMEM (Sigma-Aldrich, St. Louis, MO, USA), 1× DMEM (Gibco™, Gaithersburg, MD, USA), and 1 N NaOH (Fisher Scientific, Hampton, NH, USA) for a resulting pH of 7.4. The 50/50 w%/w% C/KTN hydrogels were prepared by combining stock collagen and KTN dissolved in neutralizing buffer of equal volume. Previously in our laboratory, we tested both 50/50 and 30/70 C/KTN hydrogels, and observed low proliferation and growth of fibroblasts in 30/70 C/KTN hydrogels, while cell growth was observed for 50/50 C/KTN hydrogels, comparable to 100% collagen hydrogels [36]. Therefore, we only tested 50/50 C/KTN against 100% collagen hydrogels. Table 1 describes the different hydrogels that were produced, with corresponding KTN concentration and total protein (collagen and keratin) concentration as w%/v%. Stock collagen and KTN/neutralizing buffer solution were combined at a 1:1 v:v ratio and mixed thoroughly with a spatula and positive displacement pipette.

**Table 1.** KTN concentration and total protein concentration for each type of hydrogel. The concentration of collagen remains constant.

| Hydrogel | Collagen Concentration | KTN Concentration | wt%/vol% |
|---|---|---|---|
| 100% Collagen | 4.0 mg/mL | 0.0 mg/ml | 4.0 |
| 50/50 C/KTN | 4.0 mg/mL | 4.0 mg/ml | 8.0 |

Samples of 40, 80, or 800 µL of each hydrogel formulation were seeded into 96-, 48-, or 6-well plates, respectively, and allowed to polymerize for 1 h. After polymerization, hydrogels were rinsed three times with 1× PBS for 5 min each. Approximately $0.012$–$0.02 \times 10^6$ cells/cm$^2$ was seeded on top of each hydrogel. Cells were also seeded as a blank sample on top of the plate itself at $0.004 \times 10^6$ cells/sample. After reaching 80–90% confluency in the plastic well sample, half of the samples were replaced with Ca-free media, and the other half was replaced with Ca+ media, to serve as the "calcium switch" used for differentiating keratinocytes, as shown in Figure 1. For the GBA assay, an extra set of samples were made and replaced with Ca+/FBS media. Tests were conducted at days 2, 4, and 7 post-calcium switch.

### 2.5. Cell Viability Assay

Cell viability of keratinocytes (NHEK and Ker-CT cells) cultured on hydrogels and plastic was determined to normalize subsequent assay results (Figure S1). The CellTiter-Blue® cell viability assay (Promega, Madison, WI, USA) was used to measure cell viability of each sample at each time point. Briefly, each sample media was replaced with 100 µL of fresh proliferation media and 20 µL of CellTiter-Blue® reagent. Cell viability was also performed on cell dilutions of $0.0025 \times 10^6$–$0.08 \times 10^6$ cells for the standard curve. Samples were incubated for 2 h, and the solution for each sample was subsequently transferred to a clean separate well, after which fluorescence intensity was measured with a plate reader at an emission of 560 nm and excitation at 590 nm. Background fluorescence intensity (blank well with media and CellTiter-Blue® reagent) was subtracted from sample readings. Cell number was determined with the standard curve.

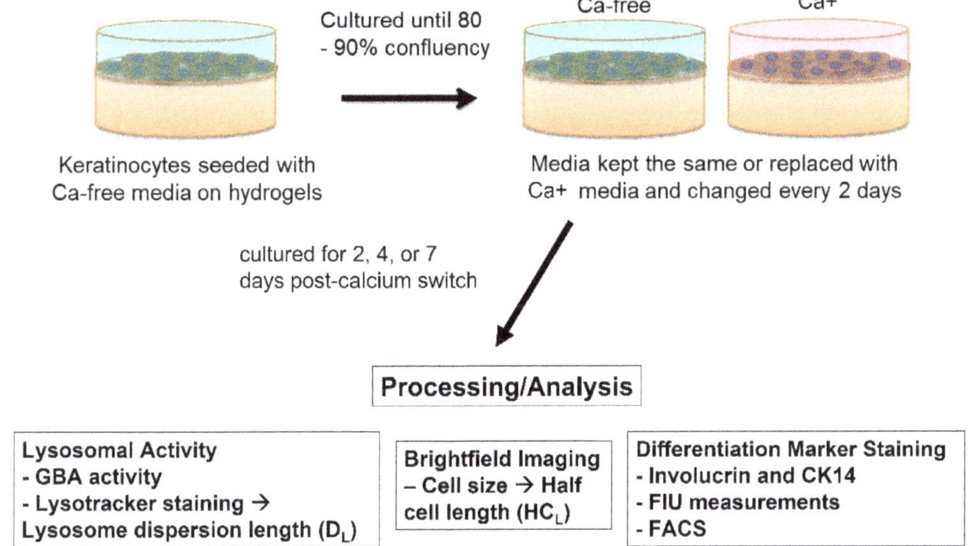



**Figure 1.** Schematic and flow chart of studies with keratinocytes seeded on either the well plate or on hydrogels. Calcium switch was utilized after reaching 80–90% confluency and then media was changed every 2 days. Half of the samples were cultured in Ca-free media. Keratinocytes were analyzed for lysosomal activity, cell size, and differentiation marker expression.

### 2.6. Glucocerebrosidase (GBA) Assay

An increase in the differentiation of keratinocytes is directly correlated with an up-regulation in lysosome activity [13,37,38]. Following a protocol by Mahanty et al., a GBA assay was conducted on both NHEK and Ker-CT cells seeded on top of the hydrogels or on a blank well plate. GBA is a lysosomal enzyme that assists with the breakdown of glucocerebroside into glucose and ceramide [39]. Complete lack of GBA results in fatal skin abnormalities with reduced barrier formation [40–42]. This GBA activity can be observed through the stratum corneum with the accumulation of lysosomal bodies, and therefore is present throughout differentiated keratinocytes [37]. For this initial study, we compared results of Ker-CT cells to primary keratinocytes to determine whether lysosomal activity was increased. Research shows that immortalized cell lines such as HaCaT cells are constantly undergoing lysosomal biogenesis compared to primary cells [43,44]. Therefore, it was critical to compare the results to primary cells to detect differences in GBA activity with increased differentiation. The assay was performed by incubating each sample with 50 µL of 3 mM 4-methylumerlliferyl β-D-glucopyranoside (MUD) dissolved in 0.2 M sodium acetate buffer at pH 4.0 for 3 h at 37 °C. After incubation, the assay was halted by adding 0.2 M glycine buffer at pH 10.8 and incubated for 5 min at room temperature. The solution was removed from each sample into a clean 96-well plate and fluorescence intensity of liberated MUD was read at Ex/Em of 365/445 nm using the Cytation3 Multi-Mode Reader (BioTek Instruments, Winooski, VT, USA). Intensity readings were normalized to the average cell number determined by the cell viability assay.

### 2.7. Lysotracker Staining

To confirm that the KTN was increasing the differentiation of the immortalized keratinocyte cell line, the dispersion length of the lysosomes was measured by lysosomal staining. Keratinocyte differentiation can lead to the dispersion of lysosomes, as increased lysosomal biogenesis occurs in response to the increased intracellular calcium levels. To avoid programmed apoptosis, membrane-bound organelles are likely to accommodate for the excess intracellular calcium, leading to lysosomal dispersion [37,45,46]. To visualize and quantify this dispersion of lysosomes, Ker-CT cells cultured in Ca-free and Ca+ media seeded on top of a plastic plate or a 100% collagen or C/KTN hydrogel were stained with LysoTracker Red DND-99 dye (Thermo Fisher Scientific, Waltham, MA, USA) according to the manufacturer's protocol. Briefly, the 1 mM probe stock solution was diluted to 75 nM in Ca-free medium. Subsequently, 200 µL of the solution was added to 48-well plates and incubated for 30 min at 37 °C. Samples were subsequently rinsed with Ca-free medium and then fixed with 4% PFA for 10 min at room temperature. Samples were then rinsed three times with $1\times$ PBS and stored at 4 °C until imaging or measurement of fluorescence intensity.

Fluorescence intensity was measured with the Cytation3 Multi-Mode Plate Reader at an Ex/Em of 577/590 nm. Intensity readings were normalized to the average cell number determined by the cell viability assay. Samples were also imaged with the Cytation3 Multi-Mode Imaging System to measure the lysosome dispersion half-cell length of the cells with brightfield (BF) imaging and fluorescence imaging at Ex/Em of 532/588 nm. The dispersion length ($D_L$) was quantified by ImageJ software by measuring the distance from the edge of the cell nucleus to the farthest lysosome visualized. The $D_L$ of 20 random cells from each sample was measured for day 2 and 4 time points. Similarly, ImageJ software was utilized to determine the half-cell length ($HC_L$) by measuring the distance from the edge of the cell nucleus to the cell periphery. The $HC_L$ of 20 random cells from each sample was measured for day 2 and 4 time points.

### 2.8. Differentiation Marker Staining

Samples at days 2 and 4 post-calcium switch were stained and the fluorescence intensity was quantified for cytokeratin 14 (CK14), involucrin, and caspase 14 differentiation markers to visualize and quantify the level of differentiation. Samples were fixed with 4% paraformaldehyde for 20 min and subsequently washed three times with $1\times$ PBS. Samples were blocked with 5% FBS for 1 h and then incubated with anti-CK14 (ab181595, Abcam, Cambridge, UK), anti-involucrin (ab68, Abcam, Cambridge, UK), and anti-caspase 14 (ab174847, Abcam, Cambridge, UK) at 1:1000, 1:1000, and 1:250 dilution, respectively, in 5% FBS overnight at 4 °C. The following day, samples were rinsed with $1\times$ PBS three times, and stained with either goat anti-rabbit AF555 (Invitrogen) or anti-mouse AF488 secondary antibody (Invitrogen) at 1:2000 dilution in 5% FBS for 1 h at room temperature.

Stained samples were imaged with the Cytation3 Imaging Plate Reader. Gain and intensity were kept constant between different samples and time points for the RFP (involucrin, CK14, LysoTracker) and the DAPI signal for post-processing analysis. Samples were imaged at 10 and $20\times$ magnification, and fluorescence intensity was measured using ImageJ software. Two methods were used to quantify the intensity of involucrin and CK14 by measuring the fluorescence intensity per cell (corrected total cell fluorescence (CTCF)) and the total fluorescence intensity per image. The corrected total fluorescence intensity per cell was calculated by measuring the integrated density (total fluorescence intensity) of the cell by creating a boundary of one cell with a total of 10 cells per image (total of three images/group, n = 30) and subtracting the integrated intensity by the area of the selected cell multiplied by the mean fluorescence of the background intensity as follows:

$$\mathrm{CTCF} = \text{Integrated Intensity} - (\text{Area of Cell} \times \text{Background Mean Fluorescence}) \qquad (1)$$

This value was then normalized to the CTCF of the DAPI intensity per cell.

With the secondary method, the total fluorescence intensity per image of either the CK14 or involucrin was measured through ImageJ software and normalized to the DAPI cell count/image of $10\times$ magnification images (n = 3).

### 2.9. Flow Cytometry

To acquire sufficient cells for flow cytometry to determine the percentage of keratinocytes expressing differentiation markers, hydrogel samples were prepared in six-well plates and polymerized as previously described. Once polymerized, Ker-CT cells were seeded on top of the hydrogels and allowed to proliferate in Ca-free media until reaching 80–90% confluency. Once reaching confluency, media was replaced with either fresh Ca-free or Ca+ media. After 4 days of culture, cells were isolated by detaching with accutase for Ca-free samples or with 7 mg/mL dispase in $1\times$ HBSS for Ca+ samples. Cells were washed with complete media and then centrifuged to isolate the cell pellet. Cells were passed through a 40 µm cell strainer, strained twice and vortexed.

Once isolated, cells were fixed in 1% paraformaldehyde (PFA) for 10 min on ice. Cells were centrifuged, rinsed in $1\times$ PBS, and treated with 0.05% Triton X for 2 min, and subsequently centrifuged and washed with $1\times$ PBS. Cells were blocked in 5% bovine serum albumin (BSA) for 30 min and subsequently incubated overnight at 4 °C with either anti-involucrin mouse monoclonal antibodies (Abcam, ab68, 1:100 dilution) or anti-cytokeratin 14 rabbit monoclonal antibodies (Abcam, ab181595, 1:190 dilution) in 5% BSA. Samples were also prepared the same way for the isotype controls and incubated with mouse $IgG_1$ isotype control (R&D Systems, MAB002) and rabbit IgG isotype controls at the same concentrations as the anti-involucrin and anti-cytokeratin 14 antibodies, respectively. The next day, cells were washed twice in 2% BSA and incubated for 1 h with goat anti-mouse Alexa Fluor 488 secondary antibody (Invitrogen, Waltham, MA, USA) at 1:2000 dilution and goat anti-rabbit phycoerythrin (PE) secondary antibody (Invitrogen, Waltham, MA, USA) at 1:1000 dilution in 5% BSA for 30 min. Cells were washed twice in 2% BSA and stored at 4 °C until analysis.

Analysis was performed with the FACSAria flow cytometer (BD Biosciences, Franklin Lakes, NJ, USA); forward and side scatter were adjusted using unstained control cells for each group to isolate and remove the cell debris from analysis, and the autofluorescence was eliminated from sample cells using the unstained controls by adjusting the signal outputs. At least 50,000 events were collected and analyzed for each marker to determine the percentage of positively stained cells.

### 2.10. Statistical Analysis

The $D_L$, $HC_L$, and mean fluorescence intensity of differentiation markers/cell are presented as average ± SEM. All other data are presented as average or weighted average ± SD. A two-way ANOVA was performed on GraphPad Prism software. Post hoc analysis was completed with a Tukey's multiple comparison test to determine any significant differences between different groups of hydrogel formulation. This included analyses between cells seeded on a plastic plate vs. 100% collagen vs. 50/50 C/KTN, differences between Ca-free and Ca+ media, and differences between time points.

## 3. Results

### 3.1. Lysosomal Activity Influenced by KTN and Increased Calcium Concentration

Cell viability and GBA activity were measured, with the latter being normalized to the cell viability at days 2, 4, and 7 post-calcium switch. Ker-CT cells showed no observable increase in GBA activity when calcium was added, as shown in Figure 2a and Table 2. This is possibly due to the immortalized cells behaving similar to cancer cells, which constantly undergo lysosomal biogenesis compared to primary cells [43]. In a previous study on HaCaT cells, an immortalized keratinocyte cell line derived from a tumor showed no increased lysosomal activity when compared to cells activated by calcium [44]. Therefore, their differentiation is not apparent with increase in GBA activity. However, significant ($p < 0.01$) increases in GBA activity were observed when KTN was added to the collagen hydrogel, as observed in Figure 2a, at days 2 and 4 for all conditions of culture media. The 50/50 C/KTN hydrogels had significantly higher GBA activity than 100% collagen hydrogels, indicating that the presence of KTN increased keratinocyte differentiation.

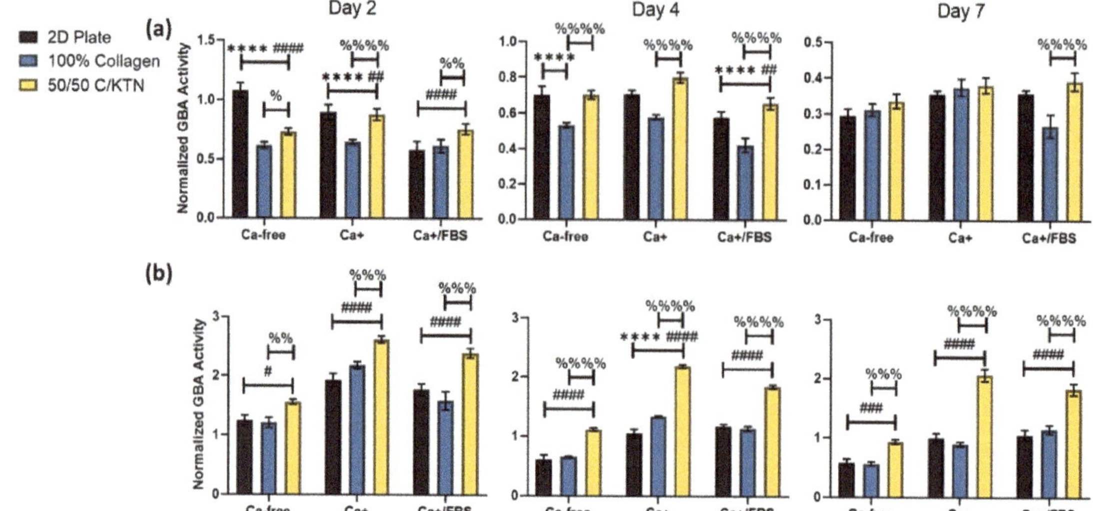

**Figure 2.** GBA activity of immortalized and primary keratinocytes influenced by KTN and media supplements. GBA activity was normalized to cell viability. Immortalized cells showed significant increase in GBA activity when KTN was added to the hydrogel, but no increase when calcium and FBS was added to the media (**a**). NHEKs showed significant increase in GBA activity when KTN was added to the hydrogel, with increases also observed with addition of calcium and FBS (**b**). GBA activity is displayed as weighted average $\pm$ SD. Two-way ANOVA was performed with n = 4 for 3 separate experiments. The different symbols (*, #, %) correspond to the comparisons and the number of symbols correspond to the p-value for each comparison: * = $p < 0.05$, ** = $p < 0.01$, *** = $p < 0.001$, **** = $p < 0.0001$ and * = 2D plate vs. 100% collagen, # = 2D plate vs. 50/50 C/KTN, % = 100% collagen vs. 50/50 C/KTN.

**Table 2.** Comparison of GBA activity between different media within each sample group. GBA activity was normalized to cell viability. For most samples, especially for primary NHEKs, addition of in $^c$calcium increased the GBA activity; addition of FBS also further increased the GBA activity. Two-way ANOVA was performed with n = 4 with 3 separate experiments. * = $p < 0.05$, ** = $p < 0.01$, *** = $p < 0.001$, **** = $p < 0.0001$, ns = not significant.

| Sample Group | Medium | *p*-Value |
|---|---|---|
| Plate, Ker-CT | Ca-free vs. Ca+ | Day 2 ($p < 0.0001$)—**** <br> Day 4 ($p > 0.9999$)—ns <br> Day 7 ($p = 0.0392$)—* |
| | Ca-free vs. Ca+/FBS | Day 2 ($p = 0.0216$)—* <br> Day 4 ($p < 0.0001$)—**** <br> Day 7 ($p = 0.0317$)—* |
| | Ca+ vs. Ca+/FBS | Day 2 ($p < 0.0001$)—**** <br> Day 4 ($p < 0.0001$)—**** <br> Day 7 ($p > 0.9999$)—ns |
| 100% Collagen, Ker-CT | Ca-free vs. Ca+ | Day 2 ($p = 0.9925$)—ns <br> Day 4 ($p = 0.2850$)—ns <br> Day 7 ($p = 0.0329$)—* |

**Table 2.** *Cont.*

| Sample Group | Medium | *p*-Value |
|---|---|---|
| 100% Collagen, Ker-CT | Ca-free vs. Ca+/FBS | Day 2 ($p > 0.9999$)—ns<br>Day 4 ($p < 0.0001$)—***<br>Day 7 ($p = 0.2709$)—ns |
| | Ca+ vs. Ca+/FBS | Day 2 ($p = 0.9859$)—ns<br>Day 4 ($p < 0.0001$)—****<br>Day 7 ($p = 0.0002$)—*** |
| 50/50 C/KTN, Ker-CT | Ca-free vs. Ca+ | Day 2 ($p = 0.0017$)—*<br>Day 4 ($p = 0.0005$)—***<br>Day 7 ($p = 0.2412$)—ns |
| | Ca-free vs. Ca+/FBS | Day 2 ($p = 0.9961$)—ns<br>Day 4 ($p = 0.3243$)—ns<br>Day 7 ($p = 0.0724$)—ns |
| | Ca+ vs. Ca+/FBS | Day 2 ($p = 0.0161$)—*<br>Day 4 ($p < 0.0001$)—****<br>Day 7 ($p = 0.9984$)—ns |
| Plate, NHEK | Ca-free vs. Ca+ | Day 2 ($p < 0.0001$)—****<br>Day 4 ($p < 0.0001$)—****<br>Day 7 ($p < 0.0001$)—**** |
| | Ca-free vs. Ca+/FBS | Day 2 ($p < 0.0001$)—****<br>Day 4 ($p < 0.0001$)—****<br>Day 7 ($p < 0.0001$)—**** |
| | Ca+ vs. Ca+/FBS | Day 2 ($p = 0.4885$)—ns<br>Day 4 ($p = 0.0526$)—ns<br>Day 7 ($p = 0.9928$)—ns |
| 100% Collagen, NHEK | Ca-free vs. Ca+ | Day 2 ($p < 0.0001$)—****<br>Day 4 ($p < 0.0001$)—****<br>Day 7 ($p = 0.0007$)—*** |
| | Ca-free vs. Ca+/FBS | Day 2 ($p = 0.0020$)—**<br>Day 4 ($p < 0.0001$)—****<br>Day 7 ($p < 0.0001$)—**** |
| | Ca+ vs. Ca+/FBS | Day 2 ($p < 0.0001$)—****<br>Day 4 ($p = 0.0006$)—***<br>Day 7 ($p = 0.0138$)—* |
| 50/50 C/KTN, NHEK | Ca-free vs. Ca+ | Day 2 ($p < 0.0001$)—****<br>Day 4 ($p < 0.0001$)—****<br>Day 7 ($p < 0.0001$)—**** |
| | Ca-free vs. Ca+/FBS | Day 2 ($p < 0.0001$)—****<br>Day 4 ($p < 0.0001$)—****<br>Day 7 ($p < 0.0001$)—**** |
| | Ca+ vs. Ca+/FBS | Day 2 ($p = 0.1154$)—ns<br>Day 4 ($p < 0.0001$)—****<br>Day 7 ($p = 0.0173$)—* |

Primary cells (NHEKs) showed a significant ($p < 0.001$) increase in GBA activity when calcium was added to the growth media, with a further rise when FBS was added (Table 2). In addition, there was significant increase for each time point and each medium when KTN was added to the hydrogel when compared to 100% collagen hydrogels (Figure 2b). This can be observed at each time point, even without the addition of calcium to the medium. This increased GBA activity is indicative of lysosomal biogenesis, showing that there is a direct relationship between KTN and increased differentiation.

In addition to the increased GBA activity, the lysosomal distribution and expression also increased when keratinocytes were seeded on top of 50/50 C/KTN hydrogels when compared to the 2D plate and 100% collagen hydrogels. Cells were stained with LysoTracker Red DND-99 dye, which labels the lysosomes of the cells, as shown in Figure 3a. The measured fluorescence intensities of these samples, when normalized to cell viability, indicate that the addition of KTN to the hydrogel significantly increased the lysosomal activity when compared to cells seeded on the plate and 100% collagen for both Ca-free and Ca+ media at days 2 ($p < 0.001$) and 4 ($p = 0.007$ (plate, Ca-free), $p < 0.0001$ (plate, Ca+ and 100% collagen, Ca+)), as shown in Figure 3b. Although $D_L$ had no significant differences between groups at day 2 ($p = 0.4780$ to $p > 0.9999$), $D_L$ did increase for 100% collagen and 50/50 C/KTN hydrogels when calcium was added to the media ($p = 0.0044$ (100% collagen), $p = 0.0002$ (50/50 C/KTN)) (Table 3). The $D_L$ also significantly increased with the addition of KTN when compared to 100% collagen at day 4 ($p = 0.0007$ (100% collagen), and further increased when calcium was included in the media compared to the cells seeded on the plate ($p < 0.0001$) (Figure 3c(i)). Individual measurements are also plotted to show the total distribution of the $D_L$, with increased distribution observed with Ca+ samples (Figure 3c(ii)).

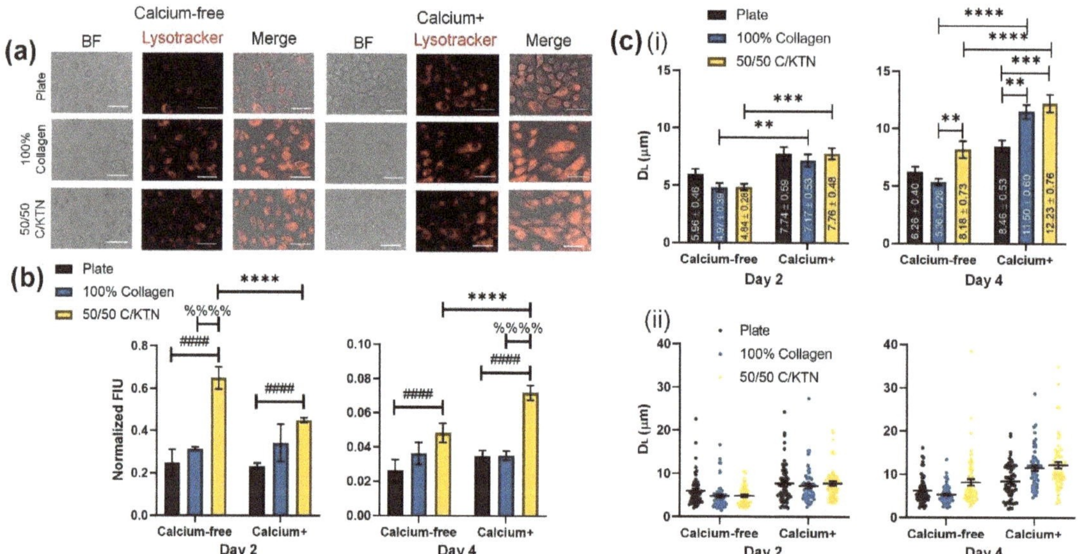

**Figure 3.** LysoTracker lysosome staining of proliferative and differentiated keratinocytes. Day 2 and day 4 keratinocytes seeded on either a plate or a hydrogel were stained with LysoTracker DND-99 dye to visualize lysosomal distribution, with day 2 shown (**a**). Fluorescence intensity units (FIUs) of stained lysosomes were measured and normalized to cell viability at days 2 and 4 post-calcium switch (**b**), with a significant increase observed for 50/50 C/KTN hydrogels compared to the plate and 100% collagen samples with and without calcium (n = 4, weighted average ± SD; * = day 2 vs. 4 for 50/50 C/KTN, # = 2D plate vs. 50/50 C/KTN, % = 100% collagen vs. 50/50 C/KTN). The $D_L$ (**c**) at day 2 showed no significant differences between groups, but increased when calcium was added to the medium (**i**). Significant increases in $D_L$ were observed with 50/50 C/KTN when compared to the plate and 100% collagen at day 4 (n = 60, average ± SEM). Each measurement of $D_L$ is also plotted to show the distribution (**ii**). Two-way ANOVA was performed for both normalized FIU and $D_L$; for all plots. The different symbols (*, #, %) correspond to the comparisons and the number of symbols correspond to the p-value for each comparison: * = $p < 0.05$, ** = $p < 0.01$, *** = $p < 0.001$, **** = $p < 0.0001$; * = 2D plate vs. 100% collagen, # = 2D plate vs. 50/50 C/KTN, % = 100% collagen vs. 50/50 C/KTN. Scale bar = 100 μm.

**Table 3.** $D_L$ of measured lysosomes tagged with LysoTracker staining. Measurements are displayed as average $\pm$ SEM with n = 60.

| Sample Group | Medium | $D_L$ (μm) |
|---|---|---|
| Plate | Ca-free | Day 2—5.96 $\pm$ 0.45<br>Day 4—6.26 $\pm$ 0.40 |
| | Ca+ | Day 2—7.74 $\pm$ 0.59<br>Day 4—8.46 $\pm$ 0.53 |
| 100% Collagen | Ca-free | Day 2—4.97 $\pm$ 0.39<br>Day 4—5.36 $\pm$ 0.28 |
| | Ca+ | Day 2—7.17 $\pm$ 0.53<br>Day 4—11.5 $\pm$ 0.60 |
| 50/50 C/KTN | Ca-free | Day 2—4.84 $\pm$ 0.28<br>Day 4—8.18 $\pm$ 0.73 |
| | Ca+ | Day 2—7.76 $\pm$ 0.48<br>Day 4—12.23 $\pm$ 0.76 |

### 3.2. Keratinocyte Size Influenced by KTN and Increased Calcium Concentration

Keratinocytes become larger during the differentiation process; therefore, the half-cell length was measured to determine whether the addition of KTN in the hydrogel with the combination of calcium in the medium had any effect in increasing cell size in the form of cell length. As shown in Figure 4A, keratinocytes at day 2 begin to appear larger with calcium in the medium. The half-cell length ($HC_L$) measured from the BF images at day 2 and 4 post-calcium switch is plotted and compared between the different groups in Figure 4B(i). Cells seeded on a plastic plate had significantly lower ($p < 0.0001$) half-cell lengths than cells seeded on both types of hydrogels, with a significant increase between Ca-free and Ca+ cells observed at day 2 ($p < 0.0001$). At day 2, the half-cell lengths of cells seeded on top of the plate were less than 19 μm (7.74 $\pm$ 0.59); thus, based on the half-cell length, these are considered proliferative cells [37]. Keratinocytes seeded on top of the hydrogels under Ca+ media had average half-cell lengths larger than 19 μm at both time points, indicating that they are in the process of differentiation. A significant increase was observed in cells from Ca-free and Ca+ media seeded on top of 100% collagen and 50/50 C/KTN hydrogels at both day 2 and 4. As shown in Figure 4B(ii), culturing in Ca+ medium led to a wider distribution of half-cell length, indicating that the cells are in the process of differentiation and proliferation.

### 3.3. Involucrin and CK14 Expression

CK14 and involucrin expression were measured by employing flow cytometry and images of stained samples. For FACS analysis, all samples and isotype controls were gated with non-stained sample. In general, efficient blocking was achieved to prevent non-specific binding, as seen by the isotype controls, although some non-specific binding was not preventable (Figure S2a). However, when compared to the samples, there was clearly more positive cells compared to each corresponding isotype control. Within the whole cell population, the proliferative and differentiated cell populations were analyzed separately, in addition to the total cell population, as seen by the forward and side scatter plot (Figure S2b). Percent positive cells are plotted in Figure S2c, with almost all cells expressing CK14, although significantly lower CK14-positive cells were observed with cells treated with the Ca+ medium. On the other hand, cells treated with the Ca+ medium had significantly higher involucrin-positive cells when compared to cells treated with the Ca-free medium for all cell populations. As expected, cells in the differentiated population also had more involucrin-positive cells than the proliferative population. In addition, 50/50 C/KTN hydrogels in the Ca+ medium had significantly higher involucrin-positive keratinocytes than 100% collagen hydrogels for all cell populations, including the total cell population.

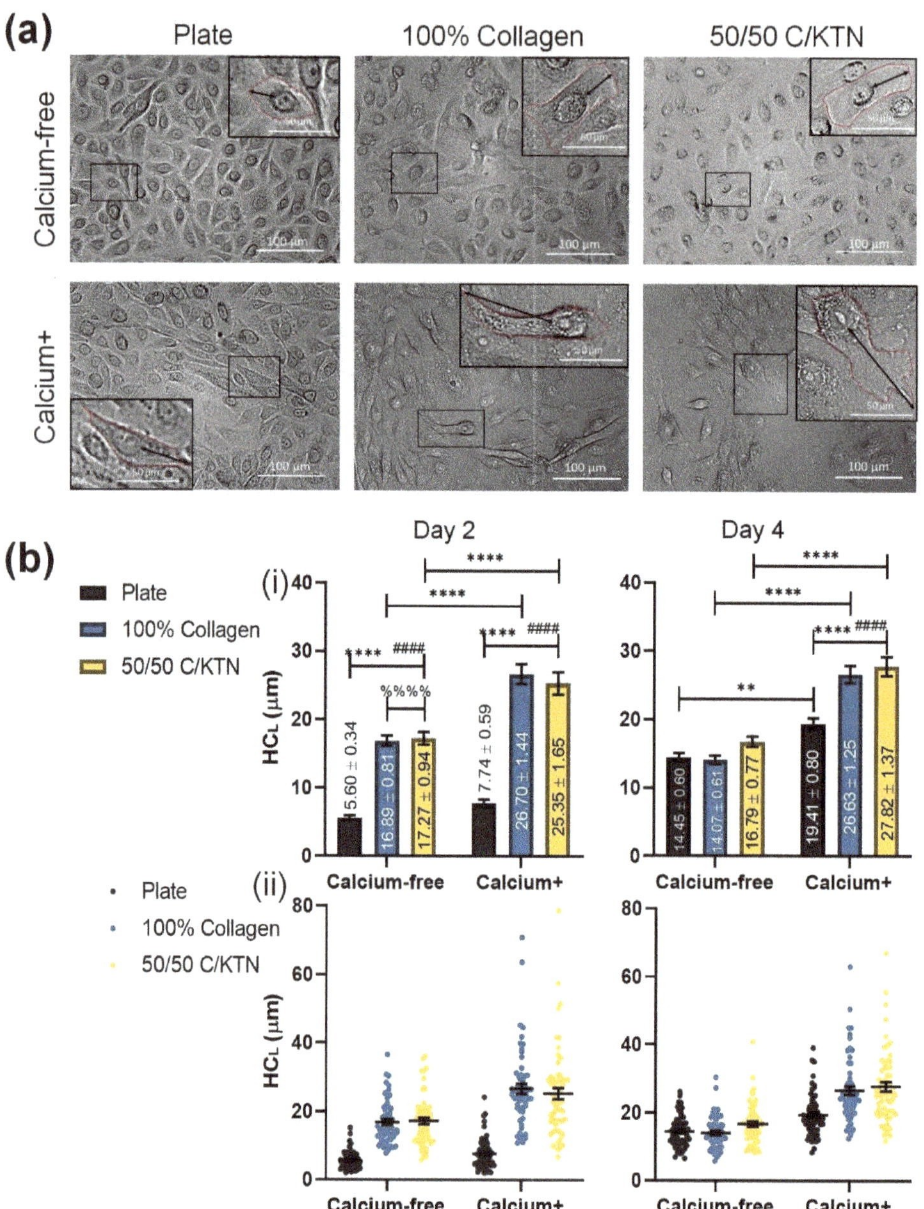

**Figure 4.** Half-cell length ($HC_L$) of keratinocytes under the influence of calcium and hydrogel composition. Day 2 and day 4 keratinocytes seeded on either a plate or a hydrogel were imaged in BF to visualize and measure the half-cell length (**a**). $HC_L$ was measured and plotted to compare different groups and media, with significant increases observed for 100% collagen and 50/50 C/KTN when compared to keratinocytes seeded on a plate; displayed as average ± SEM (**b(i)**). Each measurement of $HC_L$ is also plotted to show the distribution (**ii**). Two-way ANOVA was performed and the different symbols (*, #, %) correspond to the comparisons and the number of symbols correspond to the p-value for each comparison: with * = $p < 0.05$, ** = $p < 0.01$, *** = $p < 0.001$, **** = $p < 0.0001$; * = 2D plate vs. 100% collagen, # = 2D plate vs. 50/50 C/KTN, % = 100% collagen vs. 50/50 C/KTN; n = 60.

Images analyzed for fluorescence intensity of CK14 and involucrin showed similar results as the FACS analysis. Fluorescence images, as shown in Figure 5a, were processed using ImageJ software, and the fluorescence intensity units (FIUs) per cell and per image are plotted at days 2 and 4 post-calcium switch. From what can be observed from the images, the visually larger cells cultured in the Ca+ medium show higher fluorescence intensity of involucrin for both Ca-free and Ca+ media, with larger cells present in the Ca+ medium, as discussed earlier. Similar results were observed for both types of analyses (CTCF and total fluorescence/image), with significantly higher involucrin intensities observed for keratinocytes cultured in the Ca+ medium, and higher involucrin intensity with keratinocytes seeded on top of 50/50 C/KTN hydrogels. CK14 intensity significantly decreased with keratinocytes cultured with the Ca+ medium (Figure 5b), with significantly lower CK14 intensities observed at day 4 compared to day 2 when cultured in the Ca+ medium (Table 4).

**Table 4.** Comparison of fluorescence intensity of differentiation markers between time points within each group. Two-way ANOVA was performed with n = 4 with 3 separate experiments. * = $p < 0.05$, ** = $p < 0.01$, *** = $p < 0.001$, **** = $p < 0.0001$, ns = not significant.

| Sample Group | Media | Day 2 vs. Day 4 |
|---|---|---|
| Plate | Ca-free—CTCF | CK14 ($p = 0.0013$)—** <br> Involucrin ($p = 0.2088$)—ns |
| | Ca+—CTCF | CK14 ($p < 0.0001$)—**** <br> Involucrin ($p = 0.4185$)—ns |
| | Ca-free—total fluorescence | CK14 ($p = 0.0235$)—* <br> Involucrin ($p = 0.0349$)—* |
| | Ca+—total fluorescence | CK14 ($p = 0.0235$)—ns <br> Involucrin ($p = 0.2073$)—ns |
| 100% Collagen | Ca-free—CTCF | CK14 ($p = 0.0025$)—** <br> Involucrin ($p = 0.2609$)—ns |
| | Ca+—CTCF | CK14 ($p < 0.0001$)—**** <br> Involucrin ($p = 0.9924$)—ns |
| | Ca-free—total fluorescence | CK14 ($p = 0.0079$)—** <br> Involucrin ($p = 0.0012$)—** |
| | Ca+—total fluorescence | CK14 ($p = 0.0433$)—* <br> Involucrin ($p = 0.8726$)—ns |
| 50/50 C/KTN | Ca-free—CTCF | CK14 ($p < 0.0001$)—**** <br> Involucrin ($p = 0.0559$)—ns |
| | Ca+—CTCF | CK14 ($p < 0.0001$)—**** <br> Involucrin ($p = 0.4680$)—ns |
| | Ca-free—total fluorescence | CK14 ($p > 0.9999$)—ns <br> Involucrin ($p = 0.0272$)—* |
| | Ca+—total fluorescence | CK14 ($p = 0.6690$)—ns <br> Involucrin ($p = 0.4151$)—ns |

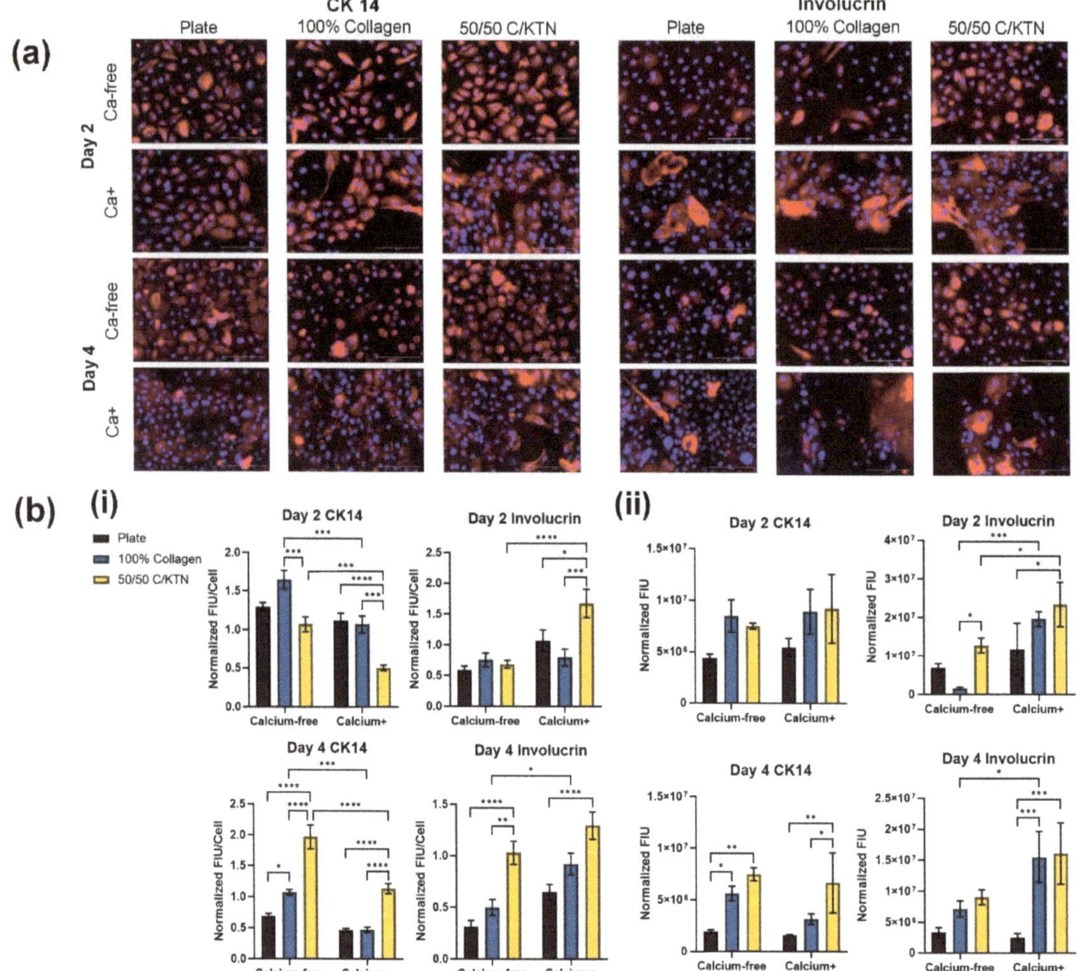

**Figure 5.** CK14 and involucrin staining and analysis of 2D monolayer samples. Keratinocytes seeded on the plate, 100% collagen hydrogels, and 50/50 C/KTN hydrogels were fixed, permeabilized, and stained for CK14 and involucrin at days 2 and post-calcium switch, with half of the samples still cultured in Ca-free medium (**a**). Images were analyzed (**b**) by CTCF of the marker (**i**) and the total fluorescence intensity per image (**ii**), which was normalized to the DAPI CTCF and the DAPI cell count, respectively. In general, CK14 was expressed significantly more in Ca-free samples when cells were analyzed individually. Involucrin expression was increased with Ca+ media with significantly higher involucrin expression observed with cells seeded on top of 50/50 C/KTN hydrogels. FIU/cell and total FIU is displayed as average ± SEM and SD, respectively. Two-way ANOVA was conducted with n = 3; * = $p < 0.05$, ** = $p < 0.01$, *** = $p < 0.001$, **** = $p < 0.0001$. Scale bar = 100 μm.

## 4. Discussion

Most organotypic skin models use collagen as the extracellular matrix (ECM) for the dermal layer, in which fibroblasts are seeded, and upon which keratinocytes are seeded to form the epidermis with exposure to the ALI. Keratin is the primary protein present in the epidermis and is produced by keratinocytes, comprising approximately 90% of the mass of the epidermis [17]. Although keratin is not present in the dermal layer, we wanted to determine whether keratin promotes the differentiation of keratinocytes when it is included in the ECM upon which the keratinocytes are seeded. Because keratin is naturally found in the epidermis, we employed a 50/50 w%/w% collagen/keratin hydrogel, which allowed the spreading and proliferation of NHDFs, as previously shown by the Rylander laboratory [36].

We hypothesized that the presence of keratin, without fibroblasts, and the use of the ALI would promote differentiation of keratinocytes with greater lysosome (GBA) activity and expression of differentiation markers, including involucrin. As demonstrated, GBA activity for immortalized (Ker-CT) and primary keratinocytes significantly increased when KTN was added to the hydrogel. In a previous study conducted by Mahanty et al., a twofold increase in GBA activity was observed with the addition of calcium in the media with a 2D model of keratinocytes seeded in a well-plate [37]. We documented similar results with the primary cells seeded on hydrogels, particularly for those seeded on 50/50 C/KTN hydrogels. We observed significant twofold increases in GBA activity with 50/50 C/KTN hydrogels compared to the 2D plate and 100% collagen hydrogels. However, immortalized keratinocytes showed no significant increases in GBA activity when calcium was added, possibly because immortalized cells behave similar to cancer cells, which constantly undergo lysosomal biogenesis as they proliferate [43,47]. Cancer and immortalized cells differ in normal metabolism to maintain uncontrolled growth and senescence, resulting in rapid depletion of cellular nutrients and damaged organelles, therefore increasing lysosome activity [47]. The rise in calcium concentration did not have any major effect in increasing lysosomal biogenesis for immortalized keratinocytes. However, the addition of KTN to the hydrogel did significantly increase the GBA activity of both primary and immortalized keratinocytes when compared to 100% collagen hydrogels, indicating that KTN increases keratinocyte differentiation. Previous studies on immortalized HaCaT keratinocytes, which are derived from a tumor, showed no increased in lysosome activity with greater calcium [44].

To confirm that the KTN was increasing the differentiation of the immortalized keratinocytes, half-cell length and dispersion length ($D_L$) and total fluorescence intensity of lysosomes was measured at days 2 and 4 post-calcium switch with controls for culture in Ca-free media. With increased keratinocyte differentiation, the $D_L$ of keratinocytes cultured on top of 100% collagen and 50/50 C/KTN hydrogels increased significantly at day 4 when compared to cells cultured on 2D plates, with overall increase in $D_L$ in Ca media. Similar results were observed by Mahanty et al., in which keratinocytes cultured in media containing calcium had increased cell size compared to keratinocytes cultured in Ca-free media [37]. This terminal differentiation of keratinocytes into corneocytes are correlated with increasing size, as keratinocytes in vivo enlarge as they progressively move up through the basal layer into the granular layer of the epidermis [28]. This may be due to the enlargement of the cytoplasm as the cell produces more proteins such as involucrin and filaggrin to prepare for terminal differentiation [48,49]. Keratinocyte differentiation can also lead to the dispersion of lysosomes as increased lysosomal biogenesis occurs in response to the increased intracellular calcium levels. To avoid programmed apoptosis, membrane-bound organelles are likely to accommodate for the excess intracellular calcium, leading to lysosomal dispersion near the cell surface instead of near the nucleus [37,45,46,50]. In addition, culture on hydrogels, a 3D surface, increased the FIU of lysosomal activity, with further significant increase in FIU with 50/50 C/KTN samples compared to 100% collagen hydrogels. This indicated an increased quantity of lysosomes in the cell. The importance of a 3D ECM structure for keratinocyte differentiation was also confirmed with cell size correlating with increased calcium concentration and culture on both 100% collagen and 50/50 C/KTN hydrogels. Keratinocytes cultured on a 2D well surface were small enough to still be considered proliferative at day 2 (<19 µm) [37,48] when cultured in both Ca-free and Ca+ media, whereas keratinocytes cultured on hydrogels in both types of media were already considered differentiated with average cell size threefold greater than those seeded on the plate. As average keratinocyte cell size increased over time and with culture on hydrogels and Ca+ media combined, the dispersion of cell size increased, indicating the presence of both proliferating and differentiating keratinocytes at different levels.

Involucrin, a terminal differentiation marker, is expressed in the cornified envelope of human skin and only expressed in enlarging keratinocytes that are in the process of terminal differentiation [49,51,52]. We were able to demonstrate in our hydrogel monolayer model that the inclusion of KTN without the influence of the ALI and the fibroblast feeder layer led to the increase in involucrin-positive cells and involucrin expression as determined by FACS and fluorescence intensity analysis, respectively. We decided to analyze two different cell populations in our forward and side scatter plot, following methods described by Sanz-Gomez et al. [53]. These two populations of proliferating and differentiating cells can be identified as proliferative basal keratinocytes, and have lower and homogenous FSC and SCC points without much dispersion, whereas the differentiating population can be identified as a heterogenous population [53]. We were able to identify these populations with keratinocytes seeded on top of 100% collagen and 50/50 C/KTN hydrogels. As expected, almost all cells in all conditions were positive for CK14 expression, a marker for keratinocyte proliferation, with a small percentage of cells staining positive for involucrin with FACS analysis. However, when analyzing the expression of CK14 by measuring fluorescence intensity, CK14 expression was significantly lower in samples when cultured in Ca+ conditions, indicating that the calcium was inducing more differentiation and less proliferation of keratinocytes compared to those cultured in Ca-free conditions. Keratinocytes cultured on 50/50 C/KTN hydrogels had significantly higher percentages of involucrin-positive cells in Ca+ culture conditions compared to 100% collagen and Ca-free culture conditions. In addition, the normalized fluorescence intensity also increased with Ca+ culture conditions, with visibly larger cells exhibiting greater fluorescence than smaller cells, validating that the larger cells are differentiating.

There are several limitations to the current study. Complete differentiation of keratinocytes into corneocytes in native human skin take on average 14 days, during which the dead cell layer of your skin turns over [17,20,54,55]. Therefore, most studies employ 14–21-day cultures of keratinocyte on the 3D surface at the ALI. Our study only investigated keratinocyte differentiation submerged in culture medium for a maximum of 7 days. Although we were unable to culture at the ALI for longer than 7 days, we were able to demonstrate that KTN was able to increase differentiation of Ker-CT cells without the influence of the ALI and longer culture. In addition, we only compared our results to primary keratinocytes for the GBA activity, in which we were unable to show increased GBA activity with Ker-CT cells with the addition of calcium to the media. We believe this is due to the Ker-CT cells possibly undergoing lysosomal biogenesis due to constant proliferation [43,44]. However, a recent study by Tito et al. demonstrated increased GBA activity with HaCaT cells when treated with *Triticum vulgare*, a plant-derived treatment known to accelerate wound healing [56]. This suggests that an appropriate inducer for increasing GBA activity due to differentiation may be needed to overcome the GBA activity reading from constant proliferation.

In relation to this study, KTN may be helpful in improving wound healing by increasing the differentiation of keratinocytes, therefore leading to increased skin renewal. Previous studies have reported the clinical uses of keratin as a therapeutic dressing to increase wound healing in chronic wounds and burn injuries [57–61]. In future studies, we aim to test KTN in the presence of an organotypic model to determine whether the addition of KTN further enhances differentiation. Further investigation into the clinical aspect of KTN in improving wound healing in relation to increasing differentiation will also be needed to understand the full potential of keratin as a therapeutic.

## 5. Conclusions

In this study, we demonstrate the presence of keratin in the ECM promoted keratinocyte differentiation without the use of a fibroblast feeder layer and the ALI, traditionally used on 3D organotypic skin models. The presence of KTN increased lysosomal activity, with increases in GBA activity and dispersion length of lysosomes throughout the cytoplasm in immortalized keratinocytes, indicating keratinocyte differentiation. More importantly, the inclusion of KTN concentration increased the expression of involucrin, a terminal differentiation marker that is normally seen in native skin, with further increased differentiation observed with calcium included in the culture media. Although it is still important to include the ALI and the fibroblast feeder layer to represent human skin, the addition of KTN in future skin models in combination with these other inducers of keratinocyte differentiation may create a more representative skin model. Furthermore, KTN may be beneficial in improving wound healing by increased keratinocyte turnover through keratinocyte differentiation.

**Supplementary Materials:** The following supporting information can be downloaded at: https://www.mdpi.com/article/10.3390/{b}ioengineering9100559/s1. Figure S1: Cell viability of keratinocytes, Figure S2: Flow cytometry analysis of CK14 and involucrin expression at day 4 post-calcium switch.

**Author Contributions:** Conceptualization, K.Z. and M.N.R.; methodology, K.Z. and N.G.; software, K.Z.; validation, K.Z., N.G., and J.S.; formal analysis, K.Z. and J.S.; investigation, K.Z.; resources, M.V.D.; data curation, K.Z.; writing—original draft preparation, K.Z.; writing—review and editing, M.N.R., M.V.D., N.G., and K.Z.; visualization, K.Z.; supervision, M.N.R.; project administration, M.N.R.; funding acquisition, M.N.R. and K.S. All authors have read and agreed to the published version of the manuscript.

**Funding:** This research was funded by the Battelle Institute Grant number US001-0000544452 and the Oak Ridge Institute for Science and Technology Burn Injury Research Fellowship USAISR-2019-0014.

**Institutional Review Board Statement:** Not applicable.

**Informed Consent Statement:** Not applicable.

**Data Availability Statement:** The data presented in this study are available in the article and the Supplementary Material.

**Acknowledgments:** The authors would like to Shanmugasundaram Natesan for providing knowledgeable advice on keratinocyte differentiation and skin modeling and Robert Christy for providing laboratory space at the U.S. Army Institute of Surgical Research.

**Conflicts of Interest:** The authors declare no conflict of interest.

# References

1. Eckert, R.L.; Efimova, T.; Dashti, S.R.; Balasubramanian, S.; Deucher, A.; Crish, J.F.; Sturniolo, M.; Bone, F. Keratinocyte survival, differentiation, and death: Many roads lead to mitogen-activated protein kinase. In *Journal of Investigative Dermatology Symposium Proceedings*; Blackwell Publishing Inc.: Oxford, UK, 2002; Volume 7, pp. 36–40.
2. Eckert, R.L.; Crish, J.F.; Robinson, N.A. *The Epidermal Keratinocyte as A Model for the Study of Gene Regulation and Cell Differentiation*; Elsevier: Amsterdam, The Netherlands, 1997; Volume 77.
3. Eckhart, L.; Lippens, S.; Tschachler, E.; Declercq, W. Cell death by cornification. *Biochim. Biophys. Acta—Mol. Cell Res.* **2013**, *1833*, 3471–3480. [CrossRef]
4. Bikle, D.D.; Xie, Z.; Tu, C.L. Calcium regulation of keratinocyte differentiation. *Expert Rev. Endocrinol. Metab.* **2012**, *7*, 461–472. [CrossRef]
5. Seo, E.Y.; Namkung, J.H.; Lee, K.M.; Lee, W.H.; Im, M.; Kee, S.H.; Geon, T.P.; Yang, J.M.; Seo, Y.J.; Park, J.K.; et al. Analysis of calcium-inducible genes in keratinocytes using suppression subtractive hybridization and cDNA microarray. *Genomics* **2005**, *86*, 528 538. [CrossRef] [PubMed]
6. Ng, D.C.; Su, M.J.; Kim, R.; Bikle, D.D. Regulation of involucrin gene expression by calcium in normal human keratinocytes. *Front. Biosci.* **1996**, *1*, 16–24. [CrossRef]
7. Rothnagel, J.A.; Greenhalgh, D.A.; Gagne, T.A.; Longley, M.A.; Roop, D.R. Identification of a Calcium-Inducible, Epidermal-Specific Regulatory Element in the 3′-Flanking Region of the Human Keratin 1 Gene. *J. Investig. Dermatol.* **1993**, *101*, 506–513. [CrossRef]
8. Huff, C.A.; Yuspa, S.H.; Rosenthal, D. Identification of control elements 3′ to the human keratin 1 gene that regulate cell type and differentiation-specific expression—PubMed. *J. Biol Chem.* **1993**, 377–384. Available online: https://pubmed.ncbi.nlm.nih.gov/7677999/ (accessed on 29 September 2022). [CrossRef]

9.    Elsholz, F.; Harteneck, C.; Muller, W.; Friedland, K. Calcium—A central regulator of keratinocyte differentiation in health and disease. *Eur. J. Dermatol.* **2014**, *24*, 650–661. [CrossRef] [PubMed]

10.   Warhol, M.J.; Roth, J.; Lucocq, J.M.; Pinkus, G.S.; Rice, R.H. Immuno-ultrastructural localization of involucrin in squamous epithelium and cultured keratinocytes. *J. Histochem. Cytochem.* **1985**, *33*, 141–149. [CrossRef]

11.   Hennings, H.; Michael, D.; Cheng, C.; Steinert, P.; Holbrook, K.; Yuspa, S.H. Calcium regulation of growth and differentiation of mouse epidermal cells in culture. *Cell* **1980**, *19*, 245–254. [CrossRef]

12.   Zamansky, G.B.; Nguyen, U.; Chou, I.N. An immunofluorescence study of the calcium-induced coordinated reorganization of microfilaments, keratin intermediate filaments, and microtubules in cultured human epidermal keratinocytes. *J. Investig. Dermatol.* **1991**, *97*, 985–994. [CrossRef] [PubMed]

13.   Monteleon, C.L.; Agnihotri, T.; Dahal, A.; Liu, M.; Rebecca, V.W.; Beatty, G.L.; Amaravadi, R.K.; Ridky, T.W. Lysosomes Support the Degradation, Signaling, and Mitochondrial Metabolism Necessary for Human Epidermal Differentiation. *J. Investig. Dermatol.* **2018**, *138*, 1945–1954. [CrossRef] [PubMed]

14.   Borowiec, A.S.; Delcourt, P.; Dewailly, E.; Bidaux, G. Optimal Differentiation of In Vitro Keratinocytes Requires Multifactorial External Control. *PLoS ONE* **2013**, *8*, 1–15. [CrossRef]

15.   Bernstam, L.I.; Vaughan, F.L.; Bernstein, I.A. Keratinocytes grown at the air-liquid interface. *Vitr. Cell. Dev. Biol.* **1986**, *22*. [CrossRef]

16.   Prunieras, M.; Regnier, M.; Woodley, D. Methods for cultivation of keratinocytes with an air-liquid interface. *J. Investig. Dermatol.* **1983**, *81*, S28–S33. [CrossRef] [PubMed]

17.   McGrath, J.A.; Eady, R.A.J.; Pope, F.M. Anatomy and organization of human skin. *Rook's Textb. Dermatol.* **2004**, *10*, 9781444317633.

18.   Ehrlich, F.; Fischer, H.; Langbein, L.; Praetzel-Wunder, S.; Ebner, B.; Figlak, K.; Weissenbacher, A.; Sipos, W.; Tschachler, E.; Eckhart, L. Differential evolution of the epidermal keratin cytoskeleton in terrestrial and aquatic mammals. *Mol. Biol. Evol.* **2019**, *36*, 328–340. [CrossRef] [PubMed]

19.   Kolarsick, P.A.J.; Kolarsick, M.A.; Goodwin, C. Anatomy and Physiology of the Skin. *J. Dermatol. Nurses. Assoc.* **2011**, *3*, 203–213. [CrossRef]

20.   Reijnders, C.M.A.; Van Lier, A.; Roffel, S.; Kramer, D.; Scheper, R.J.; Gibbs, S. Development of a Full-Thickness Human Skin Equivalent in Vitro Model Derived from TERT-Immortalized Keratinocytes and Fibroblasts. *Tissue Eng.—Part A* **2015**, *21*, 2448–2459. [CrossRef] [PubMed]

21.   Dickson, M.A.; Hahn, W.C.; Ino, Y.; Ronfard, V.; Wu, J.Y.; Weinberg, R.A.; Louis, D.N.; Li, F.P.; Rheinwald, J.G. Human Keratinocytes That Express hTERT and Also Bypass a p16INK4a-Enforced Mechanism That Limits Life Span Become Immortal yet Retain Normal Growth and Differentiation Characteristics. *Mol. Cell. Biol.* **2000**, *20*, 1436–1447. [CrossRef]

22.   Rheinwald, J.G.; Hahn, W.C.; Ramsey, M.R.; Wu, J.Y.; Guo, Z.; Tsao, H.; De Luca, M.; Catricalà, C.; O'Toole, K.M. A Two-Stage, p16INK4A- and p53-Dependent Keratinocyte Senescence Mechanism That Limits Replicative Potential Independent of Telomere Status. *Mol. Cell. Biol.* **2002**, *22*, 5157–5172. [CrossRef]

23.   Briley, B.; Shapiro, B. hTERT-Immortalized and Primary Keratinocytes Differentiate into Epidermal Structures in 3D Organotypic Culture. ATCC.org. Available online: https://www.atcc.org/resources/application-notes/htert-immortalized-and-primary-keratinocytes-differentiate (accessed on 6 April 2020).

24.   Van Drongelen, V.; Danso, M.O.; Mulder, A.; Mieremet, A.; Van Smeden, J.; Bouwstra, J.A.; El Ghalbzouri, A. Barrier properties of an N/TERT-based human skin equivalent. *Tissue Eng.—Part A* **2014**, *20*, 3041–3049. [CrossRef] [PubMed]

25.   van Drongelen, V.; Haisma, E.M.; Out-Luiting, J.J.; Nibbering, P.H.; El Ghalbzouri, A. Reduced filaggrin expression is accompanied by increased Staphylococcus aureus colonization of epidermal skin models. *Clin. Exp. Allergy* **2014**, *44*, 1515–1524. [CrossRef] [PubMed]

26.   Alloul-Ramdhani, M.; Tensen, C.P.; El Ghalbzouri, A. Performance of the N/TERT epidermal model for skin sensitizer identification via Nrf2-Keap1-ARE pathway activation. *Toxicol. Vitr.* **2014**, *28*, 982–989. [CrossRef] [PubMed]

27.   Rouse, J.G.; Van Dyke, M.E. A Review of Keratin-Based Biomaterials for Biomedical Applications. *Materials* **2010**, *3*, 999–1014. [CrossRef]

28.   Moll, R.; Divo, M.; Langbein, L. The human keratins: Biology and pathology. *Histochem. Cell Biol.* **2008**, *129*, 705. [CrossRef] [PubMed]

29.   McKittrick, J.; Chen, P.Y.; Bodde, S.G.; Yang, W.; Novitskaya, E.E.; Meyers, M.A. The Structure, Functions, and Mechanical Properties of Keratin. *JOM* **2012**, *64*, 449–468. [CrossRef]

30.   Buchanan, C.F.; Voigt, E.E.; Szot, C.S.; Freeman, J.W.; Vlachos, P.P.; Rylander, M.N. Three-dimensional microfluidic collagen hydrogels for investigating flow-mediated tumor-endothelial signaling and vascular organization. *Tissue Eng. Part C Methods* **2013**, *20*, 64–75. [CrossRef] [PubMed]

31.   Wang, S.; Taraballi, F.; Tan, L.P.; Ng, K.W. Human keratin hydrogels support fibroblast attachment and proliferation in vitro. *Cell Tissue Res.* **2012**, *347*, 795–802. [CrossRef]

32.   Arnette, C.; Koetsier, J.L.; Hoover, P.; Getsios, S.; Green, K.J. In vitro model of the epidermis: Connecting protein function to 3D structure. In *Methods in Enzymology*; Elsevier: Amsterdam, The Netherlands, 2016; Volume 569, pp. 287–308. ISBN 0076-6879.

33.   Ikuta, S.; Sekino, N.; Hara, T.; Saito, Y.; Chida, K. Mouse epidermal keratinocytes in three-dimensional organotypic coculture with dermal fibroblasts form a stratified sheet resembling skin. *Biosci. Biotechnol. Biochem.* **2006**, *70*, 610050121. [CrossRef]

34. Antoine, E.E.; Vlachos, P.P.; Rylander, M.N. Tunable collagen I hydrogels for engineered physiological tissue micro-environments. *PLoS ONE* **2015**, *10*, e0122500. [CrossRef]

35. Antoine, E.E.; Vlachos, P.P.; Rylander, M.N. Review of collagen I hydrogels for bioengineered tissue microenvironments: Characterization of mechanics, structure, and transport. *Tissue Eng. Part B Rev.* **2014**, *20*, 683–696. [CrossRef] [PubMed]

36. Zuniga, K.; Gadde, M.; Scheftel, J.; Senecal, K.; Cressman, E.; Van Dyke, M.; Rylander, M.N. Collagen/kerateine multi-protein hydrogels as a thermally stable extracellular matrix for 3D in vitro models. *Int. J. Hyperth.* **2021**, *38*, 830–845. [CrossRef]

37. Mahanty, S.; Dakappa, S.S.; Shariff, R.; Patel, S.; Swamy, M.M.; Majumdar, A.; Gangi Setty, S.R. Keratinocyte differentiation promotes ER stress-dependent lysosome biogenesis. *Cell Death Dis.* **2019**, *10*, 269. [CrossRef] [PubMed]

38. Lavker, R.M.; Gedeon Matoltsy, A. Formation of horny cells: The fate of cell organelles and differentiation products in ruminal epithelium. *J. Cell Biol.* **1970**, *44*, 501–512. [CrossRef]

39. Boer, D.E.C.; van Smeden, J.; Bouwstra, J.A.; Aerts, J.M.F. Glucocerebrosidase: Functions in and Beyond the Lysosome. *J. Clin. Med.* **2020**, *9*, 736. [CrossRef]

40. Holleran, W.M.; Takagi, Y.; Menon, G.K.; Jackson, S.M.; Lee, J.M.; Feingold, K.R.; Elias, P.M.; Takagi, Y.; Menon, G.K.; Jackson, S.M.; et al. Permeability Barrier Requirements Regulate Epidermal P-Glucocerebrosidase. Available online: https://n.d.www.jlr.org (accessed on 29 July 2020).

41. Holleran, W.M.; Ginns, E.I.; Menon, G.K.; Grundmann, J.U.; Fartasch, M.; McKinney, C.E.; Elias, P.M.; Sidransky, E. Consequences of β-glucocerebrosidase deficiency in epidermis. Ultrastructure and permeability barrier alterations in Gaucher disease. *J. Clin. Investig.* **1994**, *93*, 1756–1764. [CrossRef]

42. Holleran, W.M.; Takagi, Y.; Menon, G.K.; Legler, G.; Feingold, K.R.; Elias, P.M. Processing of epidermal glucosylceramides is required for optimal mammalian cutaneous permeability barrier function. *J. Clin. Investig.* **1993**, *91*, 1656–1664. [CrossRef]

43. Dielschneider, R.F.; Henson, E.S.; Gibson, S.B. Lysosomes as Oxidative Targets for Cancer Therapy. *Oxid. Med. Cell. Longev.* **2017**, *2017*, 3749157. [CrossRef] [PubMed]

44. Bocheńska, K.; Moskot, M.; Malinowska, M.; Jakóbkiewicz-Banecka, J.; Szczerkowska-Dobosz, A.; Purzycka-Bohdan, D.; Pleńkowska, J.; Slomiński, B.; Gabig-Cimińska, M. Lysosome alterations in the human epithelial cell line hacat and skin specimens: Relevance to psoriasis. *Int. J. Mol. Sci.* **2019**, *20*, 2255. [CrossRef] [PubMed]

45. Takahashi, H.; Aoki, N.; Nakamura, S.; Asano, K.; Ishida-Yamamoto, A.; Iizuka, H. Cornified cell envelope formation is distinct from apoptosis in epidermal keratinocytes. *J. Dermatol. Sci.* **2000**, *23*, 161–169. [CrossRef]

46. Harr, M.W.; Distelhorst, C.W. Apoptosis and Autophagy: Decoding Calcium Signals that Mediate Life or Death. *Cold Spring Harb. Perspect. Biol.* **2010**, *2*, a005579. [CrossRef]

47. Fennelly, C.; Amaravadi, R.K. Lysosomal Biology in Cancer. *Methods Mol. Biol.* **2017**, *1594*, 293. [CrossRef]

48. Barrandon, Y.; Green, H. Cell size as a determinant of the clone-forming ability of human keratinocytes. *Proc. Natl. Acad. Sci. USA* **1985**, *82*, 5390. [CrossRef] [PubMed]

49. Watt, F.M.; Green, H. Involucrin synthesis is correlated with cell size in human epidermal cultures. *J. Cell Biol.* **1981**, *90*, 738–742. [CrossRef]

50. Furuta, K.; Ikeda, M.; Nakayama, Y.; Nakamura, K.; Tanaka, M.; Hamasaki, N.; Himeno, M.; Hamilton, S.R.; August, J.T. Expression of lysosome-associated membrane proteins in human colorectal neoplasms and inflammatory diseases. *Am. J. Pathol.* **2001**, *159*, 449–455. [CrossRef]

51. Vanhoutteghem, A.; Djian, P.; Green, H. Ancient origin of the gene encoding involucrin, a precursor of the cross-linked envelope of epidermis and related epithelia. *Proc. Natl. Acad. Sci. USA* **2008**, *105*, 15481–15486. [CrossRef] [PubMed]

52. Watt, F.M.; Green, H. Stratification and terminal differentiation of cultured epidermal cells. *Nature* **1982**, *295*, 434–436. [CrossRef]

53. Sanz-Gómez, N.; Freije, A.; Gandarillas, A. Keratinocyte Differentiation by Flow Cytometry. *Methods Mol. Biol.* **2019**, *2109*, 83–92. [CrossRef]

54. Jakobsen, N.D.; Kaiser, K.; Ebbesen, M.F.; Lauritsen, L.; Gjerstorff, M.F.; Kuntsche, J.; Brewer, J.R. The ROC skin model: A robust skin equivalent for permeation and live cell imaging studies. *Eur. J. Pharm. Sci.* **2022**, *178*, 106282. [CrossRef]

55. Rikken, G.; Niehues, H.; van den Bogaard, E.H. Organotypic 3D Skin Models: Human Epidermal Equivalent Cultures from Primary Keratinocytes and Immortalized Keratinocyte Cell Lines. *Methods Mol. Biol.* **2020**, *2154*, 45–61. [CrossRef]

56. Tito, A.; Minale, M.; Riccio, S.; Grieco, F.; Colucci, M.G.; Apone, F. A Triticum vulgare Extract Exhibits Regenerating Activity During the Wound Healing Process. *Clin. Cosmet. Investig. Dermatol.* **2020**, *13*, 21. [CrossRef] [PubMed]

57. Roy, D.C.; Tomblyn, S.; Isaac, K.M.; Kowalczewski, C.J.; Burmeister, D.M.; Burnett, L.R.; Christy, R.J. Ciprofloxacin-loaded keratin hydrogels reduce infection and support healing in a porcine partial-thickness thermal burn. *Wound Repair Regen.* **2016**, *24*, 657–668. [CrossRef]

58. Ye, W.; Qin, M.; Qiu, R.; Li, J. Keratin-based wound dressings: From waste to wealth. *Int. J. Biol. Macromol.* **2022**, *211*, 183–197. [CrossRef] [PubMed]

59. Konop, M.; Sulejczak, D.; Czuwara, J.; Kosson, P.; Misicka, A.; Lipkowski, A.W.; Rudnicka, L. The role of allogenic keratin-derived dressing in wound healing in a mouse model. *Wound Repair Regen.* **2017**, *25*, 62–74. [CrossRef]

60. Tang, A.; Li, Y.; Yao, Y.; Yang, X.; Cao, Z.; Nie, H.; Yang, G. Injectable keratin hydrogels as hemostatic and wound dressing materials. *Biomater. Sci.* **2021**, *9*, 4169–4177. [CrossRef] [PubMed]

61. Roy, D.C.; Tomblyn, S.; Burmeister, D.M.; Wrice, N.L.; Becerra, S.C.; Burnett, L.R.; Saul, J.M.; Christy, R.J. Ciprofloxacin-Loaded Keratin Hydrogels Prevent Pseudomonas aeruginosa Infection and Support Healing in a Porcine Full-Thickness Excisional Wound. *Adv. Wound Care* **2015**, *4*, 457–468. [CrossRef]

 *bioengineering*

*Review*

# Extracellular Vesicles as Regulators of the Extracellular Matrix

Neil J. Patel [1], Anisa Ashraf [1] and Eun Ji Chung [1,2,3,4,5,6,*]

[1] Department of Biomedical Engineering, University of Southern California, Los Angeles, CA 90089, USA
[2] Division of Vascular Surgery and Endovascular Therapy, Department of Surgery, Keck School of Medicine, University of Southern California, Los Angeles, CA 90033, USA
[3] Mork Family Department of Chemical Engineering and Materials Science, University of Southern California, Los Angeles, CA 90089, USA
[4] Eli and Edythe Broad Center for Regenerative Medicine and Stem Cell Research, Keck School of Medicine, University of Southern California, Los Angeles, CA 90033, USA
[5] Division of Nephrology and Hypertension, Department of Medicine, Keck School of Medicine, University of Southern California, Los Angeles, CA 90033, USA
[6] Norris Comprehensive Cancer Center, Keck School of Medicine, University of Southern California, Los Angeles, CA 90033, USA
* Correspondence: eunchung@usc.edu

**Abstract:** Extracellular vesicles (EVs) are small membrane-bound vesicles secreted into the extracellular space by all cell types. EVs transfer their cargo which includes nucleic acids, proteins, and lipids to facilitate cell-to-cell communication. As EVs are released and move from parent to recipient cell, EVs interact with the extracellular matrix (ECM) which acts as a physical scaffold for the organization and function of cells. Recent work has shown that EVs can modulate and act as regulators of the ECM. This review will first discuss EV biogenesis and the mechanism by which EVs are transported through the ECM. Additionally, we discuss how EVs contribute as structural components of the matrix and as components that aid in the degradation of the ECM. Lastly, the role of EVs in influencing recipient cells to remodel the ECM in both pathological and therapeutic contexts is examined.

**Keywords:** extracellular vesicles; extracellular matrix; tissue repair; calcification; tumor microenvironment

**Citation:** Patel, N.J.; Ashraf, A.; Chung, E.J. Extracellular Vesicles as Regulators of the Extracellular Matrix. *Bioengineering* **2023**, *10*, 136. https://doi.org/10.3390/bioengineering10020136

Academic Editors: Ngan F. Huang and Brandon J. Tefft

Received: 18 December 2022
Revised: 15 January 2023
Accepted: 17 January 2023
Published: 19 January 2023

## 1. Introduction

Extracellular vesicles (EVs) are released into the extracellular environment by all cell types and function as mediators of cell-to-cell communication [1]. EVs contain cargo including lipids, nucleic acids, and proteins that are reflective of their parent cell phenotype [2]. As such, there has been significant interest in utilizing EVs derived from healthy cell sources as endogenous nanotherapeutics [3,4] and applying EVs to enhance angiogenesis, inhibit fibrosis and apoptosis, regulate the immune microenvironment, and modulate the extracellular matrix (ECM) in the context of promoting tissue regeneration [5–9].

When EVs are released from parent cells into the extracellular space, they may first interact with the surrounding ECM. The ECM is the physical scaffolding for cellular components of tissues and actively participates in cell growth, movement, and differentiation. The major components of the ECM include collagens, laminins, fibronectin, and proteoglycans that are organized into a physically crosslinked network via non-covalent interactions. The ECM interacts with cells through cell adhesion molecules (e.g., integrins, cadherins, and other transmembrane proteoglycans) which enable cells to migrate across the matrix. Importantly, cells can modify the matrix by depositing ECM components or degrading the ECM by secreting matrix-degrading enzymes such as matrix metalloproteases (MMPs). The mechanical and physical properties of the ECM have been demonstrated to be integral to the differentiation, migration, and maintenance of cells within the ECM. These proprieties include (1) mesh size, which is the distance between two crosslinks within a matrix, (2) stiffness, which is the extent to which the matrix resists deformation when stress is applied, and (3) the viscoelastic behavior, or the ability of the matrix to exhibit both viscous and elastic characteristics upon deformation [10]. Together, the components of the ECM and the mechanical characteristics of the matrix are integral to the health and function of the resident cells [11–14]. As such, EVs released by both healthy and pathological cells can act as mediators of the ECM either through direct EV–ECM interactions or by influencing cell–ECM interactions.

For the receiving cell, EVs also transport through the ECM of the target tissue to reach the recipient cells and deliver their cargo to impart a cellular response. Transport of EVs through the ECM can be a passive, diffusive process dependent on the stiffness and viscoelastic characteristics of the ECM and the deformability of the EV itself [15,16]. EVs can become structural components of the matrix, either as initiators of calcification or as bioactive signaling agents anchored to the matrix, or can modulate the ECM indirectly by inducing recipient cells to promote synthesis or degradation of the ECM [17–22].

Given its unstudied but significant role, this review will discuss both the direct and indirect influence of EVs on the ECM. First, the biogenesis and characteristics of EVs and their transport through the ECM will be delineated. Additionally, the role of EVs in modulating the ECM in cancer progression, as structural components during bone/endochondral and vascular calcification, and as ECM-bound bioactive signaling agents will be addressed. Furthermore, how EVs also indirectly modulate the ECM will be explored in both pathological and therapeutic contexts.

## 2. EV Biology

### 2.1. EV Biogenesis

EV biogenesis includes vesicle formation, cargo loading, and secretion. Exosome (50–200 nm) biogenesis begins with an early sorting endosome which then matures into a late sorting endosome and finally into a multivesicular body (MVB, Figure 1A). MVB formation is marked by the invagination of the late endosomal membrane forming intraluminal vesicles (ILVs) within the MVB. ILVs are released into the extracellular environment when the MVB merges with the cell membrane and once excreted, they are referred to as exosomes [23]. Exosomes are commonly identified by expression of the membrane-bound tetraspanin marker CD63 and syntenin-1 [24]. Exosome cargo loading is regulated by the endosomal sorting complex required for the transport (ESCRT) pathway, which is comprised of four protein complexes denoted ESCRT-0 through ESCRT-III. ESCRT-0 recruits ubiquitinated protein to the endosomal membrane after which ESCRT-I clusters the cargo and complexes with ESCRT-II [25]. The protein cargo is then sequestered into the endosome and ESCRT-II initiates ILV invagination. Lastly, ESCRT-III de-ubiquinates the protein cargo and finalizes vesicle budding to form ILVs. Additionally, RAB GTPases are associated with exosome formation and support exosome secretion by directing MVBs to the plasma membrane for fusion [26].

Beyond exosomes, microvesicles range in size from 50 to 1000 nm and are released by blebbing of the plasma membrane. Thus, they express markers of the originating cell plasma membrane as well as tetraspanins CD9 and CD81 [27,28]. In addition to surface markers, exosome and microvesicle membranes are comprised of cholesterols, sphingomyelins, and phospholipids [29]. Additionally, apoptotic bodies are large EVs (>1000 nm) released when cells undergo programmed cell death [30]. While apoptotic bodies participate in key processes in cell death, such as the removal of cell contents and delivering information from dying cells to phagocytic cells, the majority of EV biology research has been directed towards small EVs including exosomes and microvesicles which range from 50 to 200 nm in size.

### 2.2. EV Uptake

Since exosomes and microvesicles participate in cell-to-cell communication via the transfer of encapsulated messenger RNA (mRNA), microRNAs, small nuclear RNA, long non-coding RNAs, and cytosolic and membrane-associated proteins, they are often uptaken by cells to deliver their cargo [31]. Once released into the extracellular environment, multiple energy-dependent pathways have been proposed by which EVs are internalized by the target cell including clatharin-mediated endocytosis (CME), caveolin-dependent endocytosis (CDE), phagocytosis, and membrane fusion (Figure 1B) [28]. Accordingly, previous studies show that inhibition of the CME pathway by the drug chlorpromazine results in a significant decrease in EV uptake [32]. Similarly, suppressing the cavelin-1 protein, which is necessary for the CDE pathway, resulted in decreased EV endocytosis [31]. Special cases for EV uptake include phagocytic cells and within the tumor microenvironment in which EVs are uptaken primarily by phagocytosis [33]. Additionally, EV-cell membrane fusion as an entry mechanism has also been observed in acidic tumor microenvironments [34].

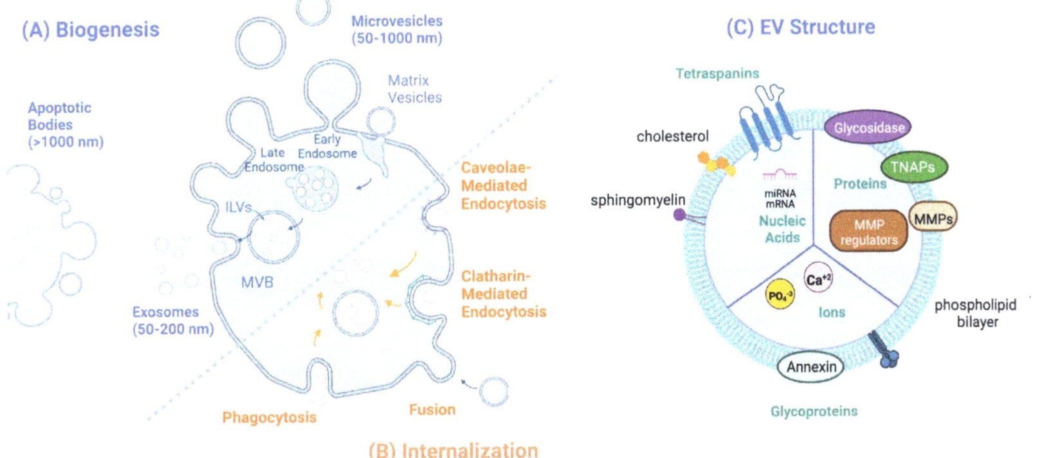


**Figure 1.** EV biology: (**A**) Routes of EV biogenesis: exosomes are formed from multivesicular bodies (MVB); apoptotic bodies are formed from cells undergoing programmed cell death; microvesicles bud off the donor cell plasma membrane and include matrix vesicles. (**B**) Routes of EV internalization include caveolae-mediated endocytosis, clathrin-mediated endocytosis, micropinocytosis, and phagocytosis, which are internalized via endosomes, and EV fusion with the plasma membrane. (**C**) EV structure: internal cargo includes nucleic acids, proteins, and ions; surface molecules include glycoproteins, tetraspanins, matrix metalloproteases (MMPs), and TNAPs.

## 3. EV Transport through the ECM

In order to participate in cell-to-cell communications, EVs must traverse the ECM from the parent cell to the recipient cell. Recent studies regarding stress relaxation of the ECM by Lenzini et al. [15,16] showed that the mechanical properties of the ECM and EVs allow EV diffusion through the extracellular space, despite the larger exosome and microvesicle diameter (50–200 nm) compared to the mesh size of the ECM (~50 nm) [15]. The authors demonstrated approximately 50% of the EVs loaded within decellularized ECM of lung tissue were released after 24 h. By using crosslinked alginate hydrogels with tunable viscoelastic properties, EVs in stiff physically crosslinked alginate hydrogels were found to have significantly higher diffusion coefficients compared to both soft viscoelastic and elastic hydrogels, indicating that the ECM undergoing stress relaxation leads to increased EV diffusion through the matrix. This was true across EVs from a variety of cell sources, implying that the mechanical properties of the matrix play a role in EV transport. In addition, depleting the membrane water channel protein aquaporin-1 on the EV surface increased the stiffness of the EV and reduced the diffusion coefficient by approximately threefold. This indicates that water permeation enables EVs to become more deformable and thus, more easily able to diffuse through the matrix [35].

## 4. Direct Influence of EVs on the ECM

EVs have been reported to directly interact with the ECM and can contribute as both a physical and bioactive structural component of the matrix. In addition, EVs can directly associate with the ECM and actively degrade the matrix with their surface-associated enzymes [5,16,36–39]. In each case, EVs play an integral role in the evolution of the ECM in both physiological and pathological states (Table 1).

### 4.1. EV-Mediated Calcification

#### 4.1.1. Bone Formation and Endochondral Calcification

Mineralization of the ECM is a physiological process during bone formation and endochondral calcification while it is pathological during vascular calcification [40]. The ECM architecture during calcification undergoes a dramatic evolution from a non-crystalline network comprised mainly of type 1 collagen fibrils to a crystalline matrix that is able to support high load and stress [41]. Matrix vesicles (MVs, 100–300 nm) are a subtype of microvesicle that are essential to the mineralization of the ECM. MVs released into the ECM from osteoblast and chondrocytes initiate the formation of hydroxyapatite ($Ca_{10}(PO_4)_6(OH)_2$) crystals and the transformation of the ECM from an entirely organic architecture to a combination of both an organic and inorganic (i.e., mineralized) structure. During the first phase of mineralization, MVs sequester $Ca^{2+}$ ions through various calcium-binding molecules concentrated on the MV surface which include the calcium binding proteins annexins II, V, and IV and the membrane phospholipid phosphatidylserine (PS) (Figures 1C and 2A) [42–44]. Simultaneously, intra- and extravesicular concentrations of $PO_4^{3-}$ ions are increased via the following: (1) MV membrane-bound tissue non-specific alkaline phosphate (TNAP), (2) the conversion of ATP to ADP via ATPases, (3) and Pit-1, a sodium-dependent phosphate transporter [44–46]. The resultant increased intravesicular $Ca^{2+}$ and $PO_4^{3-}$ concentrations leads to precipitation of hydroxyapatite crystals within the MV.

During the second phase of mineralization, hydroxyapatite crystals become large enough to break through the MV into the ECM and continue mineralization of the ECM along the length of type 1 collagen fibrils (Figure 2A). Of note, MV-mediated calcification occurs not only during endochondral calcification during development but also in the constant cycle of bone resorption and formation during adulthood. Thus, MVs are integral in the evolution of the ECM from a viscoelastic network to a fully crystalline structure in physiological calcification processes.

#### 4.1.2. Vascular Calcification

Vascular calcification is the mineralization of ECM in blood vessels and is present in diseases such as atherosclerosis, diabetes mellitus, and chronic kidney disease [47]. Vascular calcification is initiated with the transdifferentiation of VSMCs into an osteoblast-like phenotype. Under physiological conditions, VSMCs release EVs containing calcification inhibitors such as Matrix Gla protein (MGP) and fetuin-A to maintain tissue homeostasis. However, under pathogenic conditions caused by chronic inflammation or aberrant mineral metabolism, osteoblast-like VSMCs release EVs resembling MVs during bone formation [48]. These EVs can destabilize the ECM of the blood vessel by calcifying distinct regions of the vessel wall, which can lead to thrombosis or rupture of the vessel [49].

The mechanism of EV-mediated vascular calcification differs from that of MV-mediated bone or cartilage mineralization. In calcifying conditions, VSMC-derived EVs are enriched in phosphatidylserine and contain annexins I, II, V, and VI similar to MVs. However, unlike MVs, increased extracellular $Ca^{2+}$ concentration does not lead to an increase in TNAP activity in VSMC-EVs, indicating calcification nucleation is independent of TNAP [50]. Rogers et al. showed that annexin I is enriched in VSMC-derived EVs released during ectopic vascular calcification and facilitates the formation of EV aggregates within collagen fibrils of the ECM. These EV aggregates form mineralization nucleation sites on the surface of the EV which then grow in the presence of increased extracellular $Ca^{2+}$ and $PO_4^{3-}$ (Figure 2B). Importantly, the knockdown of annexin-I in osteoblast-like VSMCs inhibited the formation of calcification indicating that annexin-I-mediated EV aggregation is necessary for vascular calcification [51]. Thus, unlike bone or cartilage in which mineralization occurs within the MV, vascular calcification is mediated by the surface phospholipids and annexins of EV aggregates localized in the ECM.

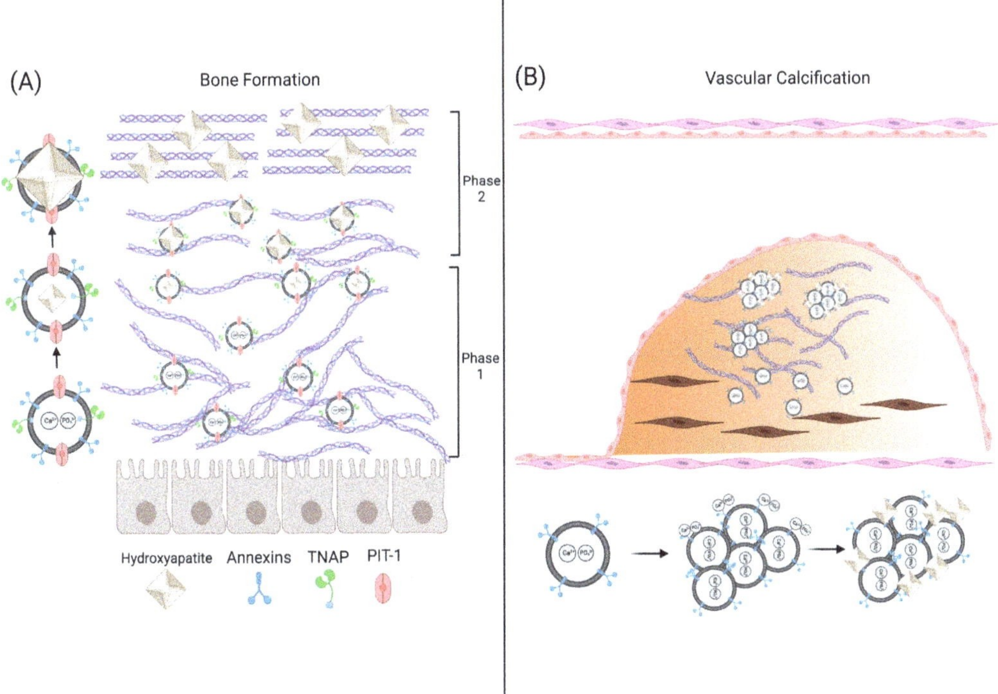



**Figure 2.** EV-mediated calcification of bone and the vasculature. (**A**) During phase 1 of mineralization during physiological bone formation, osteoblasts release matrix vesicles (MVs) which transport calcium and phosphate ions into the vesicle forming hydroxyapatite (HA) crystals. During phase 2 of mineralization, the HA crystals penetrate the MV and orient themselves with the ECM matrix, furthering crystal growth. (**B**) During vascular calcification, osteogenic-VSMC release EVs that aggregate and anchor to collagen within the ECM. Hydroxyapatite nucleation occurs on the surface of the EV aggregates.

### 4.2. Matrix-Bound Vesicles Are Integral to the Bioactive Properties of the ECM

In addition to acting as structural components during calcification, EVs embedded within the matrix are an integral part of the bioactive properties of the ECM. Recently discovered is a novel subtype of EVs anchored within the ECM called matrix-bound vesicles (MBV) which act as key biological signaling agents within the ECM [52]. Interestingly, MBVs do not contain any of the common exosome or microvesicle markers CD63, CD9, CD81, or HSP70 suggesting a different biogenesis pathway. Additionally, MBVs have tissue-specific characteristics with distinct lipid, RNA, and protein profiles indicating MBVs are an integral part of the bioactive signaling composition of the ECM and have a function in regulating tissue homeostasis [53,54].

Merwe et al. demonstrated that MBV derived from the ECM of the urinary bladder of healthy rats protected against ischemia-induced injury of retinal ganglion cells (RGC) [55]. As decellularized urinary bladder ECM (UB-ECM) had previously been shown to promote ganglion cell axon growth in coculture with microglia and astrocytes, the authors hypothesized that MBVs were a key biological agent anchored within the UB-ECM that were responsible for the increased cell growth [56]. In vitro treatment of a coculture of microglia, astrocytes, and RGC with UB-ECM-derived EVs suppressed the pro-inflammatory signaling of activated microglia and astrocytes and promoted RGC axon growth. Furthermore, intraocular administration of UB-ECM-derived EVs in an optical ischemia murine model ameliorated hypoxia-induced RGC axon degeneration by approximately 80% compared to the uninjured control. Thus, MBVs are key players responsible for the biological effects, namely promoting cell growth and regulating the immune microenvironment, exerted by the ECM.

Additionally, MBVs have been shown to be integral to ECM-dependent regulation of the immune microenvironment by modulation of immune cells within the matrix. With regards to rheumatoid arthritis, which is an autoimmune disease characterized by chronic inflammation of synovial joints, UB-ECM-derived EVs were shown to prevent acute and chronic arthritis at an efficacy comparable to the clinical standard [57]. Specifically, UB-ECM-derived EVs administered intravenously in an arthritic rat model promoted the transition of pro-inflammatory M1 macrophages to an anti-inflammatory M2 phenotype thereby decreasing inflammation in synovial joints and limiting adverse joint remodeling. Taken together, these studies underscore the indispensable function MBVs have as the bioactive structural component of the ECM.

### 4.3. EV-Mediated Modulation of the ECM

In contrast to calcification in which EVs aid in the structural development of the ECM, EVs released from cancer cells have been shown to actively degrade the ECM via enzymes such as MMPs and glycosidases present on the EV surface [5,16,36–38] (Figures 1C and 3). EVs derived from human G361 melanoma cells and HT-1080 human fibrocarcinoma cells display matrix metalloprotease MT1-MMP (MMP-14), a transmembrane protein that promotes cell migration by degrading ECM components such as collagen, fibronectin, and laminins [58,59]. Similarly, EVs derived from nasopharyngeal cancer (NPC) cells were shown to contain matrix metalloprotease 13 (MMP-13) on the EV surface. Autologous incubation of the MMP-13-expressing EVs with NPC cells in Matrigel led to increased degradation of the matrix compared to both EVs with decreased MMP-13 levels and the untreated control, resulting in increased cell invasion within the matrix [60]. In addition to EVs derived from cancer cells, human vascular endothelial cells (HVECs) also secrete EVs that present MMP-2, MMP-9, and MMP-14 on the membrane surface. Incubating Matrigel with HVEC-EVs resulted in the degradation of the matrix and a threefold increase in HVEC cell migration compared to conditions without EV treatment [61]. As evidenced by these studies, EVs have direct contact and actively degrade the ECM by their surface-bound MMPs, resulting in enhanced cell migration through the matrix.

In addition to MMPs, EVs degrade the ECM by utilizing surface-bound heparinase, a glycosidase that degrades heparan sulfate proteoglycans (HSPC) [36,62–64]. HSPC is a is a proteoglycan, which is comprised of a core glycoprotein with multiple heparan sulfate groups that can sequester and bind signaling molecules including growth factors, chemokines, and cytokines to the matrix [65]. EVs derived from myeloma cells have heparinase bound to HSPC on the EV surface [62,63]. When these myeloma-derived EVs were incubated with ECM deposited by endothelial cells, free heparan sulfate chains were detected in the culture media [63]. Since HSPCs sequester bioactive signaling molecules and anchor them to ECM, EV-mediated cleavage of HSPC can modulate the signaling properties of the ECM.

EVs have also been shown to contain the surface-bound ECM-modulating enzymes lysyl oxidase like-2 (LOXL2), elastase, and collagenase [66,67]. Specifically, Jong et al. showed that EVs from human endothelial cells grown in hypoxic conditions present LOXL2, which is responsible for collagen crosslinking, on the EV surface [66]. Incubation of LOXL2-EVs within a collagen gel increased crosslinking leading to approximately twofold higher stiffness. Thus, hypoxic endothelial cells produce EVs which directly promote collagen crosslinking in the ECM. Conversely, Genshmer et al. found that EVs from activated neutrophils contained surface-bound elastase and collagenase [67]. Incubation of these activated neutrophil-derived EVs with both elastin and collagen fibrils led to a 2.5-fold increase in the degradation of elastin and collagen compared to the inactivated neutrophil EVs. As such, EV-surface-associated enzymes can directly influence the ECM by modulating the matrix through multiple ECM targets including collagen, fibronectin, elastin, and heparan sulfate proteoglycans.

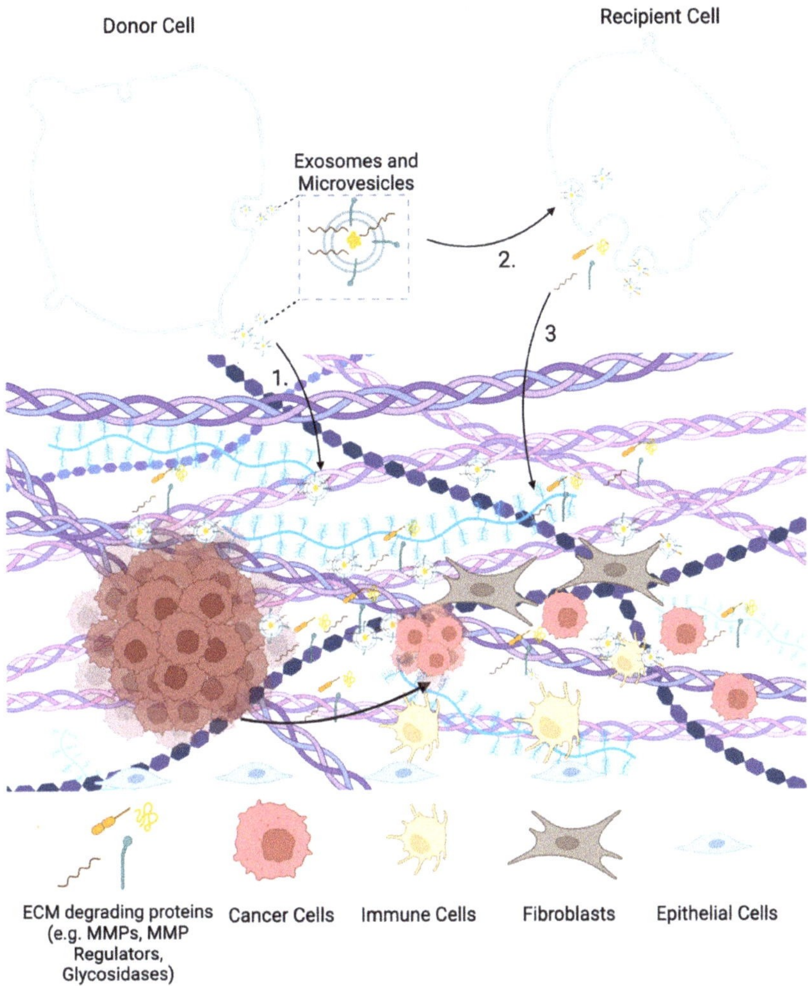


**Figure 3.** EV-mediated matrix remodeling in tumor progression. 1. Matrix metalloproteases (MMPs) and glycosidases displayed on EV surfaces promote ECM degradation. 2. EVs released from tumor cells are uptaken by recipient cells in both the tumor microenvironment and the pre-metastatic niche. Tumor-derived EVs promote the release of ECM-degrading proteins (e.g., MMPs and MMP regulators) and EVs in recipient cells. 3. Free and EV-associated ECM degrading proteins promote tumor cell invasion, metastasis, and the development of the pre-metastatic niche.

**Table 1.** Direct EV-ECM effects.

|  | EV Source | EV Effects on the ECM | Active EV Components | Refs. |
|---|---|---|---|---|
| EV-mediated calcification |  |  |  |  |
|  | Osteoblasts | Promotes calcification of the ECM during bone and endochondral calcification | Annexins II, IV, V, Pit-1, Phosphatidylserine, TNAP | [42–44] |
|  | Osteochondrogenic vascular smooth muscle cells | Promotes calcification of the vasculature during vascular calcification | Annexins I, II, IV, V, VI | [50,51] |
| Bioactive signaling Agents |  |  |  |  |
|  | Matrix-bound vesicles | Promotes anti-inflammatory signaling and cell growth in retinol ganglion cells and macrophages | N/A | [52–57] |
| Direct modulation of the ECM |  |  |  |  |
|  | G361 melanoma cells | Degrades collagen, fibronectin, and laminin; promotes cell migration | MMP-14 | [58] |
|  | HT-1080 fibro carcinoma cells | Degrades collagen, fibronectin, and laminin; promotes cell migration | MMP-14 | [59] |
|  | Nasal pharyngeal carcinoma cells | Degrades collagen; promotes cell migration | MMP-13 | [60] |
|  | Vascular endothelial cells | Degrades collagen, fibronectin, and laminin; promotes cell migration and angiogenesis | MMP-2,9,13 | [61] |
|  | Myeloma cells | Degrades heparin sulfate proteoglycans | Heparinase | [62–64] |
|  | Neutrophils | Degrades collagen and elastase | Collagenase and elastase | [66] |
|  | Endothelial cells | Promotes collagen crosslinking | Lysyl oxidase like-2 | [67] |

## 5. Indirect Influence of EVs on the ECM

While EVs can directly modulate the structural components of the matrix as discussed above, EVs can also indirectly alter the ECM by influencing recipient cells to remodel the ECM. Specifically, EVs released from tumor cells have been shown to direct recipient cells to modulate the matrix to create a favorable tumor niche [36]. In addition, EVs released from mesenchymal stem cells (MSCs) have been found to have therapeutic capability by improving ECM remodeling by fibroblasts and chondrocytes during tissue regeneration which will be further detailed below [52,54,55,57,68] (Table 2).

*5.1. Tumor-Derived EVs Mediate Cell–ECM Interactions during Tumor Progression*
5.1.1. Tumor Cell-Derived EVs Promote ECM Degradation

In Section 4.2, it was discussed that EVs derived from cancer cells have matrix-degrading enzymes present on their surface which promotes cancer cell migration and growth. However, in addition to directly interacting with the ECM, EVs secreted into the tumor microenvironment also contain cargo such as MMPs and MMP regulators that is transferred to recipient cells inducing cell-mediated remodeling of the ECM (Figure 1C).

For example, EVs derived from hypoxic nasopharyngeal carcinoma (NPC) cells contain MMP-13 as cargo, and treatment of normoxic NPC with hypoxic NPC-derived EVs resulted in a 2.5-fold increase in cellular expression of MMP-13. Additionally, hypoxic NPC-derived EV treatment of normoxic NPC cells increased cell migration and invasion by approximately 300% and 200%, respectively, compared to the non-treated control. Furthermore, EV treatment of normoxic NPC cells decreased levels of the tumor suppressor protein E-cadherin by approximately 80% compared to the non-treated control, thereby promoting tumor growth [69]. Similarly, EVs derived from aggressive myeloma patient cells, which bound to both RPMI-8226 plasmacytoma lymphocytes and human endothelial cells via surface-associated fibronectin, increased the expression of MMP-9 in the RPMI-8226 cells and promoted endothelial cell invasion. Specifically, treatment with myeloma-derived EVs promoted increased cell spreading and migration of endothelial cells through a Matrigel basement membrane by 1.5-fold, indicating that the EVs promote modulation of the matrix via MMP-9 and reprogram the healthy cells towards a more cancerous phenotype [64].

Tumor-derived EVs also contain MMP regulators such as the glycoprotein extracellular matrix metalloprotease inducer (EMMPRIN) which is a transmembrane protein that stimulates the expression of MMPs by tumor-associated fibroblasts. EVs derived from epithelioid sarcoma cells contain EMMPRIN and treatment of human fibroblasts with sarcoma-derived EVs resulted in an increased expression of MMP-2 by approximately twofold compared to both the non-treated control and EVs without EMMPRIN [70]. Another MMP regulator that has been associated with poor prognosis in multiple cancer types is TIMP-1 [71,72]. Overexpression of TIMP-1 in human lung adenocarcinoma cells increased the expression and loading of the pro-angiogenic and pro-tumorigenic microRNA-210 into EVs. These EVs induced angiogenic tube formation in human epithelial cells and promoted lung adenocarcinoma cell migration in vitro, both processes in which ECM remodeling occurs [73,74]. Overall, these studies indicate that tumor-cell-derived EVs promote recipient cells toward a cancerous phenotype in which ECM degradation is involved.

### 5.1.2. Tumor-Derived EVs Facilitate Pre-Metastatic Niche Formation

The pre-metastatic niche is a potential location of metastasis that has been primed for invasion by circulating tumor cells. According to the "seed and soil" model, metastasis requires that the "soil" (receiving microenvironment) be ideal for the "seed" (circulating tumor cells) to populate [75]. Primary tumors release EVs which actively inform distant recipient cells to modify the ECM and prime the pre-metastatic niche for tumor cell invasion and establishment of a metastatic site. The work of Lyden and colleagues unraveled the role of EVs in establishing an organotropic pre-metastatic niche [17]. They first described that melanoma-derived EVs in circulation educate bone marrow progenitor cells towards a phenotype that supports tumor growth and metastasis [17,21]. In addition, they demonstrated that integrins on the surface of tumor-derived EVs are responsible for organotropic accumulation of EVs [76]. Once localized to a specific tissue, resident cells uptake these tumor-derived EVs and are induced to remodel the ECM toward the development of a pre-metastatic niche.

The role of MMPs and other matrix-modulating proteins in the development of the pre-metastatic niche is well established, and the function of EVs in regulating the expression of these matrix-modulating proteins is now being explored [77]. Redzic et al. showed that EVs released from breast cancer, leukemia, and pancreatic cancer cell lines contain EMMPRIN [78]. When these cell lines were treated with their respective EVs, MMP-9 and IL-6, which are downstream targets of EMMPRIN, were upregulated across all cell lines. Additionally, the cells expressed a higher level of EMMPRIN and increased their secretion of EMMPRIN-containing EVs.

Furthermore, melanoma-derived EVs have also been shown to facilitate ECM deposition and promote angiogenesis to prime sentinel lymph nodes for tumor metastasis. Specifically, intravenous injection of melanoma-derived EVs into mice led to an increase in mitogen-activated protein kinase 14, laminin, collagen, and urokinase plasminogen activator, or ECM-modulating factors involved in stroma remodeling that enable invasion by circulating tumor cells. Establishing a pre-metastatic niche in sentinel lymph nodes by intravenous administration of melanoma EVs led to increased selective colocalization of circulating melanoma cells with melanoma EVs in the sentinel lymph nodes [79]. As such, tumor-derived EVs can create an organotropic pre-metastatic niche through influencing recipient cells to modulate the ECM.

### 5.2. Therapeutic Potential of EVs in Tissue Regeneration

While EVs released from pathogenic cells are involved in disease progression, EVs from healthy cells have become of interest as next-generation therapeutics. Their low immunogenicity, high biocompatibility, endogenous therapeutic cargo, and ability to efficiently transfer bioactive agents to target cells have made them an attractive therapeutic platform in recent years [80]. While there are no current EV therapeutics in clinical trials specifically related to the ECM, multiple studies have reported the therapeutic ability of EVs to promote cell-dependent modulation of the matrix during tissue regeneration.

Mesenchymal Stem Cell EV-Mediated ECM Repair

EVs derived from mesenchymal stem cells (MSC) have been shown to participate in restoring ECM during tissue regeneration. For example, EVs derived from adipose tissue MSC (AT-MSC) decreased scar formation and aided in wound healing in an in vivo model. Specifically, AT-MSC EVs were administered intravenously to a murine model with a dorsal wound. AT-MSC-EV treatment increased the ratio of type 3 collagen to type 1 collagen and MMP-3 to TIMP-1 in dermal fibroblasts, thereby aiding in matrix remodeling and limiting scar formation [81]. Interestingly, fibroblasts isolated from women with stress urinary incontinence that had significantly decreased expression of type 1 collagen, treated with AT-MSC EVs had increased expression of type 1 collagen and TIMP-1/3, and downregulation of MMP-1/2 [82]. Increased deposition of collagen along with inhibition and decreased expression of MMPs led to the restoration of ECM stiffness that is lost during stress urinary incontinence. Woo et al. demonstrated that AT-MSC EVs could also attenuate the progression of osteoarthritis, which is a chronic degenerative disease of articular cartilage. In vitro, AT-MSC EV treatment promoted chondrocyte migration and proliferation while simultaneously promoting type 2 collagen synthesis and inhibiting the expression of MMP-1, MMP-3, MMP-13, and ADAMTS-5. Additionally, intra-articular injections of AT-MSC EVs inhibited osteoarthritis progression in a murine model by protecting the cartilage matrix both by increased collagen deposition and inhibition of ECM-degrading proteases [83]. In addition to MMPs, human bone marrow (BM) MSC secretes hyaluronan-coated EVs. Hyaluronan (HA) is a glycosaminoglycan that acts as a critical bioactive factor within the ECM that plays roles in cell adhesion, motility, proliferation, and differentiation. HA-coated EVs were shown to be secreted from BM-MSC and anchored onto the ECM providing bioactive signaling cues within the matrix [84]. Taken together, these studies suggest that stem-cell-derived EVs have promise as therapeutics for tissue regeneration via modulation of the ECM.

**Table 2.** Indirect effect of EVs on the ECM.

| Function | EV Source | EV Target | Cell–ECM Interactions | Active EV Components | Refs. |
|---|---|---|---|---|---|
| Tumor EVs promote ECM degradation | | | | | |
| | Hypoxic nasal pharyngeal carcinoma cells | Normoxic nasal pharyngeal carcinoma cells | Increases cell expression of MMP-13, decreases expression of E-cadherin, promotes cell migration and invasion | MMP-13 | [69] |
| | Human myeloma cells | RPMI-8226 plasmacytoma lymphocytes and human endothelial cells | Increases MMP-9 expression, promotes cell invasion, and migration | MMP-9, fibronectin | [64] |
| | Epithelioid sarcoma cells | Fibroblasts | Increases expression of MMP-2 | EMMPRIN | [70] |
| | Human lung adenocarcinoma cells | Human epithelial cells and lung adenocarcinoma cells | Induces angiogenic tube formation and promotes cell migration | TIMP and miR-210 | [73,74] |
| Tumor EVs create a pre-metastatic niche | | | | | |
| | Human melanoma cells | Bone marrow progenitor cells | Promotes tumor growth and metastasis | N/A | [17] |
| | Breast cancer MCF-7 cells | Human fibroblasts | Increases expression of MMP-9 and IL-6 | EMMPRIN | [78] |
| | Monocytic leukemia U937 cells | Human fibroblasts | Increases expression of MMP-9 and IL-6 | EMMPRIN | [78] |
| | Pancreatic cancer L3.6pL cells | Human fibroblasts | Increases expression of MMP-9 and IL-6 | EMMPRIN | [78] |
| | Melanoma cells | Sentinel lymph nodes | Increases in mitogen-activated protein kinase 14, laminin, collagen, and urokinase plasminogen activator | N/A | [79] |
| Mesenchymal stem cell EV-mediated ECM repair | | | | | |
| | Adipose tissue MSCs | Dermal fibroblasts | Increases the ratio of type 3 collagen to type 1 collagen and the ratio of MMP-3 to TIMP-1 | N/A | [81,82] |
| | Adipose tissue MSCs | Chondrocytes | Promotes type 2 collagen synthesis and inhibits the expression of MMP-1, MMP-3, MMP-13, and ADAMTS-5 | N/A | [83] |
| | Bone marrow MSCs | Bone marrow MSCs | Promotes cell adhesion and motility | Hyaluronan | [84] |

## 6. Conclusions

During pathological processes such as tumor progression, EVs function via both the direct (i.e., EV-surface-associated enzymes) and indirect mechanisms (i.e., influencing recipient cells) to aid in ECM degradation, tumor progression, and the development of the pre-metastatic niche. Similarly, in physiological processes such as bone formation and tissue regeneration, EVs from osteoblasts and MSCs can directly evolve the structure of the ECM through calcification or by directing recipient cells to remodel the ECM during tissue repair, respectively. More knowledge regarding EV–ECM interactions during pathological processes such as cancer will allow for new therapeutic and diagnostic approaches targeting

EV-mediated tumor progression and metastases. Similarly, characterizing the effect of EVs on the ECM during physiological processes such as tissue regeneration presents the opportunity to create novel or augment current therapeutic approaches while overcoming translational limitations in tissue repair.

**Author Contributions:** Conceptualization, N.J.P. and E.J.C.; writing-original draft preparation, N.J.P., A.A. and E.J.C.; writing-review and editing, N.J.P., A.A. and E.J.C.; funding acquisition, E.J.C. All authors have read and agreed to the published version of the manuscript.

**Funding:** The authors would like to acknowledge the financial support from the National Science Foundation Graduate Student Fellowship awarded to N.J.P., and USC Women in Science and Engineering (WiSE), Ming Hsieh Institute for Research on Engineering-Medicine for Cancer, and NSF EAGER (DMR 2132744) awarded to E.J.C.

**Institutional Review Board Statement:** Not applicable.

**Informed Consent Statement:** Not applicable.

**Data Availability Statement:** No new data was created or analyzed. Data sharing is not applicable to this article.

**Conflicts of Interest:** The authors declare no competing interests.

## References

1. van Niel, G.; Carter, D.R.F.; Clayton, A.; Lambert, D.W.; Raposo, G.; Vader, P. Challenges and directions in studying cell–cell communication by extracellular vesicles. *Nat. Rev. Mol. Cell Biol.* **2022**, *23*, 369–382. [CrossRef] [PubMed]
2. Dai, J.; Su, Y.; Zhong, S.; Cong, L.; Liu, B.; Yang, J.; Tao, Y.; He, Z.; Chen, C.; Jiang, Y. Exosomes: Key players in cancer and potential therapeutic strategy. *Signal Transduct. Target. Ther.* **2020**, *5*, 145. [CrossRef] [PubMed]
3. Murphy, D.E.; de Jong, O.G.; Brouwer, M.; Wood, M.J.; Lavieu, G.; Schiffelers, R.M.; Vader, P. Extracellular vesicle-based therapeutics: Natural versus engineered targeting and trafficking. *Exp. Mol. Med.* **2019**, *51*, 1–12. [CrossRef] [PubMed]
4. Wiklander, O.P.B.; Brennan, M.; Lötvall, J.; Breakefield, X.O.; El Andaloussi, S. Advances in therapeutic applications of extracellular vesicles. *Sci. Transl. Med.* **2019**, *11*, eaav8521. [CrossRef]
5. Nawaz, M.; Shah, N.; Zanetti, B.R.; Maugeri, M.; Silvestre, R.N.; Fatima, F.; Neder, L.; Valadi, H. Extracellular Vesicles and Matrix Remodeling Enzymes: The Emerging Roles in Extracellular Matrix Remodeling, Progression of Diseases and Tissue Repair. *Cells* **2018**, *7*, 167. [CrossRef]
6. Kalfon, T.; Loewenstein, S.; Gerstenhaber, F.; Leibou, S.; Geller, H.; Sher, O.; Nizri, E.; Lahat, G. Gastric Cancer-Derived Extracellular Vesicles (EVs) Promote Angiogenesis via Angiopoietin-2. *Cancers* **2022**, *14*, 2953. [CrossRef] [PubMed]
7. Brigstock, D.R. Extracellular Vesicles in Organ Fibrosis: Mechanisms, Therapies, and Diagnostics. *Cells* **2021**, *10*, 1596. [CrossRef]
8. Xie, F.; Zhou, X.; Fang, M.; Li, H.; Su, P.; Tu, Y.; Zhang, L.; Zhou, F. Extracellular Vesicles in Cancer Immune Microenvironment and Cancer Immunotherapy. *Adv. Sci.* **2019**, *6*, 1901779. [CrossRef] [PubMed]
9. Sanwlani, R.; Gangoda, L. Role of Extracellular Vesicles in Cell Death and Inflammation. *Cells* **2021**, *10*, 2663. [CrossRef]
10. Padhi, A.; Nain, A.S. ECM in Differentiation: A Review of Matrix Structure, Composition and Mechanical Properties. *Ann. Biomed. Eng.* **2020**, *48*, 1071–1089. [CrossRef]
11. Bonnans, C.; Chou, J.; Werb, Z. Remodelling the extracellular matrix in development and disease. *Nat. Rev. Mol. Cell Biol.* **2014**, *15*, 786–801. [CrossRef]
12. Humphrey, J.D.; Dufresne, E.R.; Schwartz, M.A. Mechanotransduction and extracellular matrix homeostasis. *Nat. Rev. Mol. Cell Biol.* **2014**, *15*, 802–812. [CrossRef] [PubMed]
13. Mouw, J.K.; Ou, G.; Weaver, V.M. Extracellular matrix assembly: A multiscale deconstruction. *Nat. Rev. Mol. Cell Biol.* **2014**, *15*, 771–785. [CrossRef] [PubMed]
14. Yue, B. Biology of the extracellular matrix: An overview. *J. Glaucoma* **2014**, *23*, S20–S23. [CrossRef] [PubMed]
15. Lenzini, S.; Bargi, R.; Chung, G.; Shin, J.-W. Matrix mechanics and water permeation regulate extracellular vesicle transport. *Nat. Nanotechnol.* **2020**, *15*, 217–223. [CrossRef]
16. Chaudhuri, O.; Gu, L.; Darnell, M.; Klumpers, D.; Bencherif, S.A.; Weaver, J.C.; Huebsch, N.; Mooney, D.J. Substrate stress relaxation regulates cell spreading. *Nat. Commun.* **2015**, *6*, 6365. [CrossRef]
17. Peinado, H.; Alečković, M.; Lavotshkin, S.; Matei, I.; Costa-Silva, B.; Moreno-Bueno, G.; Hergueta-Redondo, M.; Williams, C.; García-Santos, G.; Ghajar, C.M.; et al. Melanoma exosomes educate bone marrow progenitor cells toward a pro-metastatic phenotype through MET. *Nat. Med.* **2012**, *18*, 883–891. [CrossRef]
18. Costa-Silva, B.; Aiello, N.M.; Ocean, A.J.; Singh, S.; Zhang, H.; Thakur, B.K.; Becker, A.; Hoshino, A.; Mark, M.T.; Molina, H.; et al. Pancreatic cancer exosomes initiate pre-metastatic niche formation in the liver. *Nat. Cell Biol.* **2015**, *17*, 816–826. [CrossRef]

19. Fong, M.Y.; Zhou, W.; Liu, L.; Alontaga, A.Y.; Chandra, M.; Ashby, J.; Chow, A.; O'Connor, S.T.; Li, S.; Chin, A.R.; et al. Breast-cancer-secreted miR-122 reprograms glucose metabolism in premetastatic niche to promote metastasis. *Nat. Cell Biol.* **2015**, *17*, 183–194. [CrossRef]
20. Nogués, L.; Benito-Martin, A.; Hergueta-Redondo, M.; Peinado, H. The influence of tumour-derived extracellular vesicles on local and distal metastatic dissemination. *Mol. Asp. Med.* **2018**, *60*, 15–26. [CrossRef]
21. Peinado, H.; Zhang, H.; Matei, I.R.; Costa-Silva, B.; Hoshino, A.; Rodrigues, G.; Psaila, B.; Kaplan, R.N.; Bromberg, J.F.; Kang, Y.; et al. Pre-metastatic niches: Organ-specific homes for metastases. *Nat. Rev. Cancer* **2017**, *17*, 302–317. [CrossRef] [PubMed]
22. Siveen, K.S.; Raza, A.; Ahmed, E.L.; Khan, A.Q.; Prabhu, K.S.; Kuttikrishnan, S.; Mateo, J.M.; Zayed, H.; Rasul, K.; Azizi, F.; et al. The role of extracellular vesicles as modulators of the tumor microenviornment, metastasis and drug resistance in colorectal cancer. *Cancers* **2019**, *11*, 746. [CrossRef]
23. Gurung, S.; Perocheau, D.; Touramanidou, L.; Baruteau, J. The exosome journey: From biogenesis to uptake and intracellular signalling. *Cell Commun. Signal.* **2021**, *19*, 47. [CrossRef] [PubMed]
24. Kugeratski, F.G.; Hodge, K.; Lilla, S.; McAndrews, K.M.; Zhou, X.; Hwang, R.F.; Zanivan, S.; Kalluri, R. Quantitative proteomics identifies the core proteome of exosomes with syntenin-1 as the highest abundant protein and a putative universal biomarker. *Nat. Cell Biol.* **2021**, *23*, 631–641. [CrossRef]
25. Henne, W.M.; Stenmark, H.; Emr, S.D. Molecular mechanisms of the membrane sculpting ESCRT pathway. *Cold Spring Harb. Perspect. Biol.* **2013**, *5*, a016766. [CrossRef] [PubMed]
26. Blanc, L.; Vidal, M. New insights into the function of Rab GTPases in the context of exosomal secretion. *Small GTPases* **2018**, *9*, 95–106. [CrossRef]
27. Ståhl, A.L.; Johansson, K.; Mossberg, M.; Kahn, R.; Karpman, D. Exosomes and microvesicles in normal physiology, pathophysiology, and renal diseases. *Pediatr. Nephrol.* **2019**, *34*, 11–30. [CrossRef]
28. Mulcahy, L.A.; Pink, R.C.; Carter, D.R.F. Routes and mechanisms of extracellular vesicle uptake. *J. Extracell. Vesicles* **2014**, *3*, 24641. [CrossRef]
29. Skotland, T.; Sagini, K.; Sandvig, K.; Llorente, A. An emerging focus on lipids in extracellular vesicles. *Adv. Drug Deliv. Rev.* **2020**, *159*, 308–321. [CrossRef]
30. Battistelli, M.; Falcieri, E. Apoptotic Bodies: Particular Extracellular Vesicles Involved in Intercellular Communication. *Biology* **2020**, *9*, 21. [CrossRef]
31. Dellar, E.R.; Hill, C.; Melling, G.E.; Carter, D.R.F.; Baena-Lopez, L.A. Unpacking extracellular vesicles: RNA cargo loading and function. *J. Extracell. Biol.* **2022**, *1*, e40. [CrossRef]
32. Escrevente, C.; Keller, S.; Altevogt, P.; Costa, J. Interaction and uptake of exosomes by ovarian cancer cells. *BMC Cancer* **2011**, *11*, 108. [CrossRef] [PubMed]
33. Simons, K.; Ehehalt, R. Cholesterol, lipid rafts, and disease. *J. Clin. Investig.* **2002**, *110*, 597–603. [CrossRef] [PubMed]
34. Parolini, I.; Federici, C.; Raggi, C.; Lugini, L.; Palleschi, S.; De Milito, A.; Coscia, C.; Iessi, E.; Logozzi, M.; Molinari, A. Microenvironmental pH is a key factor for exosome traffic in tumor cells. *J. Biol. Chem.* **2009**, *284*, 34211–34222. [CrossRef]
35. Clarke-Bland, C.E.; Bill, R.M.; Devitt, A. Emerging roles for AQP in mammalian extracellular vesicles. *Biochim. Et Biophys. Acta (BBA)—Biomembr.* **2022**, *1864*, 183826. [CrossRef] [PubMed]
36. Sanderson, R.D.; Bandari, S.K.; Vlodavsky, I. Proteases and glycosidases on the surface of exosomes: Newly discovered mechanisms for extracellular remodeling. *Matrix Biol.* **2019**, *75–76*, 160–169. [CrossRef] [PubMed]
37. Reiner, A.T.; Tan, S.; Agreiter, C.; Auer, K.; Bachmayr-Heyda, A.; Aust, S.; Pecha, N.; Mandorfer, M.; Pils, D.; Brisson, A.R.; et al. EV-Associated MMP9 in High-Grade Serous Ovarian Cancer Is Preferentially Localized to Annexin V-Binding EVs. *Dis. Mrk.* **2017**, *2017*, 9653194. [CrossRef]
38. Dolo, V.; D'Ascenzo, S.; Violini, S.; Pompucci, L.; Festuccia, C.; Ginestra, A.; Vittorelli, M.L.; Canevari, S.; Pavan, A. Matrix-degrading proteinases are shed in membrane vesicles by ovarian cancer cells in vivo and in vitro. *Clin. Exp. Metastasis* **1999**, *17*, 131–140. [CrossRef]
39. Sung, B.H.; Ketova, T.; Hoshino, D.; Zijlstra, A.; Weaver, A.M. Directional cell movement through tissues is controlled by exosome secretion. *Nat. Commun.* **2015**, *6*, 7164. [CrossRef]
40. Murshed, M.; Schinke, T.; McKee, M.D.; Karsenty, G. Extracellular matrix mineralization is regulated locally; different roles of two gla-containing proteins. *J. Cell Biol.* **2004**, *165*, 625–630. [CrossRef]
41. Lin, X.; Patil, S.; Gao, Y.G.; Qian, A. The Bone Extracellular Matrix in Bone Formation and Regeneration. *Front. Pharm.* **2020**, *11*, 757. [CrossRef] [PubMed]
42. Veschi, E.A.; Bolean, M.; Strzelecka-Kiliszek, A. Bandorowicz-Pikula, J.; Pikula, S.; Granjon, T.; Mebarek, S.; Magne, D.; Ramos, A.P.; Rosato, N.; et al. Localization of Annexin A6 in Matrix Vesicles During Physiological Mineralization. *Int. J. Mol. Sci.* **2020**, *21*, 1367. [CrossRef] [PubMed]
43. Rojas, E.; Arispe, N.; Haigler, H.T.; Burns, A.L.; Pollard, H.B. Identification of annexins as calcium channels in biological membranes. *Bone Miner.* **1992**, *17*, 214–218. [CrossRef]
44. Golub, E.E. Role of matrix vesicles in biomineralization. *Biochim. Biophys. Acta* **2009**, *1790*, 1592–1598. [CrossRef]
45. Suzuki, A.; Ghayor, C.; Guicheux, J.; Magne, D.; Quillard, S.; Kakita, A.; Ono, Y.; Miura, Y.; Oiso, Y.; Itoh, M.; et al. Enhanced Expression of the Inorganic Phosphate Transporter Pit-1 Is Involved in BMP-2–Induced Matrix Mineralization in Osteoblast-Like Cells. *J. Bone Miner. Res.* **2006**, *21*, 674–683. [CrossRef]

46. Anderson, H.C. Molecular biology of matrix vesicles. *Clin. Orthop. Relat. Res.* **1995**, *314*, 266–280. [CrossRef]

47. Wu, M.; Rementer, C.; Giachelli, C.M. Vascular calcification: An update on mechanisms and challenges in treatment. *Calcif Tissue Int* **2013**, *93*, 365–373. [CrossRef]

48. Krohn, J.B.; Hutcheson, J.D.; Martínez-Martínez, E.; Aikawa, E. Extracellular vesicles in cardiovascular calcification: Expanding current paradigms. *J. Physiol.* **2016**, *594*, 2895–2903. [CrossRef] [PubMed]

49. Ruiz, J.L.; Weinbaum, S.; Aikawa, E.; Hutcheson, J.D. Zooming in on the genesis of atherosclerotic plaque microcalcifications. *J. Physiol.* **2016**, *594*, 2915–2927. [CrossRef] [PubMed]

50. Kapustin, A.N.; Davies, J.D.; Reynolds, J.L.; McNair, R.; Jones, G.T.; Sidibe, A.; Schurgers, L.J.; Skepper, J.N.; Proudfoot, D.; Mayr, M.; et al. Calcium Regulates Key Components of Vascular Smooth Muscle Cell–Derived Matrix Vesicles to Enhance Mineralization. *Circ. Res.* **2011**, *109*, e1–e12. [CrossRef]

51. Rogers, M.A.; Buffolo, F.; Schlotter, F.; Atkins, S.K.; Lee, L.H.; Halu, A.; Blaser, M.C.; Tsolaki, E.; Higashi, H.; Luther, K.; et al. Annexin A1-dependent tethering promotes extracellular vesicle aggregation revealed with single-extracellular vesicle analysis. *Sci. Adv.* **2020**, *6*, eabb1244. [CrossRef] [PubMed]

52. Huleihel, L.; Hussey, G.S.; Naranjo, J.D.; Zhang, L.; Dziki, J.L.; Turner, N.J.; Stolz, D.B.; Badylak, S.F. Matrix-bound nanovesicles within ECM bioscaffolds. *Sci. Adv.* **2016**, *2*, e1600502. [CrossRef] [PubMed]

53. Turner, N.J.; Quijano, L.M.; Hussey, G.S.; Jiang, P.; Badylak, S.F. Matrix Bound Nanovesicles Have Tissue-Specific Characteristics That Suggest a Regulatory Role. *Tissue Eng. Part A* **2022**, *28*, 879–892. [CrossRef] [PubMed]

54. Hussey, G.S.; Pineda Molina, C.; Cramer, M.C.; Tyurina, Y.Y.; Tyurin, V.A.; Lee, Y.C.; El-Mossier, S.O.; Murdock, M.H.; Timashev, P.S.; Kagan, V.E.; et al. Lipidomics and RNA sequencing reveal a novel subpopulation of nanovesicle within extracellular matrix biomaterials. *Sci. Adv.* **2020**, *6*, eaay4361. [CrossRef]

55. van der Merwe, Y.; Faust, A.E.; Sakalli, E.T.; Westrick, C.C.; Hussey, G.; Chan, K.C.; Conner, I.P.; Fu, V.L.N.; Badylak, S.F.; Steketee, M.B. Matrix-bound nanovesicles prevent ischemia-induced retinal ganglion cell axon degeneration and death and preserve visual function. *Sci. Rep.* **2019**, *9*, 3482. [CrossRef]

56. Zhang, L.; Zhang, F.; Weng, Z.; Brown, B.N.; Yan, H.; Ma, X.M.; Vosler, P.S.; Badylak, S.F.; Dixon, C.E.; Cui, X.T.; et al. Effect of an inductive hydrogel composed of urinary bladder matrix upon functional recovery following traumatic brain injury. *Tissue Eng. Part A* **2013**, *19*, 1909–1918. [CrossRef]

57. Crum, R.J.; Hall, K.; Molina, C.P.; Hussey, G.S.; Graham, E.; Li, H.; Badylak, S.F. Immunomodulatory matrix-bound nanovesicles mitigate acute and chronic pristane-induced rheumatoid arthritis. *Npj Regen. Med.* **2022**, *7*, 13. [CrossRef]

58. Page-McCaw, A.; Ewald, A.J.; Werb, Z. Matrix metalloproteinases and the regulation of tissue remodelling. *Nat. Rev. Mol. Cell Biol.* **2007**, *8*, 221–233. [CrossRef]

59. Hakulinen, J.; Sankkila, L.; Sugiyama, N.; Lehti, K.; Keski-Oja, J. Secretion of active membrane type 1 matrix metalloproteinase (MMP-14) into extracellular space in microvesicular exosomes. *J. Cell. Biochem.* **2008**, *105*, 1211–1218. [CrossRef]

60. You, Y.; Shan, Y.; Chen, J.; Yue, H.; You, B.; Shi, S.; Li, X.; Cao, X. Matrix metalloproteinase 13-containing exosomes promote nasopharyngeal carcinoma metastasis. *Cancer Sci.* **2015**, *106*, 1669–1677. [CrossRef]

61. Taraboletti, G.; D'Ascenzo, S.; Borsotti, P.; Giavazzi, R.; Pavan, A.; Dolo, V. Shedding of the Matrix Metalloproteinases MMP-2, MMP-9, and MT1-MMP as Membrane Vesicle-Associated Components by Endothelial Cells. *Am. J. Pathol.* **2002**, *160*, 673–680. [CrossRef] [PubMed]

62. Thompson, C.A.; Purushothaman, A.; Ramani, V.C.; Vlodavsky, I.; Sanderson, R.D. Heparanase Regulates Secretion, Composition, and Function of Tumor Cell-derived Exosomes. *J. Biol. Chem.* **2013**, *288*, 10093–10099. [CrossRef] [PubMed]

63. Bandari, S.K.; Purushothaman, A.; Ramani, V.C.; Brinkley, G.J.; Chandrashekar, D.S.; Varambally, S.; Mobley, J.A.; Zhang, Y.; Brown, E.E.; Vlodavsky, I.; et al. Chemotherapy induces secretion of exosomes loaded with heparanase that degrades extracellular matrix and impacts tumor and host cell behavior. *Matrix Biol.* **2018**, *65*, 104–118. [CrossRef]

64. Purushothaman, A.; Bandari, S.K.; Liu, J.; Mobley, J.A.; Brown, E.E.; Sanderson, R.D. Fibronectin on the Surface of Myeloma Cell-derived Exosomes Mediates Exosome-Cell Interactions. *J. Biol. Chem.* **2016**, *291*, 1652–1663. [CrossRef]

65. Sarrazin, S.; Lamanna, W.C.; Esko, J.D. Heparan sulfate proteoglycans. *Cold Spring Harb. Perspect. Biol.* **2011**, *3*, a004952. [CrossRef]

66. de Jong, O.G.; van Balkom, B.W.; Gremmels, H.; Verhaar, M.C. Exosomes from hypoxic endothelial cells have increased collagen crosslinking activity through up-regulation of lysyl oxidase-like 2. *J. Cell Mol. Med.* **2016**, *20*, 342–350. [CrossRef]

67. Genschmer, K.R.; Russell, D.W.; Lal, C.; Szul, T.; Bratcher, P.E.; Noerager, B.D.; Abdul Roda, M.; Xu, X.; Rezonzew, G.; Viera, L.; et al. Activated PMN Exosomes: Pathogenic Entities Causing Matrix Destruction and Disease in the Lung. *Cell* **2019**, *176*, 113–126.e115. [CrossRef]

68. Hussey, G.S.; Dziki, J.L.; Lee, Y.C.; Bartolacci, J.G.; Behun, M.; Turnquist, H.R.; Badylak, S.F. Matrix bound nanovesicle-associated IL-33 activates a pro-remodeling macrophage phenotype via a non-canonical, ST2-independent pathway. *J. Immunol Regen. Med.* **2019**, *3*, 26–35. [CrossRef]

69. Shan, Y.; You, B.; Shi, S.; Shi, W.; Zhang, Z.; Zhang, Q.; Gu, M.; Chen, J.; Bao, L.; Liu, D.; et al. Hypoxia-Induced Matrix Metalloproteinase-13 Expression in Exosomes from Nasopharyngeal Carcinoma Enhances Metastases. *Cell Death Dis.* **2018**, *9*, 382. [CrossRef] [PubMed]

70. Aoki, M.; Koga, K.; Hamasaki, M.; Egawa, N.; Nabeshima, K. Emmprin, released as a microvesicle in epithelioid sarcoma, interacts with fibroblasts. *Int. J. Oncol.* **2017**, *50*, 2229–2235. [CrossRef]

71. Song, G.; Xu, S.; Zhang, H.; Wang, Y.; Xiao, C.; Jiang, T.; Wu, L.; Zhang, T.; Sun, X.; Zhong, L.; et al. TIMP1 is a prognostic marker for the progression and metastasis of colon cancer through FAK-PI3K/AKT and MAPK pathway. *J. Exp. Clin. Cancer Res.* **2016**, *35*, 148. [CrossRef] [PubMed]

72. Gong, Y.; Scott, E.; Lu, R.; Xu, Y.; Oh, W.K.; Yu, Q. TIMP-1 promotes accumulation of cancer associated fibroblasts and cancer progression. *PLoS ONE* **2013**, *8*, e77366. [CrossRef] [PubMed]

73. Dang, K.; Myers, K.A. The role of hypoxia-induced miR-210 in cancer progression. *Int. J. Mol. Sci.* **2015**, *16*, 6353–6372. [CrossRef] [PubMed]

74. Cui, H.; Seubert, B.; Stahl, E.; Dietz, H.; Reuning, U.; Moreno-Leon, L.; Ilie, M.; Hofman, P.; Nagase, H.; Mari, B.; et al. Tissue inhibitor of metalloproteinases-1 induces a pro-tumourigenic increase of miR-210 in lung adenocarcinoma cells and their exosomes. *Oncogene* **2015**, *34*, 3640–3650. [CrossRef] [PubMed]

75. Paget, S. The distribution of secondary growths in cancer of the breast. *Lancet* **1889**, *133*, 571–573. [CrossRef]

76. Hoshino, A.; Costa-Silva, B.; Shen, T.-L.; Rodrigues, G.; Hashimoto, A.; Tesic Mark, M.; Molina, H.; Kohsaka, S.; Di Giannatale, A.; Ceder, S.; et al. Tumour exosome integrins determine organotropic metastasis. *Nature* **2015**, *527*, 329–335. [CrossRef]

77. Quintero-Fabián, S.; Arreola, R.; Becerril-Villanueva, E.; Torres-Romero, J.C.; Arana-Argáez, V.; Lara-Riegos, J.; Ramírez-Camacho, M.A.; Alvarez-Sánchez, M.E. Role of Matrix Metalloproteinases in Angiogenesis and Cancer. *Front. Oncol.* **2019**, *9*, 1370. [CrossRef]

78. Redzic, J.S.; Kendrick, A.A.; Bahmed, K.; Dahl, K.D.; Pearson, C.G.; Robinson, W.A.; Robinson, S.E.; Graner, M.W.; Eisenmesser, E.Z. Extracellular Vesicles Secreted from Cancer Cell Lines Stimulate Secretion of MMP-9, IL-6, TGF-β1 and EMMPRIN. *PLoS ONE* **2013**, *8*, e71225. [CrossRef]

79. Hood, J.L.; San, R.S.; Wickline, S.A. Exosomes Released by Melanoma Cells Prepare Sentinel Lymph Nodes for Tumor Metastasis. *Cancer Res.* **2011**, *71*, 3792–3801. [CrossRef]

80. Klyachko, N.L.; Arzt, C.J.; Li, S.M.; Gololobova, O.A.; Batrakova, E.V. Extracellular Vesicle-Based Therapeutics: Preclinical and Clinical Investigations. *Pharmaceutics* **2020**, *12*, 1171. [CrossRef]

81. Wang, L.; Hu, L.; Zhou, X.; Xiong, Z.; Zhang, C.; Shehada, H.M.A.; Hu, B.; Song, J.; Chen, L. Exosomes secreted by human adipose mesenchymal stem cells promote scarless cutaneous repair by regulating extracellular matrix remodelling. *Sci. Rep.* **2017**, *7*, 13321. [CrossRef] [PubMed]

82. Liu, X.; Wang, S.; Wu, S.; Hao, Q.; Li, Y.; Guo, Z.; Wang, W. Exosomes secreted by adipose-derived mesenchymal stem cells regulate type I collagen metabolism in fibroblasts from women with stress urinary incontinence. *Stem. Cell Res. Ther.* **2018**, *9*, 159. [CrossRef] [PubMed]

83. Woo, C.H.; Kim, H.K.; Jung, G.Y.; Jung, Y.J.; Lee, K.S.; Yun, Y.E.; Han, J.; Lee, J.; Kim, W.S.; Choi, J.S.; et al. Small extracellular vesicles from human adipose-derived stem cells attenuate cartilage degeneration. *J. Extracell. Vesicles* **2020**, *9*, 1735249. [CrossRef] [PubMed]

84. Arasu, U.T.; Kärnä, R.; Härkönen, K.; Oikari, S.; Koistinen, A.; Kröger, H.; Qu, C.; Lammi, M.J.; Rilla, K. Human mesenchymal stem cells secrete hyaluronan-coated extracellular vesicles. *Matrix Biol.* **2017**, *64*, 54–68. [CrossRef] [PubMed]

 *bioengineering*

*Review*

# Engineering Cell–ECM–Material Interactions for Musculoskeletal Regeneration

Calvin L. Jones, Brian T. Penney and Sophia K. Theodossiou *

Department of Mechanical and Biomedical Engineering, Boise State University, 1910 University Dr MS2085, Boise, ID 83725, USA
* Correspondence: sophiatheodossiou@boisestate.edu

**Abstract:** The extracellular microenvironment regulates many of the mechanical and biochemical cues that direct musculoskeletal development and are involved in musculoskeletal disease. The extracellular matrix (ECM) is a main component of this microenvironment. Tissue engineered approaches towards regenerating muscle, cartilage, tendon, and bone target the ECM because it supplies critical signals for regenerating musculoskeletal tissues. Engineered ECM–material scaffolds that mimic key mechanical and biochemical components of the ECM are of particular interest in musculoskeletal tissue engineering. Such materials are biocompatible, can be fabricated to have desirable mechanical and biochemical properties, and can be further chemically or genetically modified to support cell differentiation or halt degenerative disease progression. In this review, we survey how engineered approaches using natural and ECM-derived materials and scaffold systems can harness the unique characteristics of the ECM to support musculoskeletal tissue regeneration, with a focus on skeletal muscle, cartilage, tendon, and bone. We summarize the strengths of current approaches and look towards a future of materials and culture systems with engineered and highly tailored cell–ECM–material interactions to drive musculoskeletal tissue restoration. The works highlighted in this review strongly support the continued exploration of ECM and other engineered materials as tools to control cell fate and make large-scale musculoskeletal regeneration a reality.

**Keywords:** extracellular matrix (ECM); musculoskeletal tissue engineering; musculoskeletal regeneration; skeletal muscle; tendon; cartilage; bone; atomic force microscopy (AFM)

Citation: Jones, C.L.; Penney, B.T.; Theodossiou, S.K. Engineering Cell–ECM–Material Interactions for Musculoskeletal Regeneration. *Bioengineering* **2023**, *10*, 453. https://doi.org/10.3390/bioengineering10040453

Academic Editor: Stuart Goodman

Received: 28 January 2023
Revised: 23 March 2023
Accepted: 28 March 2023
Published: 7 April 2023



## 1. Introduction

Engineering cell-extracellular matrix (ECM) interactions to achieve desired regenerative outcomes are promising approaches in musculoskeletal tissue engineering. Musculoskeletal tissue injuries are a common and increasing clinical problem in the US and around the world [1]. Injuries and diseases of the skeletal muscles, tendons, cartilage, and bones have widespread impacts on the activities and quality of daily life, as musculoskeletal tissues enable movement and daily function. Current musculoskeletal injury treatments rely on rehabilitative or surgical interventions that may not restore normal function, resulting in long-term disability, with associated economic losses in terms of lost wages, decreased productivity, and high healthcare costs [1]. Regenerative medicine promises to improve the prognosis for musculoskeletal pathologies. The last fifteen years have substantially deepened the knowledge of how cells and biomaterial scaffolds can restore functionality to musculoskeletal tissues [2]. In this review, we highlight recent and noteworthy studies that have engineered the relationship between cells, the ECM, and natural or synthetic biomaterials to selectively exploit desired interactions with musculoskeletal cells and tissues.

Engineering the ECM to facilitate desired cell–cell and cell–material interactions is a relatively new approach in tissue engineering. In the last decade, biomaterial fabrication methods have progressed to enable precise control over the mechanical, chemical, and biological properties of the extracellular microenvironment [3]. Thanks to this enhanced control of scaffold properties, tailoring the microenvironment to support targeted cell growth and differentiation is a viable tissue engineering strategy. Scaffolds with highly controlled porosity [4,5], pH-responsiveness [6,7], stiffness [8–11], and biochemistry [12] are increasingly used to drive cellular behavior towards the regeneration of large-scale musculoskeletal tissues and musculoskeletal tissue interfaces [13–15]. Additional components of the extracellular environment, such as growth factors and extracellular vesicles, also direct cell behavior, often in combination with the ECM; engineered approaches focusing on these aspects of the microenvironment are discussed elsewhere [16]. In this review, we focus on approaches targeting or utilizing the ECM, though we note that several of the highlighted studies incorporate extracellular vesicles and growth factors alongside their ECM focus.

The cell- and tissue-derived extracellular matrix (ECM) has been extensively characterized for uses in tissue engineering, and it has been reviewed elsewhere [2,17]. To generate ECM for scaffolds, various methods have been developed for extracting ECM from human [18,19] and animal [20–22] tissues. A persistent concern is the rejection of animal ECM-containing scaffolds in human recipients, though this review highlights studies that indicate that the animal ECM can be used safely and successfully in human patients [23]. Human-derived ECM, such as ECM generated from human acellular dermal matrix, also has potential uses in bioscaffolds that enhance musculoskeletal regeneration [24]. Human-sourced ECM has recently been isolated from the decellularized human chorion matrix [19], transforming the placenta from an organ that is commonly discarded as medical waste to a rich source of ECM starting material for tissue engineering applications in several tissues, including the musculoskeletal system. Moving forward, genetic engineering techniques, such as CRISP-Cas9, may expand the potential uses of animal-derived ECM in humans. Additionally, though it is less common than ECM originating from animal or human tissues, cell-derived ECM has been generated from cells cultured for the purpose of producing ECM with specific, desirable qualities, with varied but consistent success [2,25]. In the concluding sections of this review, we briefly discuss how genetic engineering may enable the generation of highly specialized and useful ECM from genetically-modified cells possessing desirable ECM characteristics. We predict that ECM and ECM-derived biomaterials will continue to be valuable tools for directing and engineering cell functions.

The value of engineering the ECM to foster specific cell–ECM interactions extends beyond its use in regenerative approaches and into its widespread potential applications in directing cell function throughout entire tissue and organ systems [26]. Engineering the ECM to resemble various musculoskeletal pathologies can be used to model injury and disease, enhancing the search for treatments and cures. To treat musculoskeletal disease, a deeper understanding of the pathologies is needed. Engineering the ECM to resemble specific conditions, or to mimic particular stages of development or aging, has numerous uses in musculoskeletal research. Throughout this review, we use the term "engineering" in reference to the ECM to denote research approaches that extract, augment, modify, pattern, or otherwise adapt features of the ECM and purposefully impact cell- or tissue-level experimental outcomes. We pay special attention to emerging treatments for musculoskeletal fibrosis, a persistent ECM-mediated problem both within and outside the musculoskeletal system. Our goal is to illustrate promising ideas in the hopes of fostering novel discussions and collaborations that move the field of musculoskeletal tissue engineering forward.

## 2. Survey of Engineered Cell–ECM–Material Interactions in Musculoskeletal Tissues

### 2.1. Skeletal Muscle

Selected approaches utilizing cell–ECM–material interactions to regenerate skeletal muscle are summarized in Table 1. Skeletal muscle composes nearly 40% of the weight of the average human body [27], enables locomotion and fine motor skills, and participates in multiple homeostatic processes [28]. Skeletal muscle ECM supports the function of muscle cells by providing three-dimensional scaffolding for their function and growth, as well as facilitating interactions with the various collagens, glycoproteins, proteoglycans, and elastins that give skeletal muscleits mechanical and biological properties [29]. The ECM regulates skeletal muscle growth, maintenance, response to injury, and repair. Importantly, the structure of skeletal muscle ECM directly contributes to the tissue's mechanical capabilities.

The highly organized, hierarchical structure of skeletal muscle has made it a particularly challenging tissue to engineer, as its structure provides unique mechanical properties. Replicating muscle structure and the complex multi-tissue interfaces required for muscle function, such as neuromuscular junctions (NMJs) and tendons, has proven elusive. While muscle cells from human and non-human sources readily grow and differentiate in vitro, and they can be successfully implanted in numerous animal models, the resulting muscle tissue often lacks the force production, growth, and vascularization capacities of native muscle. A significant limitation of several regenerative approaches for restoring skeletal muscle is the inability to replicate the structural and spatial components of healthy muscle tissue. This problem is especially pronounced in research attempting to reverse volumetric muscle loss (VML) [30]. VML results from extensive trauma to the muscles that overwhelms their intrinsic regenerative capacity [31]. VML is characterized by a complete breakdown of the muscle ECM and its associated structures [31]. Restoring that structure in lab-grown muscle may be the key to addressing not only VML, but other traumatic and wasting diseases characterized by widespread muscle tissue damage.

### 2.1.1. ECM-Based Approaches for Skeletal Muscle Regeneration

Engineered approaches replicating the organized ECM of muscle have seen some success in terms of the force production of the regenerated muscle [32] and have improved the formation of NMJs. The restoration of highly oriented and structured muscle and NMJ tissue may require engineering the ECM with appropriate templates or sacrificial scaffolds until cells can differentiate and deposit their own matrix. Sacrificial templates enhanced ECM deposition in a rat model of a tibialis anterior defect, where engineered ECM scaffolds with parallel microchannels were subcutaneously implanted, and the template was then removed and decellularized (Figure 1). Compared to controls, rats that received the scaffolds had extensive neotissue formation in the grafting area, as well as cell infiltration, blood vessel formation, and new ECM deposition [33]. Neo-muscle tissue had acetylcholine receptors and nerve fiber contacts resembling early NMJ formation, suggesting this engineered ECM microchannel model is a promising in vitro and in vivo platform.

Although decellularized ECM has been shown to aid tissue regeneration, it may require additional materials for effective tissue delivery. A recent study coupled the use of ECM with extracellular vesicles (EVs), cell-secreted nanoparticles that convey intercellular signals to propagate tissue repair. Decellularized muscle ECM was combined with human EVs derived from Wharton Jelly mesenchymal stromal cells (MSC EVs) to boost tissue regeneration in a murine VML model [34]. Thirty days after the removal of the tibialis anterior, ECM plus MSC EV treated mice had significantly higher myosin heavy chain (MHC) expression compared to controls treated with muscle ECM in phosphate-buffered saline (PBS). Notably, the EV-treated mice showed gains in muscle function compared to the control group, underscoring the potential of ECM and EVs as aids in tissue-specific regeneration and functional restoration.

Xenogeneic ECM (ECM derived from animal tissue) scaffolds were tested as platforms for skeletal muscle restoration in another complex injury model [35]. A scaffold composed of small intestinal submucosa ECM (SIS-ECM) was used to repair resections in canine vastus lateralis and vastus medialis muscles. At 1, 2, 3, and 6 months post-injury, muscle function was not restored [35]. However, the SIS–ECM appeared to promote integration of soft and bony tissues, suggesting it may be a useful tool in engineering the ECM after injury to promote an integrative cellular response. Other uses of xenogeneic ECM in skeletal muscle tissue engineering have involved porcine bladder tissue in murine models of VML, and they have shown improved perivascular stem cell mobilization and accumulation, with increased skeletal muscle cell proliferation at the injury sites [23]. The porcine bladder ECM supported the formation of stimulus-responsive skeletal muscle cells and tissues in mice, which resulted in functional improvement in three implanted human patients. Notably, all patients showed signs of de novo formation of muscle tissue and vasculature. In the same study, an accompanying murine VML model found no evidence of new skeletal muscle tissue proliferation in the untreated controls. Conversely, animals treated with ECM formed islands of desmin and myosin heavy chain-positive (MHC+) striated skeletal muscle throughout the defect and scaffold placement areas, consistent with an ECM-mediated constructive remodeling effect [23]. EMG analysis further showed that ECM treated mice displayed muscle activation characteristics closer to those of uninjured mice, rather than untreated controls. Taken together, these studies suggest xenogeneic ECM has the potential to improve functional muscle regeneration in human patients.

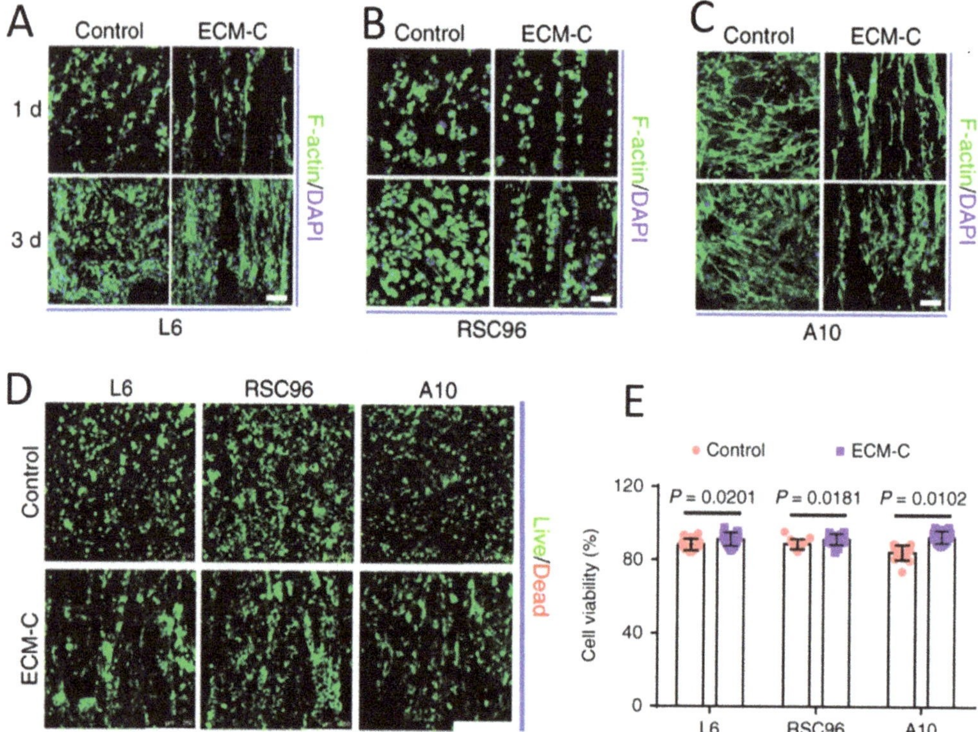


**Figure 1.** Materials composed of cells and ECM show promise for recreating the complex, hierarchical structure of functional skeletal muscle tissue. ECM–cell composite scaffolds effectively guide cells towards the spatial organization required for muscle, nerve, and blood vessel formation. (**A–C**) Cellular guiding effects of ECM–cellular (ECM–C) and control scaffolds. Skeletal actin fibers and nuclei of (**A**) L6, (**B**) RSC96, and (**C**) A10 cells were respectively stained with fluorescein isothiocyanate (FITC) conjugated phalloidin and DAPI at one and three days. (**D,E**) Live/dead staining and cell viability of L6, RSC96, and A10 cells on different scaffolds at seven days. Live cells are stained green and dead cells are stained red (*n* = 15). Figure adapted with permission from Zhu et al. 2019 [33].

### 2.1.2. ECM-Based Personalized Disease Models

Engineered muscle ECM is a crucial component of recently developed and highly accurate in vitro disease models, which are of significant interest as personalized test platforms for novel muscle-targeting therapeutics. Collagen VI- related dystrophies (COL6-RDs) are rare congenital neuromuscular dystrophies where alterations in one of the three collagen VI genes change the collagen's incorporation into the ECM, severely impacting the mechanical coherence of the matrix [36]. COL6-RDs result in muscle weakness, proximal joint contractures, and distal hyperlaxity. To bridge the gap between the specific ECM features of COL6-RD patients and the known clinical phenotype, recent work engineered a novel, personalized preclinical model of COL6-RDs using cell-derived matrices (CDMs). These CDMs were developed using the forearm skin fibroblasts of patients with COL6-RD and healthy donors. Disease markers were significantly increased in CDMs from COL6-RD patients compared to controls (CDMs derived from healthy patients). Additionally, higher Collagen VI and fibronectin alignment, length, width, and straightness were observed in control CDMs compared to patient-derived CDMs, suggesting the model successfully captured the matrix characteristics of COL6-RD [36]. This and other studies summarized elsewhere [37] hint towards a future of personalized disease models for identifying the most effective treatment for individual patients. Engineering the ECM to accurately replicate specific disease morphologies will lay the groundwork for novel surgical, rehabilitative, and pharmacological treatments of currently incurable conditions.

The future of engineered cell–ECM interactions in skeletal muscle tissue engineering extends to personalized injury models. ECM–cell materials can potentially recreate the tissue structure that is destroyed during VML, with the goal of restoring the large quantities of the specific muscle tissue (for example, the quadriceps or the gastrocnemius) lost in VML, thus reversing the associated disability. Recreating the highly organized mechanical and biological structure of muscle tissue remains a significant challenge. Both natural and engineered ECM, as a standalone biomaterial or in combination with other scaffolds, is uniquely suited to providing a culture environment that mimics native muscle. Though much exploration is still needed regarding the role of ECM in directing cell behavior in degenerative or muscle wasting diseases, improved in vitro and in vivo models of engineered ECM–cell interactions will advance our knowledge of which mechanical, biochemical, and genetic factors synergize to initiate skeletal muscle pathologies.

To prevent or reverse skeletal muscle atrophy, a viable strategy may be targeted hypertrophy using ECM-based materials to stimulate muscle cells. Matricellular proteins play key roles in ECM–cell communication and are necessary to understand in engineering ECM–cell interactions for musculoskeletal regeneration. A recent study indicated that matricellular protein cellular communication network factor 2 (CCN2), also known as connective tissue growth factor (CTGF), is induced during a time course of overload-driven skeletal muscle hypertrophy in mice and may be a useful addition to current strategies. Genetically engineered mouse models for myofiber-specific CCN2 gain- and loss-of-function were subjected to mechanical stimuli via muscle overload. Testing showed that myofiber-specific deletion of CCN2 blunted the muscle's hypertrophic response to overload without interfering with ECM deposition. Conversely, CCN2 overexpression was efficient in promoting overload-induced aberrant ECM accumulation without affecting myofiber growth [38]. The study results, following these genetic alterations, suggest the existence of independent ECM and myofiber stress adaptation reposes, both of which appear to be mediated by CCN2. Furthermore, the study proposed that CCN2 acts by regulating focal adhesion kinase (FAK)-mediated transduction of overload-induced extracellular signals, including interleukin 6 (IL6), as well as their downstream impact on global protein synthesis in skeletal muscle [38]. These findings highlight the crucial role of the muscle-derived ECM factor CCN2 in enabling mechanically induced, hypertrophic muscle growth, and they suggest that providing native CCN2 may be a significant advantage of ECM-based regenerative approaches.

**Table 1.** Recent and noteworthy studies focused on cell–ECM–material interactions in skeletal muscle.

| Tissue | Model | Material | Main Findings |
|---|---|---|---|
| Skeletal Muscle | ECM combined with extracellular vesicles (EVs) and mesenchymal stem cells (MSCs) in a murine VML model. | Decellularized ECM and extracellular vesicles (EVs). | Muscle regeneration was enhanced after 30 days in mice treated with ECM and EVs. Higher MHC and gains in muscle function compared to control groups [34]. |
| | ECM scaffolds with parallel microchannels (ECM-C) by subcutaneous implantation of sacrificial templates followed by template removal and decellularization. | ECM scaffolds with parallel microchannels (ECM-C). | Compared to controls, rats that received the scaffolds had extensive neo-tissue formation in the grafting area, as well as cell infiltration, blood vessel formation, and new ECM deposition, which were not observed in the controls. Neo-muscle tissue had acetylcholine receptors and nerve fiber contacts, resembling early neuromuscular junction formation [33]. |
| | Unilateral resection of the distal third of the vastus lateralis and medial half of the distal third of the vastus medialis in dogs; defects replaced with scaffolds composed of small intestinal submucosa extracellular matrix (SIS-ECM). | Scaffolds composed of small intestinal submucosa extracellular matrix (SIS-ECM). | SIS-ECM promoted integration of soft and bony tissues, suggesting it may be a useful tool in engineering the ECM after injury to promote an integrative response in the cells [35]. |
| | Xenogeneic porcine urinary bladder ECM scaffolds used as a surgical treatment for volumetric muscle loss in both a preclinical rodent model and human male patients. | Xenogeneic porcine urinary bladder ECM Scaffolds. | Porcine bladder ECM supported the formation of stimulus-responsive skeletal muscle cells and tissues in mice, and functional improvement was observed in three implanted human patients. ECM-treated mice showed muscle activation [23]. |
| | Preclinical model of collagen VI- related dystrophies (COL6-RDs) using cell-derived matrices (CDMs) developed using the forearm skin fibroblasts of both patients with (COL6-RD), as well as from healthy donors without neuromuscular disease. | Cell-derived matrices (CDMs) developed using the forearm skin fibroblasts of both patients with (COL6-RD), and from healthy donors without neuromuscular disease. | Disease markers were significantly increased in CDMs from COL6-RD patients compared to controls (CDMs derived from healthy patients). Higher collagen VI and fibronectin alignment, length, width, and straightness were observed in control CDMs compared to patient-derived CDMs [36]. |
| | Decellularized canine placentas and murine skeletal muscle ECM placed in male Wistar rats with pockets at the posterior limbs. | Decellularized canine placentas and murine skeletal muscles. | Higher percentage of proliferative PCNA+ cells three days after implantation in placenta-derived matrices, compared to muscle derived matrices. Higher percentage of $CD163^{+high}$ macrophages in muscle-derived ECM; higher percentage of $CD163^{+low}$ macrophages found in placenta-derived ECM 3- and 15-days post-implantation [39]. |

### 2.1.3. Interactions between ECM and the Immune System in Skeletal Muscle

Immune cell–ECM interactions are also critical for muscle growth and regeneration. Despite their emerging potential in regenerative approaches, decellularized allogeneic and xenogeneic ECM scaffolds may cause an aggressive inflammatory response when implanted. To combat this, a recent study compared decellularized canine placentas and murine skeletal muscles as scaffolds for regenerating skeletal muscles in a rat model [39]. Muscle pockets were created at the posterior limbs of male Wistar rats, into which the muscle and placenta derived ECM were implanted. Three days after implantation, a higher percentage of proliferative cells (PCNA$^+$) was seen in placenta-derived matrices compared to muscle derived matrices [39]. Additionally, higher percentages of CD163$^{+high}$ macrophages were observed in the muscle derived ECM group, whereas CD163$^{+low}$ macrophages were found at higher percentages in the placenta-derived ECM group three- and fifteen-days post implantation. These findings indicate that the local inflammatory response mediated by the implantation of the placenta-derived ECM was similar to that of the allogeneic muscle ECM, suggesting that placenta-derived ECM may minimize inflammation. Placental tissue is an abundant biomaterial that is often discarded as medical waste. Deriving ECM scaffolds from the placenta is desirable, as it is easier to procure than skeletal muscle, and decellularization produces large quantities of tissue. Future decellularized ECM approaches may source placental and other currently underutilized tissues, as research showing the source of the ECM to be consequential continues to emerge.

Decellularized ECM-based approaches may overstimulate the immune system by using excessive ECM material. Insufficient ECM can compromise musculoskeletal tissue function, but excessive ECM deposition that mimics or instigates fibrosis is detrimental, as it drives replacement of the resident muscle fibers with collagenous scar tissue, leading to atrophy and poor mechanical function. Unwanted infiltration of fibrotic tissue is commonly seen in implantation of allogeneic ECM scaffolds and in acute surgical injury. Anticipation and prevention of the physiological mechanisms responsible for fibrosis are needed within regenerative approaches. To this end, a recent study targeted cyclooxygenase-2 (COX-2), the rate-limiting enzyme in synthesis of prostaglandin and a positive regulator in pathophysiological processes, such as inflammation and oxidative stress [40]. Injured muscles in both human patients and mouse models overexpressed COX-2 compared to non-damaged muscle regions. COX-2 was also upregulated in human and murine fibroblasts following TGF-$\beta$ stimulation [40]. The same study investigated how celecoxib, a COX-2 inhibitor, impacted fibrogenesis in human patients. Celecoxib-mediated COX-2 inhibition was anti-fibrotic via inhibition of fibroblast differentiation, proliferation, and migration, as well as inactivation of TGF-$\beta$-dependent signaling, non-canonical TGF-$\beta$ pathways, and suppression of both reactive oxygen species (ROS) formation and oxidative stress [40]. Similarly, celecoxib-mediated inhibition of COX-2 in mouse models resulted in decreased tissue fibrosis and increased skeletal muscle fiber preservation, reflected by less ECM formation and myofibroblast accumulation with decreased p-ERK1/2, p-Smad2/3, TGF-$\beta$R1, VEGF, NOX2, and NOX4 expression. This work suggests that pharmacologically targeting the COX-2/PDK1/AKT signaling pathway may be a beneficial co-intervention to block immune rejections when using ECM scaffolds in skeletal muscle.

### 2.1.4. Other ECM-Based Approaches in Skeletal Muscle

Other ECM proteins, such as SPARCL1, have recently been targeted due to their roles in myogenic differentiation. SPARCL1 was quantified in transfected C2C12 mouse muscle myoblasts and in a mouse tibialis anterior injury model, where in vivo results suggested that SPARCL1 is associated with muscle damage repair in mice, and in vitro data showed that SPARCL1 binds to bone morphogenetic protein 7 (BMP7) by regulating BMP and transforming growth factor (TGF)-$\beta$ cell signaling [41], both pathways that ultimately promote myogenic differentiation in C2C12 cells. Using CRISPR/Cas9, the same study showed that SPARCL1 activated BMP/TGF-$\beta$ to promote the differentiation of C2C12 cells, and BMP7 molecules interacted directly with SPARCL1. Overall, SPARCL1 influenced the expression of BMP7 and activity of the BMP/TGF-β signaling pathway, while SPARCL1 activation was accompanied by BMP7 inhibition in C2C12 cells, confirming that SPARCL1 affects BMP7 expression and can promote C2C12 cell differentiation through the BMP/TGF-β pathway.

In addition to signals from ECM proteins, muscle cell differentiation is dependent on the correct assembly of the ECM itself, which is in turn regulated by matrix-metalloproteinases (MMPs). Reversion-inducing-cysteine-rich protein with kazal motifs (RECK) is a membrane-anchored protein that negatively regulates the activity of different MMPs. RECK's role in skeletal muscle differentiation, regeneration, and fibrosis have only recently been explored, but one study showed that, during myogenic differentiation of C2C12 myoblasts and satellite cells on isolated muscle fibers, RECK was transiently upregulated, and upregulation alternated with periods of RECK reduction [42]. Additionally, when RECK levels were reduced, C2C12 myoblasts were likely to enter their differentiation program with an accelerated differentiation process. RECK- deficient (RECK±) mice models were also compared to WT controls. In vivo results indicated that the transient upregulation of RECK occurs during skeletal muscle regeneration and is accelerated in RECK± mice. The RECK± mice had diminished fibrosis in response to chronic muscle damage compared to WT controls, suggesting that RECK acts as a myogenic repressor during muscle formation and regeneration [42]. Attenuating the effects of RECK in the ECM may prove useful for engineered and regenerative muscle therapies.

### 2.1.5. ECM-Cell Interactions and Muscle Fibrosis

Beyond acute injuries, muscle fibrosis can often result from muscular dystrophies. Though there are many different pathologies that directly lead to muscular dystrophy, research focus has recently shifted to patients with defects in fukutin-related protein (FKRP). FKRP is a glycosyltransferase, with the only identified function of transferring ribitol-5-phosphate to α-dystroglycan (α-DG) [43]. Interestingly, this modification is crucial for ECM attachment. A recent study showed that FKRP directs sialylation of fibronectin, a process essential for collagen recruitment to the muscle basement membrane [43]. The study findings indicated that FKRP regulates the major muscle–ECM linkages essential for fiber survival, and thus warrants further investigation as a disease axis for several muscular dystrophies. Understanding how the FKRP-dependent glycosylation of fibronectin regulates muscle pathology may prove useful, not only for understanding muscular dystrophies, but also as a new pathway to explore in ECM-mediated muscle regeneration.

A related study specifically targeted Duchenne muscular dystrophy (DMD), which is characterized by increased muscle stiffness alongside a buildup of collagenous fibrotic tissue and other ECM materials. It was previously unknown if, and by which mechanisms, collagen organization changes with the progression of DMD in diaphragm muscle tissue, and it was difficult to predict how collagen organization influences the mechanical properties of the ECM. C57BL/10ScSn-Dmdmdx/J (mdx) and wild type (WT) mice models were compared and assessed for collagen fibers straightness and alignment after three months and six months. Collagen was significantly straighter and more aligned in the WT mice at both timepoints, though collagen fibers retained a transverse orientation relative to the muscle fibers in both models. Image-based finite element mechanical models predicted an increase in the transverse relative to longitudinal (muscle fiber direction) stiffness, with the stiffness ratio (transverse/longitudinal) increased in the mdx model compared to the WT at three and six months. This study highlights that the changes in diaphragm ECM structure and mechanical properties during the disease progression in the mdx muscular dystrophy mouse may be the underlying mechanism driving changes to the diaphragm muscle function that are severe complications of DMD. These results suggest that exploiting cell–ECM–material interactions to increase collagen fiber organization and alignment may be a viable strategy for engineering muscle grafts or materials to treat dystrophies and alleviate the ECM disorganization that fibrosis causes.

Muscular fibrosis can also be the result of other pathologies, such as denervation (denervation-induced fibrosis), which follows loss of nervous system stimulation or connectivity to a skeletal muscle. A recent study showed that mice subjected to a unilateral sciatic nerve transection accumulated ECM protein, such as collagen and fibronectin, in the denervated hindlimb, together with increased levels of the profibrotic factors TGF-$\beta$ and connective tissue growth factor (CTGF/CCN2). Mice that were hemizygous for CTGF/CCN2 or mice treated with a blocking antibody against CTGF/CCN2 had a reduced accumulation of ECM proteins after denervation, compared to control mice. Additionally, these mice had no changes in fibro/adipogenic progenitors (FAPs). The same study found that ECM proteins and CTGF/CCN2 levels were increased two to four days after denervation; however, TGF-$\beta$ signaling did not increase until one to two weeks post-denervation. Blocking TGF-$\beta$ did not decrease fibronectin or CTGF levels four days following denervation, suggesting that CTGF/CCN2 is not upregulated by canonical TGF-$\beta$ signaling early after denervation. Further investigation is necessary to understand the other factors involved in the early fibrotic response following skeletal muscle denervation. Understanding these signals can lead to strategic ECM-mediated approaches to simultaneously regenerate muscle and reduce fibrosis in denervated areas.

Primary myelofibrosis (PMF) is a type of myeloproliferative neoplasm (MPN) characterized by a buildup of fibrotic tissue in the bone marrow. Given the negative side effects of current treatment options, such as the JAK2 inhibitor, as well as ruxolitinib, which also does not always have high efficacy, recent research has turned toward the possibility that ECM and bone marrow (BM) microenvironments may have an important role in the development of PMF. Lysyl oxidase (LOX), an enzyme that plays a key role in the ECM by facilitating the cross-linking of collagen and elastin fibers, has recently been targeted as an area for research, as it has been shown to be upregulated in megakaryocytes (MKs) of PMF mice and in PMF patients. This upregulation alludes to the possibility that LOX has a role in the progression of BM fibrosis. As such, LOX has been identified as a potential novel therapeutic target for PMF, leading to the development of small molecule LOX inhibitors, PXS-LOX 1 and PXS-LOX 2, which have shown that they may slow the progression of PMF in preclinical studies. Furthermore, these inhibitors prove to be potential PMF therapeutic agents, as they can target the dysregulation of the ECM via LOX inhibition. Including LOX inhibitors into ECM-derived tissue engineering may be useful in preventing potential osteo-fibrotic diseases, such as PMF.

The ECM ultimately provides two avenues for guiding myogenic cell behavior: ECM–material interactions for improved scaffold fabrication, as well as ECM–cell interactions for enhanced control of cellular outcomes. Understanding ECM proteins and their role in cellular communication is proving to be a crucial element in EMC-mediated muscle regeneration. New materials that incorporate native or engineered ECM, or that mimic key aspects of the ECM, are supporting unprecedented control over cell fate following cell–material interactions. As myogenic materials continue to improve in terms of cost, reproducibility, and immunogenicity prevention, we expect that their uses and relevance in both clinical and research settings will expand. We now shift our focus to cell–ECM–material interactions in cartilage tissue engineering.

### 2.2. Cartilage

Selected approaches utilizing cell–ECM–material interactions to regenerate cartilage are summarized in Table 2. Cartilage tissue engineering aims to slow, halt, or reverse the progression of osteoarthritis (OA), an inflammatory disease that impacts daily function, and it is likely to affect over 75 million Americans and over 300 million patients globally by 2030 [1,44]. Changes to the body's biochemical and mechanical environment initiate and drive OA. The ECM is critical in the context of OA, as alterations in ECM mechanics, particularly stiffness, have been shown to activate macrophages (MFs) [45,46]. Briefly, OA is associated with immune cell activation, including MFs [47], in response to stiffness changes in the ECM [48,49]. Further, ECM stiffening with alterations to the chondrocytes (cartilage cells) occurs in articular cartilage during OA [50–53]. In a recent proof-of-concept study, the ECM mechanical properties were tuned to reduce pro-inflammatory macrophage phenotypes, highlighting the therapeutic potential of keeping ECM mechanics within a desired stiffness range [46]. Given the lack of treatments that can restore normal articular function, once clinically-identifiable OA sets in, as well as the early promise of mechanics-based anti-inflammatory interventions in engineered cartilage ECM [45], and there is significant interest in engineering articular cartilage ECM with mechanical and biochemical properties that can treat or reverse OA.

Currently, invasive and noninvasive methods are used to treat early to moderate OA. Available treatments, including pharmaceuticals and physical therapy, do not adequately address the problem of tissue degeneration, but instead reduce pain and slow OA progression. However, existing treatments do not meaningfully improve the underlying clinical problem of tissue deterioration. A recently developed method that directly addresses cartilage damage is targeted cellular therapy. While its promising results are reviewed elsewhere [54], a key takeaway from approaches utilizing targeted cell delivery is that is that chondrocytes can undergo rapid dedifferentiation and lose their ability to produce hyaline cartilage and form fibrocartilage during in vitro culture, likely due to the lack of ECM-supplied spatial and mechanical cues. Engineered cartilage ECM that can provide chondrocytes with the necessary signals is an important and ongoing area within OA research.

Changes in the nanoscale structure of cartilage ECM lead to changes in macroscale tissue behavior that may initiate and propagate OA, but these changes can also potentially be deliberately induced to prevent or reverse OA [46,55]. The concept of harnessing the cartilage ECM mechanics to prevent macrophage-mediated initiation of OA has only recently been suggested as a proof-of-concept study that showed that certain stiffnesses could be protective against MF activation in inflammatory OA [45]. Generally, stiffer environments elicit a pro-inflammatory M1 phenotype, while softer environments induce polarization to a pro-remodeling M2 phenotype [53,56]. This relationship between MFs and ECM dynamics could be exploited towards reversing OA or developing novel treatments for MF-mediated OA. More studies are needed to understand the role small changes in the ECM mechanical environment plays in initiation and propagation of OA. Taken together, recent studies illustrate how the ECM can be engineered to enhance cell participation in articular cartilage tissue repair or otherwise support cell delivery [54].

### 2.2.1. Cartilage ECM as an Engineered Material

The ECM has outsized role in cartilage homeostasis and dysregulation. The specialized ECM found in cartilage is a crucial component of the tissue structure and function, accounting for 90% of the tissue's dry weight, a higher proportion than what is found in most other tissues, except for tendon and perhaps bone [57]. Chondrocytes account for only 2–5% of total cartilage tissue volume [58]. Due to the unique composition of its ECM, cartilage withstands the high loads experienced during human locomotion [59]. Numerous ECM-mediated environmental factors influence chondrocyte metabolic activities, including composition of the matrix and the secretion of soluble mediators. Growth factors and cytokines are secreted in response to mechanical loading, which is also transmitted to the cells via the matrix [59,60]. Since articular cartilage is avascular, the ECM plays a disproportionately large role in regulating chondrocyte metabolism compared to tissues with higher vascularization.

Cartilage ECM further supports the specialized mechanical properties of the tissues that make articular cartilage engineering a challenge [61]. Articular cartilage has a low coefficient of friction (COF), which is challenging to replicate in engineered constructs. Due to its influence on the COF of engineered cartilage [62], the ECM has emerged as a possible material for lowering the COF of in vitro cartilage constructs. Decellularized ECM extracted from the superficial zone cartilage of juvenile bovine femoral condyles increased the compressive and tensile stiffness of self-assembled articular cartilage constructs prepared from juvenile bovine stifle joints. Importantly, the addition of ECM decreased both the COF and glycosaminoglycan content, and it increased the collagen content of the constructs [63]. This approach demonstrated that the ECM may address the longstanding challenge of mimicking the COF of cartilage. As the COF has been shown to be higher in hydrogel-based biomaterials used to treat articular cartilage defects [64], other materials, such as those fabricated from ECM, may be more appropriate if COF is the main concern. Indeed, ECM scaffolds for cartilage regeneration have generated considerable interest and are reviewed thoroughly elsewhere [3]. More work is needed to understand how to maintain a native cartilage-like COF in vitro and in vivo, particularly over long periods of time. Engineered ECM may be an optimal approach for recreating the appropriate COF.

Human ECM from allogeneic sources [65] or cadaveric donors has also shown promise as an engineered material for cartilage regeneration. An acellular three-dimensional interconnected porous scaffold derived from human cartilage ECM improved cell adhesion, proliferation, and differentiation of bone marrow-derived mesenchymal stem cells (BMSCs) into chondrocyte-like cells over 21 days [66]. These chondrocyte-like BMSCs were transferred onto another cartilage ECM-derived porous scaffold and implanted subcutaneously into nude mice. Over four weeks, these scaffolds produced cartilage-like tissue with high levels of sulfated proteoglycans and collagen type II, illustrating that human ECM can support chondrocyte maturation [66]. Though human ECM is in short supply, small amounts may be sufficient to initiate the proliferation and maturation of chondrocytes needed to trigger endogenous regeneration of cartilage.

Moving forward, cartilage ECM may be purposefully generated from animal or human cells to avoid the need for a donor altogether [67]. A recent study tested decellularized human bone marrow mesenchymal stem cell-derived extracellular matrix (hBMSC-ECM) as a culture substrate for chondrocyte expansion in vitro, as well as a scaffold for chondrocyte-based cartilage repair [68]. Chondrocytes were grown on both tissue culture plastic (TCP) and decellularized hBMSC-ECM. Chondrocytes deposited on the hBMSC-ECM showed significantly increased proliferation rates and effectively maintained the chondrocyte phenotype when compared to the group cultured on TCP. Additionally, chondrocytes in the ECM group had improved chondrogenic differentiation profiles compared to the TCP group controls. To test the in vivo compatibility of the ECM, a three-dimensional chondrocyte impregnated hBMSC-ECM (cell/ECM) culture system was implanted into SCID mice. Fourteen days post-implantation, prominent cartilage formation was observed in the cell/ECM group. These findings suggest that the stem cell-ECM combinatorial approach is a promising avenue for in vitro and eventual in vivo cartilage regeneration.

In addition to serving as the main scaffold material, cartilage ECM can augment other commonly used scaffolds, including polyurethane. A notable advantage of recently developed waterborne polyurethane-ECM (WPU-ECM) scaffolds is that they can be three-dimensionally printed at temperatures suitable for cell survival [69]. These scaffolds successfully achieved hierarchical macro-microporous structures. Following ECM addition, WPU scaffolds had optimal porosity, hydrophilicity, and bioactivity. These scaffolds also enhanced cell distribution, adhesion, and proliferation compared to control WPU-only scaffolds. Most importantly, the WPU–ECM scaffold could facilitate the production of glycosaminoglycan (GAG) and collagen, as well as the upregulation of cartilage-specific genes, which is difficult to achieve in other cartilage tissue engineering platforms. In vivo studies in rabbits showed that WPU-ECM scaffold combined with a microfracture procedure successfully regenerated hyaline cartilage six months after implantation [69]. Overall, although ECM was not the main component of these scaffolds, the combination of ECM with WPU significantly increased scaffold functionality, creating a favorable microenvironment for cell adhesion, proliferation, differentiation, and ECM production. Given the difficulties in sourcing large quantities of ECM, the ability to functionalize readily available scaffold materials, such as polyurethane, with smaller quantities of ECM, while still achieving significant regeneration and repair of the target tissue, is highly desirable.

### 2.2.2. Cartilage ECM to Modulate Fibrosis

Finally, we examine recent progress in engineering the cartilage ECM to control the progression of fibrosis. OA progression results in increased deposition of fibrotic tissue throughout the joint. As cartilage tissue is primarily ECM, with few chondrocytes, blood vessels, lymph nodes, or nerves, fibrotic tissue overwhelms intrinsic repair responses. The quantity and density of the cartilage ECM further prevents normal healing due to the reduced diffusion of stem cells and nutrients through the tightly packed extracellular and pericellular spaces [70]. In response to the challenges that the ECM poses to healing, studies have attempted to treat fibrosis through the replacement or modification of the cartilage ECM. Drawing inspiration from studies outside of the musculoskeletal system that inhibited fibrotic YAP signaling by using collagenase to soften the ECM in uterine fibroids [71], pro-fibrotic pathways were successfully inhibited using the anti-fibrotic agent nintedanib to protect chondrocytes from fibrosis due to ECM degradation [72]. Other methods for repairing or preventing fibrosis in cartilage have incorporated MSCs, using implanted cells to increase the production of functional ECM and fill in cartilage lesions to replace fibrotic regions [73]. By recalibrating the balance between chondrocyte-generated, functional ECM and fibrotic ECM, such approaches seek to shield the underlying articular tissue from further damage and overloading. Despite their promise, techniques seeking to regenerate the ECM struggle to adequately direct migration of implanted cells, suggesting crucial signals are still missing. Future approaches should engineer the ECM to better enable targeted migration of implanted cells to the desired cartilage areas.

The mere presence of cartilage ECM in engineered constructs promotes further excretion of ECM from chondrocytes. This can be a useful tool in cartilage reparative techniques, as the implanted ECM stimulates the chondrocytes' reparative response. A survey of various scaffolds for cartilage reconstruction found that growing MSCs in extracted ECM scaffolds resulted in higher chondrocyte proliferation and cartilage ECM production compared to a pure collagen scaffold, and this was conducted without the need for extrinsic growth factors [74]. The chondrogenic effect of the matrix may be due to the natural ECM material's retention of GAGs, particularly chondroitin sulfate and aggrecan [74]. These proteins have been shown to induce chondrogenesis in in vitro cell populations [75]. As more approaches utilize scaffolds incorporating natural ECM components, improved control over chondrocyte behavior and targeted, chondrogenic regeneration will be achieved.

Another key aspect of cartilage ECM is the basement membrane, which is responsible for cellular adhesion to the articular interstitial matrix. The basement membrane supports cell signaling, mechanotransduction, and acts as a protective barrier for the cells. Cartilage basement membrane consists primarily of Collagen IV (Col IV), laminin, nidogen, and perlecan. The proportions of these components in cartilage vary due to aging and overall tissue health. Perlecan is of particular interest, as it is a developmental regulator of articular cartilage, binds with a variety of ligands in the ECM and pericellular matrix, and is involved in mechanotransduction and intracellular communication. While the understanding of how perlecan and other basement membrane components contribute to healthy articular cartilage is limited, a recent study found that adequate perlecan in both the basement membrane and pericellular matrix greatly enhanced cartilage repair and structural integrity of the tissue [76]. Perlecan increases around cartilage injury sites, but its role in articular cartilage repair mechanisms remains understudied. More work is needed to understand how this promising ECM component may support articular cartilage engineering.

Overall, substantial challenges remain in engineering cartilage ECM to prevent or reverse OA and other cartilage diseases. Promising future approaches build upon a strong and expanding foundation of ECM-material combinations that can recreate crucial mechanical and biochemical aspects of healthy and diseased articular cartilage. Unique approaches using bioinks [77,78] (Figure 2) and ECM from organisms outside the mammalian kingdom are also promising [79], adopting research directions that are "outside the box" of traditional tissue engineering strategies. For the first time, emerging techniques, such as three-dimensional printing and electro-writing for scaffold fabrication, can recreate some of the differences observed in natural cartilage layers [77]. An ongoing challenge will be engineering cartilage ECM to possess the unique properties of native tissue, namely, the mechanics and low COF. As new materials emerge and are incorporated into combinatorial scaffold-ECM approaches, musculoskeletal tissue engineering will progress towards functional regeneration of articular cartilage.

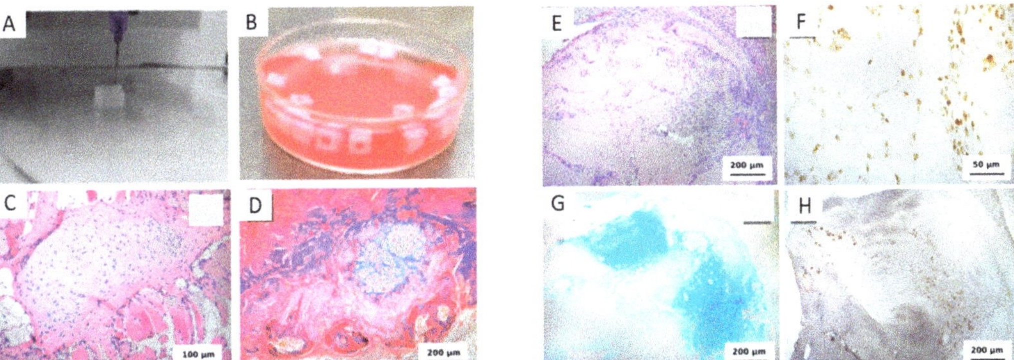

**Figure 2.** Recent approaches to regenerate articular cartilage have used cartilage ECM and ECM-component ink, in combination with cells. (**A**,**B**) The printing process uses high concentration collage and chondrocyte ink, and the results in scaffolds retain their three-dimensional shape (as shown in (**B**)). (**C**) Hematoxylin and eosin staining of the cartilage tissue that formed after 12 days in culture. Cartilage formation is observed within the striated muscle tissue (20× magnification). (**D**) Mason's trichrome staining shows bone tissue formation (in the lower section of the image; 10× magnification). (**E**) Five days after implantation, (**H**,**E**) staining shows cellularity within the scaffold (10× magnification). The scaffold in five days after the implantation. (**F**) PCNA-positive nuclei of chondroblasts and proliferating cells in the connective tissue capsule (40× magnification). (**G**) Glycosaminoglycans staining in blue shows cells are producing components of native cartilage matrix (10× magnification). (**H**) Staining for type II collagen (brown) shows additional components of the cartilage matrix are being produced by the metabolically active cells (10× magnification). Taken together, three-dimensional printing of cartilage with custom bioinks is a recent and viable strategy for cartilage tissue engineering studies. Adapted from Beketov et al. (2021) with permission under a Creative Commons CC BY 4 license [78].

*2.3. Tendon*

Selected approaches utilizing cell–ECM–material interactions to regenerate tendon are summarized in Table 3. A tendon is a collagenous musculoskeletal tissue that connects muscle to bone to enable normal movement and daily activity. The tendon is relatively acellular and comprised mainly of Collagen 1 (Col1) that is secreted by tenocytes (tendon cells) during early embryonic and post-natal development. A mature tendon has limited regenerative capacity. Damage to the collagenous tendon ECM is associated with long-term pathology and diminished daily function. Due to its low cellularity, the tendon heals poorly, and this is often accomplished via fibrotic scar formation that prevents normal mechanical function. The precise role of the ECM in tendon development, homeostasis, and disease is multifaceted and remains incompletely understood, but tendon ECM is known to support various mechanical and cellular functions [80]. A critical component of the tendon ECM is its hierarchical organization of Col1 fibers. Disruptions to the structure of these fibers are associated with tendon dysfunction and disease [81]; an important consideration when engineering the tendon ECM is the three-dimensional organization of the matrix components. Non-collagenous components of the tendon ECM must also be considered, as they have numerous functions in aging and disease [82,83].

A recent study addressed the need for hierarchical structure in engineered tendon by developing engineered and aligned autologous ECM scaffolds (Figure 3) [84]. Precision-designed templates were subcutaneously implanted in rats to generate decellularized autologous ECM from rats. The resulting scaffolds had highly aligned microchannels, which persisted after the templates and cellular components were removed. Once implanted, the scaffolds promoted rapid cell infiltration, modulated the MF response, supported collagen-rich ECM synthesis, and resulted in physiological tissue remodeling in Achilles tendon defects [84]. Three months post-operatively, the mechanical strength of the tenocyte-populated 'neo-tendons' was comparable to that of the pre-injury tendons. This study was one of the first to utilize subcutaneous ECM engineering, and it lays the groundwork for similar approaches in other animals and human patients, where autologous ECM might also be grown subcutaneously.

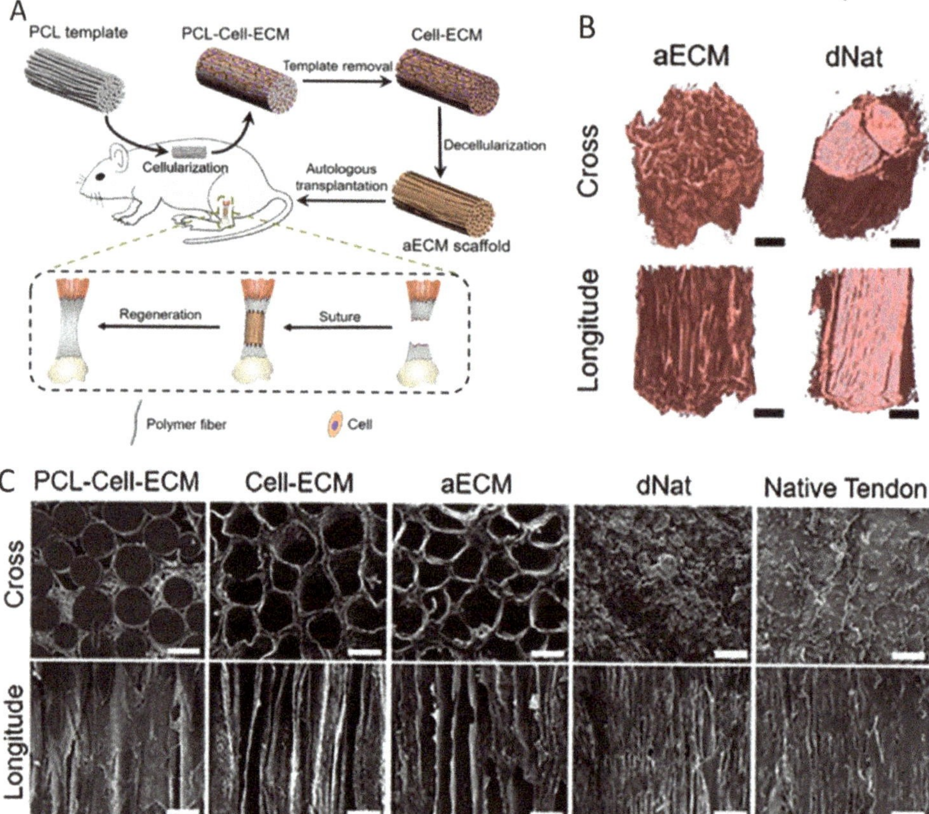


**Figure 3.** Polycaprolactone (PCL)–cell–ECM scaffolds support tendon regeneration. (**A**) Schematic of the scaffolds and implantation procedure. (**B**) Micro-CT scans of the transverse and longitudinal position in the aECM and dNat scaffolds. (**C**) PCL–cell–ECM, cell–ECM, aECM, decellularized native tendon (dNat) scaffold and native tendon microstructures were visualized using SEM cross- (top panels) and longitude- (bottom panels) sections. The figure was adapted with permission from Li et. al. 2019 [85].

Overuse injuries are common in tendons. Tendon overuse gradually degrades the ECM and alters its crucial mechanotransductive properties, causing pain and disability [85]. Mechanotransduction is the conversion of mechanical signals to cellular signals; damage to the ECM results in warped signals reaching the cells, in turn impacting the cellular response to the loads caused by daily movement. Continuous microdamage accumulation in the tendon ECM, combined with the tissue's poor regenerative capabilities, contributes to tendinopathy. Engineered tendon ECM must account for the wide variety of mechanical stimuli different tendons experience (for example, force-transferring Achilles tendons versus limb-positioning digital flexor and extensor tendons) and their reaction to these loading profiles [86,87]. Tendon tissue engineering is further complicated by the poor understanding of how distinct tendons may follow different developmental pathways, and they may vary in certain cellular processes, such as ECM homeostasis and remodeling [88]. While cell–EMC interactions are generally mediated by structures including cilia, integrins, connexins, and the cytoskeleton to support homeostasis, adaptation, or repair, it is possible, and perhaps likely, that these structures are not consistent across all tendon types. Attempts to engineer tendon ECM for use in regenerative studies must consider the unique functions of the target tendons. Such approaches have yet to be explored, but they are a promising future direction for tendon tissue engineering.

### 2.3.1. Engineered Tendon ECM to Understand Development and Disease

Mechanical force regulates tendon ECM organization, as well as tendon growth factor signaling through the transforming growth factor beta (TGFβ) family [89], indicating that tailoring the ECM to modulate the mechanical signals received by tendon cells is a crucial consideration of tendon tissue engineering. Work in zebrafish showed that, in addition to being transmitted via the ECM, mechanical forces can change the composition of the tendon ECM [89]. The feedback between tendon cells and ECM composition in response to mechanical loading supports the need to utilize ECM in tendon tissue engineering as a means of delivering desired signals to the cells. Multiscale mechanotransduction via the ECM has been proposed as a central focus of future tendon therapeutics [88].

An engineered tendon ECM can also be used to direct stem cell differentiation towards the tendon phenotype. The extracellular compartment is responsible for tendon morphogenesis during early development, in part due to the spatial cues it provides [90]. ECM elasticity [91] and proteoglycan content [92,93] have also been extensively implicated in tendon development and disease, but have yet to be considered as tunable and targetable components of engineered tendon ECM. As the tendon ECM is highly specialized [94], tendon tissue engineering may benefit by exploiting this specialization and incorporating specific ECM components in both in vitro and in vivo work. Future work that designs an ECM consisting of selected ECM components and other molecules found to be critical to tendon differentiation, such as cadherins and connexins [95], can elucidate the unknown cellular and mechanical processes that underlie tendon disease.

Beyond its mechanical properties, tendon ECM has been engineered to support tissue function via enhancement of its structural and compositional functions. Solubilized tendon ECM was recently used to induce tenogenesis (differentiation towards tendon) in human adipose-derived stem cells [96]. Extracted tendon ECM yielded consistent protein compositions that enhanced tenogenic gene expression in the cells via pathways related to ECM-associated processes. These findings support the continued use of solubilized ECM-based materials and approaches for enhancing musculoskeletal differentiation from stem cells. Such approaches may prove especially useful for induced pluripotent stem cells (iPSCs), in which consistent control of differentiation has proved challenging. ECM derived from specific tissues may provide cues that are otherwise difficult to incorporate within in vitro studies. In addition to providing specific matrix components, engineered ECM may also supply degradation cues, in response to which the cells can initiate remodeling. Such an approach may approximate the degradation–regeneration cycle seen with exercise [97]. Mimicking these cues in vitro via targeted ECM modifications can establish what amount of exercise is protective versus damaging to tendons in various stages of development or healing.

**Table 2.** Recent and noteworthy studies focused on cell–ECM–material interactions in cartilage.

| Tissue | Model | Material | Main Findings |
|---|---|---|---|
| Cartilage | Rat bone marrow-derived mesenchymal stem cells (rBMSCs) cultured with cryo-ground decellularized cartilage ECM. | Cryo-ground decellularized cartilage ECM. | Chemically decellularized cartilage (DCC) particles significantly outperformed TGF-β in chondroinduction of the rBMSCs. Collagen II gene expression was more than an order of magnitude greater compared to controls [74]. |
| | Porcine methacryl-modified solubilized and devitalized cartilage (MeSDVC) hydrogels. | Cryo-ground decellularized cartilage ECM methacrylated with glycidyl methacrylate (GM) and methacrylic anhydride (MA). | Methacrylation of the ECM increased printability of the MeSDVC hydrogels by creating paste-like consistency. Hydrogel stiffness increased to physiologically useful ranges [98]. |
| | BMSCs grown in dual-stage crosslinked hyaluronic acid-based bioink that was covalently linked to transforming growth factor-beta 1 (TGF-β1). | Hyaluronic acid (HA) bioink with covalently bonded TGF-β1. | Tethered TGF-β1 maintained functionality post three-dimensional printing and generated high quality cartilaginous tissues without exogenous growth factors [99]. |
| | BMSCs grown in porcine photocrosslinkable methacrylated cartilage ECM-based hydrogel bioink (cECM-MA). | Decellularized MA-methacrylated cartilage ECM bioink. | BMSCs were viable post-printing and underwent chondrogenesis in vitro, generating tissue rich in sulphated glycosaminoglycans and collagens [100]. |
| | Rat chondrocytes grown in genipin-crosslinked gelatin scaffolds with varying porosity. | Genipin-crosslinked gelatin scaffolds. | Chondrocytes proliferated and readily generated ECM with pore sizes of 250 and 500 μm [101]. |
| | hMSCs grown in tunicate exoskeleton-derived dECM. | Tunicate dECM. | Tunicate ECM was decellularized while retaining the honeycombed-shaped microstructure that improved metabolic activity, cell proliferation, and chondrogenic differentiation in hMSCs [79]. |
| | Rat chondrocytes grown in high concentration collagen bioprinted hydrogel scaffolds. | An amount of 4% collagen hydrogel bioink. | Subcutaneous implantation of the bioprinted scaffold resulted in cartilage-like tissue formation in rats as early as one week post implantation [78]. |
| | BMSCs grown in polyethylene glycol diacrylate (PEGDA) and ECM electro-written hydrogel. | High porosity PEDGA and porcine-derived ECM electro-written scaffold. | Electro-written PEDGA and ECM scaffold induced chondrogenesis and had anti-inflammatory effects [79]. |
| | Adipose-derived stem cells (ADSCs) grown in cartilage dECM and waterborne polyurethane (WPU) scaffolds, using low-temperature deposition manufacturing (LDM). | Cartilage dECM and WPU. | Hierarchical macro-microporous dECM- WPU scaffolds regenerated hyaline cartilage in a rabbit articular cartilage microfracture model [69]. |

**Table 2.** *Cont.*

| Tissue | Model | Material | Main Findings |
|---|---|---|---|
| | Mouse chondrocytes in human bone marrow-derived MSC-ECM (hBMSC). | hBMSC-ECM. | In vivo subcutaneous implantation of hBMSC-ECM scaffold in mice improved chondrocyte proliferation and development of a bioactive matrix [68]. |
| | Decellularized allogeneic hyaline cartilage graft (dLhCG) for porcine knee repair. | Decellularized pure hyaline-like cartilaginous ECM. | dLhCG resulted in superior efficacy in articular cartilage repair, surpassing living autologous chondrocyte-based cartilaginous engraftment repair methods [65]. |
| | Self-assembled articular cartilage constructs grown in bovine femoral condyle superficial zone cartilage ECM. | Bovine femoral condyle superficial zone cartilage ECM. | Extracted cartilage ECM reduced friction coefficients of the self-assembled articular cartilage constructs [63]. |

### 2.3.2. Engineered Tendon ECM as a Repair Material

As tendon ECM provides essentially all the mechanical functions of tendon tissue, replicating its structure and function is of particular importance to regenerative and tissue engineered approaches for treating tendon injuries. The acellular matrix has been explored as a means of augmenting rotator cuff repairs, where disruption to the gradated tendon-mineralized fibrocartilage-bone matrix is associated with complex recoveries and long-term loss of function. Human acellular dermal matrix (ADM) used as a patch was shown to be a viable option based on results in canine large rotator cuff defect models [24]. More recently, bovine decellularized tendon matrix (DTM) used in a rabbit model of Achilles tendon injury successfully prevented post-surgical adhesion, a common complication following surgical tendon repairs [98]. DTM acts as an anti-adhesion membrane, with the added benefit of being completely degraded after 12 weeks of subcutaneous implantation. This same study pioneered the use of tandem mass tag (TMT) labeling proteomics to analyze the protein compositions of native tendon, acellular tendon, and DTM, so as to quantitatively show the significant bioactivity and regenerative potential of the bovine DTM, and this is likely for other DTM-type materials. Recent advances in decellularization techniques, combined with micro-sectioning to preserve tissue viability and bioactivity with reduced reliance on chemical saturation, are expanding the pool of viable ECM sources. The emerging concept of using engineered ECM materials not only as a component of scaffolds, but also as a post-surgical treatment to augment existing repair techniques, will expand the potential applications of engineered ECM in the coming decades.

Another promising approach in tendon repair is the use of ECM to address post-operative fibrosis. Tendon surgery can lead to fibrosis of the affected or implanted tissue due to the graft adhering to surrounding native tissues. This adherence is the most common post-surgery complication in tendon repair [99], but decellularized ECM shows promise for preventing this outcome. A recent clinical trial showed that freeze-dried amniotic membrane could prevent adhesion following surgical repair of flexor tendons injured in zone II [99]. Adhesion was significantly lower in patients with the amnion membrane wrapped around the ends of the tendon compared to control patients who received implants made of poly-DL-lactic acid (PDLLA). Immune responses were also minimal in the amnion group, suggesting that this method may be used to prevent post-surgical tendon adherence in the future, as well as in larger tendons than the flexors of the hands. Another recent clinical trial also successfully used ADM during hand flexor reparative surgery as a tendon scaffold and tendon adhesion preventative [100]. Twelve months post-operatively, patients treated with the ADM had significantly reduced adhesion and increased tendon functionality compared to controls treated without the implant. A third trial using ADM found improved range of motion of the proximal and distal interphalangeal joints six months post operatively versus control groups treated conventionally [101]. Taken together, these and other studies show that matrix implants can aid in preventing adhesion and halting further tissue and joint destruction following invasive reparative surgeries. More work is needed to characterize the cellular interactions and behaviors that these implants facilitate within the tendons. Once the cellular pathways are better understood, new matrix-based materials may be developed that stimulate desired cellular behaviors and suppress unwanted outcomes, such as activation of the cellular pathways leading to fibrosis and adhesion.

**Table 3.** Recent and noteworthy studies focused on cell–ECM–material interactions in tendon.

| Tissue | Model | Material | Main Findings |
|--------|-------|----------|---------------|
| Tendon | Acellular dermal matrix (ADM) tendon scaffold affixed to hand flexor tendon post-operation. | Decellularized dermal ECM. | Addition of ADM post operation reduced tendon adhesion and improved long term functionality of the flexor tendon [100]. |
| | Decellularized bovine tendon ECM used as an anti-adhesion membrane. | Decellularized tendon matrix (DTM). | DTM improved tendon repair in rabbits by reducing adhesion and cellular proliferation, as well as improving healed tendon quality [98]. |
| | Human adipose-derived stem cells (hASCs) grown in urea-extracted bovine decellularized tendon matrix (DTM). | Urea-extracted decellularized tendon matrix (DTM). | Urea-extracted DTM increased hASC proliferation and tenogenic differentiation, and it also induced unique tenogenic gene expression profiles [96]. |
| | Rat tendon self-repair with implanted decellularized autologous extracellular matrix (aECM) scaffolds with highly aligned microchannels. | aECM scaffolds with aligned microchannels created through poly ($\varepsilon$-caprolactone) (PCL) microfiber bundle templates. | Subcutaneously implanted aECM scaffolds with aligned microchannels increased cellular infiltration and proliferation in the damaged tendon, resulting in improved restoration of rat tendon post-injury [84]. |
| | Human acellular dermal matrix graft for canine tendon repair. | Decellularized dermal ECM. | Within 12 weeks of implantation, the graft restored tendon functionality and mimicked autologous tendon both histologically and mechanically [24]. |

### 2.3.3. Engineering Cell Interactions with the Non-Collagenous Components of the Tendon ECM

It is worthwhile to consider the interactions between the cells and the non-collagenous components of tendons. While collagens and their hierarchical structural formations make up 70 to 90% of the total tissue volume in tendon, its less abundant components also provide crucial functions. These include proteoglycans, glycosaminoglycans, and non-collagenous proteins that are incompletely understood and poorly characterized, particularly with respect to their roles in mechanotransduction and their contributions to the mechanical properties of tendon [88,102]. These non-collagenous components are important for mediating cellular interactions with the matrix. The pericellular matrix directly surrounding tendon cells is comprised of mostly non-collagenous proteins that transfer mechanical stimuli to the cells in order to regulate mechanosensitive expression of genes [102]. To date, no approaches have directly engineered the makeup of the tendon ECM to quantify how specific cellular interactions with specific non-collagenous proteins affect tendon development, homeostasis, injury, or repair. There is significant potential for future approaches to manipulate the non-collagenous components of tendon ECM with the goal of directing specific cellular responses for tendon regeneration.

Engineered cell–ECM–material interactions may also benefit tendon wound repair. The tendon has limited healing capacity, but prior work has shown that biglycan and fibromodulin, two non-collagenous components of the tendon ECM, can enhance the wound healing capacity of tendon progenitor cells [103]. Recent work has shown that interactions with both the collagenous and non-collagenous components of the tendon ECM supports proliferation and wound healing properties of tendon stem/progenitor cells [104], while a new study also demonstrated the potential of the hippo-pathway downstream effector, yes-associated protein (Yap), as a signaling target using exosomes for targeted wound repair [105]. Notably, hippo and Yap are known to be mediated by mechanical and biochemical signals from the ECM. The results of this study indicate that exosomes stimulating these pathways could be a way to bypass the need for a healthy ECM to drive wound healing, in the case that there is extensive damage to the tendon ECM following injury or disease. More work must be performed regarding the use of exosomes in moderating and enhancing certain cell-ECM interactions.

Ultimately, though the tendon may be the least explored tissue in the context of engineered ECM approaches for regeneration, we see rapid advancements in the use of natural and engineered ECM and ECM-composite materials for uses in tendon tissue engineering. As ongoing research elucidates the impact of the ECM in tendon development, maintenance, and injury, engineered ECM-material approaches will provide useful interventions for improved tendon healing and regeneration. Tendon ECM is responsible for many of the tissues' mechanical and biochemical properties, and expanding the capabilities of engineered ECM to drive cell behavior is a critical and ongoing area of investigation.

### 2.4. Bone

Selected approaches utilizing cell–ECM–material interactions to regenerate bone are summarized in Table 4. Skeletal tissues serve important functions related to overall health, most notably providing the structure and protection necessary for the internal organs to perform their vital functions. Bone ECM has additional roles as a reservoir of calcium and other inorganic ions [106]. The cells housed within the bone matrix are major and active regulators of calcium homeostasis, and they secrete important hormones in response to cues from the ECM [106]. Bone is a regenerative tissue that is continuously remodeled throughout the lifespan by the osteoblasts on the bone surface that deposit new matrices, and the osteoclasts resorb the matrices. Osteocytes, the terminally differentiated cells embedded within the bone, maintain bone homeostasis and orchestrate many endocrine and paracrine functions [107]. Much of the interest in the bone matrix is driven by the search for improved treatments for osteoporosis, a bone-mass loss and demineralization disease that impacts hundreds of millions of patients globally [108]. Osteoporosis, more common in women over 50 due to menopause, reducing bone density, leads to decreased bone mass that can result in frequent fractures, particularly of the spine, arms, and hips [109]. Understanding how the bone ECM contributes to disease prevention, progression, and potential prognosis [110] is an urgent clinical need. To this end, several approaches have used bone-derived ECM scaffolds to understand how the bone matrix impacts cell behavior and bone regeneration.

### 2.4.1. Engineered Bone ECM-Mimicking Scaffolds

The well-defined composition and structure of bone ECM facilitates biofabrication of bone ECM scaffolds for uses in tissue engineering. Mimicking the porosity, permeability, and tortuosity of bone ECM is relatively straightforward and may enhance cell migration, as show in a recent computational study [111]. Another study integrated distinct biofabrication strategies to develop a multiscale porous scaffold that was both mechanically functional at the time of implantation, and that facilitated rapid vascularization, while providing stem cells with appropriate cues to enhance differentiation into osteoblasts [112]. A polycaprolactone (PCL) scaffold was integrated with decellularized bone ECM, producing osteoinductive filaments for three-dimensional printing. The addition of bone ECM to the PCL not only increased the mechanical properties of the resulting scaffold, but it also improved cellular attachment and enhanced osteogenesis of mouse mesenchymal stem cells (MSCs). The use of three-dimensional printing, particularly for bone ECM with its stable, interconnected, and somewhat predictable structure, will likely feature prominently in ongoing studies seeking to enhance cellular bone deposition.

Oriented ECM scaffolds have also been fabricated to imitate the material and structural properties of the natural bone growth plate, utilizing bone marrow stromal cells (BMSCs) to prevent bone bridge formation and growth plate injuries [113]. Growth plate injuries can disrupt normal bone development and lead to irregularities in bone length and shape after healing [114]. Limited treatments exist for growth plate injuries, especially if intervention is not rapid [115,116]. A recent study induced a growth plate injury in rabbits. Animals were treated with engineered, oriented ECM scaffolds and autogenous BMSCs, ECM scaffolds only, or injured but not treated with a scaffold or cells. At 16 weeks, the tibial defects (simulated growth plate injuries) treated with the ECM scaffolds and cells were filled with a neogenetic growth plate. Rabbits treated with only the ECM scaffolds still had physical defects, but the growth plate displayed some closure, with bone trabeculae and fibrous tissue growth. Controls showed some closure of the growth plate. Overall, BMSCs successfully adhered to and distributed within in the oriented scaffold in the group treated with both the ECM scaffold and the cells [117]. Radiological testing further showed that the scaffold and cell treatment decreased angular deformities and length discrepancy of the tibia when compared to other groups. The addition of BMSCs within the ECM scaffolds also promoted the regeneration of neogenetic chondrocytes during the repair of the injured growth plates and prevented the formation of bone bridges. Although total prevention of angular deformities and length discrepancies was not achieved, ECM-derived growth plate scaffolds combined with BMSCs have exciting potential applications in growth plate repair.

ECM from non-musculoskeletal tissues has also been successfully used to induce in situ bone regeneration. Porous polycaprolactone (PCL)/decellularized small interesting submucosa (SIS) scaffolds were fabricated using cryogenic free-form extrusion, followed by surface modification with aptamer and $PlGF-21_{23-144}$peptide-fused bone morphogenetic protein 2 (pBMP2) [118]. Rats were used to model a critically sized calvarial defects, into which a scaffold was implanted. At four- and eight-weeks post-operatively, defects implanted with the PCL/SIS-BMP2-Apt and PCL/SIS-pBMP2-APT scaffolds had substantial mineralized tissue, which was not seen in defects implanted with PCL/SIS and PCL/SIS-Apt groups. Micro-CT showed that the bone volume/tissue volume percentage (BV/TV%) and bone mineral density (BMD) values for defects implanted with PCL/SIS-Apt were significantly higher than those in the control group, suggesting that scaffolds with aptamers stimulated bone regeneration in vivo [119]. At eight weeks post-operatively, the bones in the injury sites with the PCL/SIS-pBMP2-Apt scaffold had completely bridged the defect. Bone formation was not seen in the PCL/SIS group, where only connective tissue formed. Rats implanted with the PCL/SIS-pBMP2-Apt scaffolds also showed evidence of angiogenesis. The combined use of BMP2 and Apt-19 (Aptamer 19) within a biomimetic PCL/SIS scaffold has promising applications in bone regeneration, despite the ECM source (the SIS) being outside the musculoskeletal system. Using SIS in musculoskeletal regenerations may be advantageous compared to other types of ECM, as large quantities of SIS are readily available.

Another promising scaffold for bone regeneration was recently developed using polylactic glycolic acid, bone ECM, and magnesium hydroxide. Magnesium is an emerging scaffold material that resists the loading from human activities and has potential uses as an orthopedic implant due to its porosity and robustness [120]. To mediate the negative effects associated with PLGA (i.e., low mechanical properties and acidic byproducts), a porous PLGA (P) scaffold was created and combined with magnesium hydroxide,

(MH, M) bone-extracellular matrix (bECM, E), and polydeoxyribonucleotide (PDRN, P) to improve anti-inflammatory ability, osteoconductivity, and introduce pro-osteogenic and pro-angiogenic effects, as well as osteoclast inhibition [121]. PLGA (control), PME, and PMEP groups were assessed for biocompatibility and cellular activity. PME and PMEP groups had significantly increased biocompatibility compared to the PLGA group, as both scaffolds increased the population of calcein AM positive human bone-marrow mesenchymal stem cells (hBMSCs), i.e., live cells, at one, three, and seven days post-implantation [117]. The angiogenic properties of the PDRN were confirmed using human umbilical vein endothelial cells (HUVECs) treated with PDRN, since PDRN treatment formed a significant number of branch points and longer vessel lengths in the HUVECs. Quantitative real-time PCR (qRT-PCR) determined the that the PME scaffold reduced inflammation markers in the 3D hBMSC scaffolds at seven and twenty-one days, since expression of interleukin-6 (IL-6) and interleukin 1β (IL-1β) were lower compared to the PLGA scaffolds. The PMEP scaffold further reduced inflammatory markers when compared to PME scaffolds. Osteogenic capacity was also highest in the PMEP scaffolds. These findings show that mMH has the potential to both enhance the mechanical properties and neutralize acidification of PLGA scaffolds, with widespread implications for bone tissue engineering, as PLGA has potential to be an excellent material for bone regeneration if its mechanical properties and acidification are improved. bECM improved osteogenesis by effectively providing natural calcium and phosphate. The vascularization driven by the PDRN underscores the potential applications of these engineered, ECM–cell–material-based scaffolds.

### 2.4.2. Bone Scaffold Functionalization with ECM

In addition to serving as a scaffold material, ECM can functionalize existing scaffolds to enhance osteogenic potential for bone regeneration applications. As a recent example, ECM was functionalized onto the surface of multi-channel biphasic calcium phosphate granules (MCG) to increase osteogenesis and regeneration. The collagen, sulfated glycosaminoglycan, and trace amounts of growth factors, such as BMP2, vascular endothelial growth factor (VEGF), transforming growth factor TGFβ, and fibroblast growth factor (FGF), were found in the ECM supported proliferation of MC3T3-E1 cells over seven days [118]. Protein adsorption and osteogenic properties were improved on ECM functionalized MCG scaffolds compared to controls, and ECM functionalized scaffolds enhanced bone regeneration in a rabbit model of a femoral head defect. While this particular study functionalized MCG scaffolds only, ECM can be formatted into powders, gels, sponges, bioinks, and other formats to improve existing biomaterial scaffold systems used for musculoskeletal regeneration.

In a unique formatting approach, bone ECM was configured into demineralized bone paper (DBP) as a material to direct osteoblasts to deposit structural mineralized bone tissue [119]. DBP effectively stimulated the trabecular osteoid, directed rapid, and structural mineralization by osteoblasts, and it contained the microenvironment necessary to support bone remodeling. Notably, this study focused on the trabecular bone niche, rather than the cortical bone niche explored in many other approaches, an important detail in the context of diseases, such as osteoporosis that severely impact the trabecular bone of the vertebrae [108]. Compared to control cells cultured in TCP, which is commonly used in osteogenic in vitro experiments, cells grown in the DBP displayed significantly higher mineralization, collagen alignment, and elongated morphology that was also aligned with the underlying lamellar structure of the demineralized bone [119] (Figure 4). As one of the first studies to create a bone organoid for understanding the mineralization of trabecular bone, this work lays the foundation for future inquiries using a new generation of in vitro bone models.

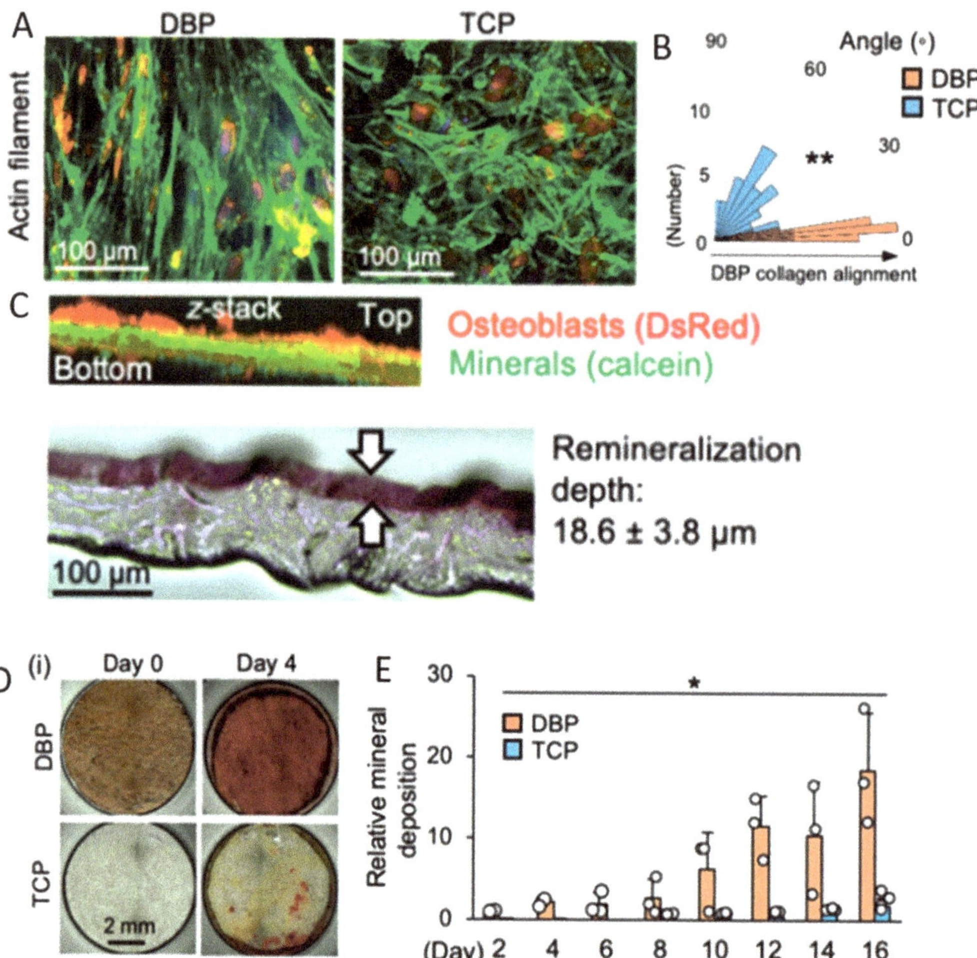

**Figure 4.** Osteoblasts grown on demineralized bone paper (DBP) display enhanced regenerative capacities compared to osteoblasts grown on tissue culture polystyrene (TCP). (**A**) Morphology of OBs grown on vertically sectioned DBP and TCP for one week. (**B**) Immunofluorescent staining of actin filaments. (**C**) Circular histogram of cell alignment angles ($n = 100$). (**C**) Top: $z$-staked cross-sectional image of DBP will osteoblasts growing. Bottom: Cross section of 100-μm-thick DBP stained with alizarin red after three-week culture of osteoblasts ($n = 3$). (**D**) Mineral deposition by osteoblasts on DBP and TCP, showing alizarin red mineral stain on days zero and four. (**E**) Time-course measurement of mineral deposition by osteoblasts for 16 days ($n = 3$). * indicates $p$-value $< 0.05$; ** indicates $p$-value $< 0.01$. Figure adapted with permission under a Creative Commons Open Access License from Park et al., 2021 [120].

Other notable in vitro bone models work towards elucidating the mechanisms necessary for proper bone–ECM interactions (alignment and osteogenic differentiation). It was recently established that many ECM-associated proteins contain thrombospondin type 1 repeat (TSR) motifs that play critical roles in cellular processes, such as cell–cell attachment, extracellular matrix (ECM) remodeling, cell proliferation, and apoptosis [120]. TSRs are modified with O-linked fucose that is added in the endoplasmic reticulum to folded TSRs by the enzyme protein O-fucosyltransferase-2 (POFUT2). POFUT2 likely promotes efficient trafficking of substrates. Mouse embryos that lack POFUT2 die early in development, emphasizing the importance of the modification for TSR-protein function. Using Prrx1-Cre recombinase, a recent study investigated the impact of POFUT2 knockout on the secretion of POFUT2 substrates and on ECM properties in vivo. Loss of POFUT2 in the limb mesenchyme caused significant shortening of the limbs, long bones, and tendons, and it stiffened the joints, resembling the deformities seen in musculoskeletal dysplasias in humans and mice with ADAMTS or ADAMTSL mutations [120]. In the mouse model, limb shortening was evident by day 14.5, and loss of O-fucosylation led to an accumulation of fibrillin 2 (FBN2), decreased BMP and IHH signaling, and increased TGF-β signaling. The hypertrophic zone decreased in size, with lower levels of collagen X (ColX). Surprisingly, there were minimal effects of the POFUT2 knockout on secretion of two POFUT2 substrates, CCN2 or ADAMTS17, in the developing bone. Conversely, CCN2, ADAMTS6, and ADAMTS10 (POFUT2 substrates important for bone development) had decreased secretion from POFUT2-null HEK293T cells in vitro. Overall, this study suggests that there are cell-type specific responses to the POFUT2 mutation, and O-fucose modification on TSRs may stretch further than the promotion-efficient trafficking of POFUT2 substrates, with the potential to influence their function in the extracellular environment [120]. Further studies that characterize the mechanisms behind O-fucose modification on TSRs may provide valuable insight to how ECM-associated protein modifications affect bone tissue regeneration.

*2.5. Tuning ECM Mechanical Signals for Bone Tissue Engineering*

While this review has already discussed the widely accepted premise that ECM stiffness is important in the regulation of stem cell differentiation, how cells sense stiffness cues and adapt their metabolic activities remains less understood. Recent research explores the role of mitochondrial phosphoenolpyruvate carboxykinase (PCK2) in enhancing osteogenesis in three-dimensional ECM via glycolysis in the context of stem cell osteogenesis [121]. The three-dimensional trabeculae network of normal and osteoporotic bone with different microstructure and stiffness were mimicked via encapsulation of bone marrow derived mesenchymal stem cells (BMMSCs) in methacrylate gelatin (GelMA) hydrogels. The osteoporotic mice models were established by ovariectomy and compared with sham controls, where three-dimensional BMMSCs/GelMA complexes were inserted into surgically induced femoral defects. PCK2 promoted osteogenesis in three-dimensional ECM with tunable stiffness in vitro and in vivo [121]. PCK2 enhanced the rate-limiting metabolic enzyme pallet isoform phosphofructokinase (PFKP) in three-dimensional ECM, and it further activated the AKT/extracellular signal-regulated kinase 1/2 (ERK1/2) cascade, which directly regulates osteogenic differentiation of MSCs. The newfound complexity of the crosstalk between cell mechanics and metabolism may inform ongoing approaches seeking to harness kinase cascades for bone regeneration. Elucidating such mechanisms provides a unique direction for ECM-mediated bone tissue regeneration, as well as new therapeutic strategies for osteoporosis.

A similar study used a mouse model of trauma-induced heterotopic ossification (HO) to examine how cell-extrinsic forces impact mesenchymal progenitor cell (MPC) fate. Mice were given a partial thickness scald burn injury to produce a burn/tenotomy HO model. Post injury, mice had their hind limbs immobilized. Single-cell (sc) RNA sequencing of the injury site revealed an early increase in MPC genes associated with pathways of cell adhesion and ECM-receptor interactions, as well as MPC trajectories to cartilage and bone [122]. Interestingly, active mechanotransduction was observed after the injury, with increased focal adhesion kinase signaling and nuclear translocation of transcriptional coactivator TAZ. TAZ inhibition mitigates HO, but joint immobilization was shown to decrease mechanotransductive signaling, which completely inhibited HO [122]. As HO is a debilitating complication following both soft and hard musculoskeletal tissue injury, there is significant interest in inhibiting HO without also inducing permanent bone loss or adipogenesis within musculoskeletal tissues. This study further identified decreases in collagen alignment and increases in adipogenesis as a result of immobilization, highlighting the need for a better understanding of the delicate balance between HO and adipogenesis following bone injury. Overall, this study suggests that joint immobilization after injury causes decreased ECM alignment, altered MPC mechanotransduction, and changes in genomic architecture, favoring adipogenesis over osteogenesis. Understanding the importance of mobilization for correct fiber alignment in the ECM is important for both bone tissue regeneration and the long-term management of other musculoskeletal tissues injuries.

Faulty skeletal muscle repair can also result in HO. A recent study used fibrodysplasia ossificans progressiva (FOP), a disease of progressive HO caused by an activin receptor type-1 (ACVR1) mutation, to elucidate how ACVR1 affects skeletal muscle repair [123]. Primary FOP human muscle stem cells (Hu-MuSCs) isolated from cadaveric skeletal muscle had increased ECM marker expression, as well as skeletal muscle-specific impaired engraftment and regeneration ability. Human iPSC-derived muscle stem/progenitor cells (iMPCs) single-cell transcriptome analyses from FOP showed uncharacteristically increased ECM and osteogenic marker expression compared to control iMPCs [124]. These findings suggest that ACVR1 activation in iMPCs or Hu-MuSCs may contribute to HO by changing the local tissue environment, and blocking this activation can prevent HO in ECM and non-ECM based regenerative approaches. Alternatively, genetically modified cells with dysfunctional ACVR1 could be useful starting platforms for growing scaled-up muscle tissues in vitro that can be subsequently implanted into VML defects.

Bone ECM stiffness has also been implicated in pathologies of the spine. Intervertebral disc degeneration (IDD) is thought to be initiated by the mechanical stimulation provided by the ECM stiffness. ECM defines the mechanical microenvironment of the nucleus pulposus (NP). The mechanosensitive ion channel Piezo1 mediates mechanical transduction within the disc, and a recent study explored the function of Piezo1 in human NP cells subjected stiffness levels comparable to those found in the ECM. The expression of Piezo1 and the ECM elasticity modulus increased in degenerative NP tissues, and stiff ECM activated the Piezo1 channel and increased intracellular $Ca^{2+}$ levels [124]. Increased intracellular reactive oxygen species (ROS) levels and factors that contribute to endoplasmic reticulum (ER) stress were also observed as a result of Piezo1 activation. Additionally, oxidative stress-induced senescence and apoptosis in human NP cells were correlated with stiffer ECM [125]. By inhibiting Piezo1, senescence and apoptosis were alleviated in the stiff ECM environment. Correlatively, Piezo1 silencing in vivo improved the condition of IDD and decreased the elastic modulus of rat NP tissues. Manipulating Piezo1 may be beneficial in mediating the effects of stiffening ECM that could occur with ECM mediated tissue regeneration techniques.

Moving forward, as the use of scaffolds containing cell-derived ECM continues to be a viable strategy for mediating bone regeneration, a mechanistic understanding of the benefits of ECM compared to other scaffolding techniques is needed. A recent study used ECM-loaded three-dimensional printed gelatin (Gel), sodium alginate (SA), and 58s bioglass (58sBG) gels seeded with either rat aortic endothelial cells (RAOECs) or rat bone mesenchymal stem cells (RBMSCs) were used to explore the mechanisms underlying the advantages of ECM-containing scaffolds. In vitro osteogenic differentiation results showed that scaffolds coated with ECM significantly increased the expression of osteogenic and angiogenic genes compared to uncoated control scaffolds, likely due to the presence of native ECM proteins [125]. In vivo experimentation showed that ECM-loaded scaffolds implanted into mandibular defects effectively promoted bone healing compared to non-ECM scaffolds. Both RAOECs–ECM and RBMSCs–ECM scaffolds greatly enhanced bone formation as a result of multiple factors, namely, the increased expression of RUNX2, OCN, BMP2, CD31, and VEGF [125]. More studies are needed that exploit the ECM specifically to enhance expression of these factors in an osteogenic manner, given they may be the mechanism behind the improved cellular outcomes seen with ECM scaffolds.

Looking towards the future, with large proportions of the population aging in the United States, China, Japan, and globally, the clinical need for regenerative bone therapies will continue to grow exponentially. We have highlighted unique strategies for exploiting the existing cellular ability to remineralize and regenerate bone, without the need for invasive surgical grafts or drug combinations with side effects that reduce the patients' quality of life. As research continues to elucidate the mechanisms by which bone ECM supports and regulates bone function, paracrine and endocrine signaling, and homeostasis, engineered ECM approaches will provide novel and exciting platforms for regenerative bone therapies.

**Table 4.** Recent and noteworthy studies focused on cell–ECM–material interactions in bone.

| Tissue | Model | Material | Main Findings |
|---|---|---|---|
| Bone | Polycaprolactone (PCL) scaffold integrated with decellularized bone ECM seeded with mouse mesenchymal stem cells (MSCs). | Polycaprolactone (PCL) scaffold integrated with decellularized bone ECM. | The addition of bone ECM to the PCL increased the mechanical properties of the resulting scaffold, increased cellular attachment, and enhanced osteogenesis of mouse mesenchymal stem cells (MSCs) [112]. |
| | Growth plate injury was induced in rabbits and treated with engineered oriented ECM scaffolds and autogenous BMSCs, ECM scaffolds only, or injured but not treated with a scaffold or cells. | Engineered oriented ECM scaffolds. | BMSCs successfully adhered to and distributed within the oriented scaffold in the group treated with both the ECM scaffold and the cells. The ECM scaffold and BMSCs generated functional tissue-engineered cartilage superior to the other groups, and the scaffold and cell treatment decreased angular deformities and length discrepancy of the tibia when compared to other groups. Addition of BMSCs within the ECM scaffolds promoted regeneration of neogenetic chondrocytes during the repair of the injured growth plates and prevented the formation of bone bridges [113]. |
| | Critical-sized calvarial defect in a rat model; porous polycaprolactone (PCL)/decellularized small interesting submucosa (SIS) scaffolds injected into defect. Scaffolds were fabricated using cryogenic free-form extrusion and surface modification with aptamer and PlGF-2123-144 peptide-fused bone morphogenetic protein 2 (pBMP2). | Porous polycaprolactone (PCL)/decellularized small interesting submucosa (SIS) scaffolds. | Four- and eight-weeks post-op, defects implanted with the PCL/SIS-BMP2-Apt and PCL/SIS-pBMP2-APT scaffolds had substantial mineralized tissue not seen in defects implanted with PCL/SIS and PCL/SIS-Apt groups. Significantly higher bone volume/tissue volume percentage and bone mineral density for defects implanted with PCL/SIS-Apt compared to controls. Eight weeks post-op, the bones in the injury site with PCL/SIS-pBMP2-Apt scaffold had completely bridged the defect, and angiogenesis occurred in rats implanted with the PCL/SIS-pBMP2-Apt scaffolds [116]. |
| | Porous PLGA (P) scaffold combined with magnesium hydroxide (MH, M), bone-extracellular matrix (bECM, E), and polydeoxyribonucleotide (PDRN, P). | Polylactic glycolic acid, magnesium hydroxide, and bone ECM. | PME and PMEP groups displayed significantly increased biocompatibility compared to the PLGA group, and both scaffolds had an increased population of calcein-AM positive human bone-marrow mesenchymal stem cells (hBMSCs), i.e., live cells at one, three, and seven days post-implantation [117]. |

**Table 4.** *Cont.*

| Tissue | Model | Material | Main Findings |
|---|---|---|---|
|  | ECM functionalized onto the surface of multi-channel biphasic calcium phosphate granules (MCG), seeded with MC3T3-E1 cells, and implanted into a rabbit femoral head defect model. | ECM functionalized onto the surface of multi-channel biphasic calcium phosphate granules (MCG). | Protein adsorption and osteogenic properties were improved on ECM functionalized MCG scaffolds compared to controls. ECM functionalized scaffolds enhanced bone regeneration in a rabbit model of a femoral head defect [118]. |
|  | Bone ECM was configured into demineralized bone paper (DBP) as a material to direct osteoblasts to deposit structural mineralized bone tissue, and it was seeded with osteoblasts from DsRed reporter mice. | ECM configured into demineralized bone paper (DBP). | DBP effectively stimulated the trabecular osteoid, directed rapid and structural mineralization by osteoblasts, and contained the microenvironment necessary to support bone remodeling. Compared to control cells cultured on TCP, cells grown in the DBP displayed significantly higher mineralization, collagen alignment, and elongated morphology that was aligned with the underlying lamellar structure of the demineralized bone [119]. |
|  | Osteoporotic mice models (established by ovariectomy) with three-dimensional complexes of encapsulated bone marrow derived mesenchymal stem cells (BMMSCs) in methacrylate gelatin (GelMA) hydrogels, and they inserted it into surgically induced femoral defects. | Methacrylate gelatin (GelMA) hydrogels. | Mitochondrial phosphoenolpyruvate carboxykinase (PCK2) promoted osteogenesis in three-dimensional ECM with tunable stiffness in vitro and in vivo. PCK2 enhanced the rate-limiting metabolic enzyme pallet isoform phosphofructokinase (PFKP) in three-dimensional ECM, and it further activated AKT/extracellular signal-regulated kinase 1/2 (ERK1/2) cascades to regulate osteogenic differentiation of MSCs [121]. |
|  | ECM-loaded three-dimensional printed gelatin (Gel), sodium alginate (SA), and 58s bioglass (58sBG) gels were seeded with either rat aortic endothelial cells (RAOECs) or rat bone mesenchymal stem cells (RBMSCs), and they were implanted into rat mandibular defects. | ECM-loaded three-dimensional printed gelatin (Gel), sodium alginate (SA), and 58s bioglass (58sBG) gels. | Scaffolds coated with ECM significantly increased the expression of osteogenic and angiogenic genes. ECM-scaffolds promoted bone defect healing in vivo compared to the pure scaffold. RAOECs–ECM scaffolds and RBMSCs–ECM scaffolds enhanced bone formation, likely via increased expression of RUNX2, OCN, BMP2, CD31, and VEGF [125]. |

## 3. Future Directions

Enhanced understanding of micro- and nanoscale mechanical properties of engineered ECM materials are considered here.

Significant research has already been conducted to understand the mechanics and topography of ECM, as well as how these properties relate to cellular changes at a macroscopic level [126–129]. There is much opportunity to improve understanding of the ECM at the microscopic and nanoscale levels, which are the scales encountered by developing, differentiating, injured, and healing cells. Assessing these properties at the micro- and nanolength scales is critical, as several types of ECM exhibit mesh-like nanoscale structures with fiber diameter and pore size governing the topographical landscape [129]. Although there is conflicting, tissue-dependent data, most research has concluded that ECM scaffolding with pore sizes between 100 and 500 μm is optimal for cellular proliferation [130,131]. The fiber diameters in ECM specifically can range from 2 nm (fibronectin) to 500 nm (collagen) [132].

In order for the engineering of both natural and synthetic scaffolds to progress, the study of in vivo ECM at the micro and nanoscale levels is crucial for the accurate replication of feature height, porosity, fiber diameter, and mechanical properties of native ECM in engineered scaffolds and other biomimetic materials.

Atomic force microscopy (AFM) is a promising imaging modality for the study of the topography, mechanical properties, and other ECM characteristics at the micro- and nanoscopic levels. AFM is capable of precisely resolving the nano-topography of diverse surfaces and mapping the spatial distribution of their physicochemical properties to include charge density and potential in a variety of hard and soft materials, including hydrogels [133,134]. This knowledge is invaluable, as it associates surface charge with protein and cellular adhesion and interactions [135–137]. Furthermore, cells cultured in hydrogels frequently remodel their local environment, especially when immune cells are present as is the case for many disease model co-cultures, and these localized mechanical alterations are lost in macroscale mechanical testing [138].

AFM can quantify the stiffness of natural ECM and synthesized ECM scaffolds, which allows for correlations between the nanomechanical characteristics of surfaces and their impact on cellular interactions and behaviors. Cell behavior can then be compared to what has been observed in the context of the material's bulk characteristics. AFM also offers a method of mechanical testing that is less destructive than standard techniques (i.e., tensile testing and microhardness). AFM can probe multiple mechanical properties simultaneously to produce accurate measurements that include stiffness, maximum indentation, indentation at maximum force, hysteresis, hardness, compliance, adhesion force, detachment distance, detachment energy, and Young's modulus. As precise attributes of the ECM are engineered to elicit specific responses from cells or create individualized patient disease models, techniques, such as AFM, may prove invaluable in determining the exact properties of various ECM scaffolds.

While AFM stands as an incredibly powerful tool to analyze ECM at the microscopic and nanoscopic level, several experimental limitations remain. The long imaging time required for a high-resolution image, potential damages exerted by the probe tip, imaging artifacts, and quantitative interpretation of the results have constrained AFM's utility in tissue engineering research. However, AFM has already been successfully used before to obtain quality concerning engineered graft surfaces [139]. This same study showed that loading type significantly influenced mechanical and histological outcomes of engineered cartilage surfaces, particularly in regards to the coefficient of friction at multiple length scales. This type of AFM-based analysis can indicate which types of loading may favor regeneration versus maintenance versus injury in cartilage [139]. It is within reason to expect similar studies in other musculoskeletal tissues to yield equally insightful results. As the demand for highly accurate disease models increases, we foresee increased use for AFM and similar techniques that can offer insight into the multiscale mechanical properties of engineered scaffolds, particularly those that incorporate biological components, such as ECM.

### 3.1. Genetically Engineered ECM and ECM-Based Bioinks

In addition to using highly sensitive mechanical testing, such as AFM, to inform engineered ECM and biomaterial scaffold design, we expect the role of genetic engineering to expand into the fabrication of custom, genetically-modified ECM for use in biomaterial scaffolds. As highlighted extensively in this review, a critical role of ECM is mechanotransduction. It stands to reason that engineered ECM from genetically modified cells/organisms that contains a specific array of mechanotransduction proteins may be a useful and desirable component of newly developed scaffolds. Techniques, such as CRISPR-based genome editing, may support the fabrication of custom ECMs that can replicate features of diseases that, until now, have been difficult or impossible to mimic in vitro. Alongside genetic engineering, novel composite materials, such as ECM-silk and ECM-titanium constructs, and ECM-based bioinks [78,80,101,140], may provide platforms with previously unachievable mechanical, biochemical, and genetic properties for use in regenerative therapies. Bioinks in particular are seeing increases in popularity and potential applications.

Until recently, most printers were unable to produce the slightly heterogeneous morphologies observed in native tissues. Recent gains in printer accuracy, porosity, and random structure generation are supporting the fabrication of biomaterials that mimic the native structure of small soft tissues, including musculoskeletal tissues. While ink availability has limited the application of this technology, new inks are rapidly emerging and enable multilayer printing while also incorporating growth factors, such as TGFβs and BMPs [100,141]. Advances in the rheological properties and printability of ECM-based materials via addition of methacrylation reactions or components continue to improve the usability of custom and commercial bioinks targeted for musculoskeletal tissue engineering applications [98]. Incorporating ECM into these inks may provide the missing cues needed to drive in situ tissue regeneration. These printed ECM and growth factor-augmented materials may seamlessly integrate with the patient's own tissues and be further modifiable post-implantation, degradable via the body's immune cells [142], and highly bio- and cytocompatible. We expect a prominent role for ECM as a component of the bioinks and biomaterials of the future.

### 3.2. Engineered ECM for Understanding and Treating Musculoskeletal Fibrosis

An underexplored application of engineered ECM is in understanding fibrotic disease, both in regards to musculoskeletal tissues and in the rest of the body. Significant interest in controlling ECM mechanics, and particularly ECM stiffening, has emerged in research related to idiopathic pulmonary fibrosis (IPF). IPF is a progressive scarring disease characterized by extracellular matrix accumulation and altered mechanical properties of lung tissue. Recent studies support the hypothesis that these compositional and mechanical changes create a progressive feed-forward loop, in which enhanced matrix deposition and tissue stiffening contribute to fibroblast and myofibroblast differentiation and activation, which further perpetuates matrix production and stiffening [143]. As the mechanotransduction pathways that sense and respond to the biomechanical properties of tissues are present across tissue types, work in one disease model may yield useful information to others. The relationship between stiffening and disease observed in IPF can be extended to musculoskeletal tissues, and indeed emerging evidence increasingly implicates localized ECM stiffening to disease initiation in the cartilage [45,50], tendons [144], and muscles [145–147]. Despite recent data, few studies have focused on musculoskeletal fibrosis specifically, largely due to the lack of ECM and ECM-composite materials that can accurately recreate the mechanical profile of fibrotic tissue.

Additional evidence of the critical role of matrix mechanics in fibrotic diseases comes from the field of vascular regeneration. Novel engineered ECM and polyethylene glycol-based materials were recently utilized to understand the pathophysiology of pulmonary artery hypertension (PAH), a progressive disease of the lung vasculature that is characterized by elevated pulmonary blood pressure, remodeling of the pulmonary arteries, and ultimately right ventricular failure. Therapeutic interventions for PAH are limited in part by the lack of in vitro screening platforms that accurately reproduce dynamic arterial wall mechanical properties. In this study, the ECM of the pulmonary arteries was engineered to mimic disease progression. The engineered ECM and polyethylene glycol-based model allowed for unprecedented recreation of the mechanical properties of both PAH and healthy tissue [148–150], providing a new in vitro tool for understanding how fibrotic pathogenesis initiates. The phototunable constructs developed to mimic the mechanical properties of normal and diseased arteries could be adapted for use in musculoskeletal applications, providing an existing and validated platform for interrogating the role of the ECM in musculoskeletal health and disease.

### 3.3. Limitations

As with any literature summary, we intend for this review to be a brief survey of each of these commonly studied tissues, rather than a comprehensive guide to every engineered ECM-based approach in the field of musculoskeletal regeneration. This review fills a gap in the existing literature by uniting a diverse array of experimental approaches under the common theme of musculoskeletal regeneration. While highlighting every recent and exciting use of engineered ECM-cell interactions in the context of musculoskeletal regeneration is outside of the scope of this review, and we show the expansive promise of ECM in musculoskeletal tissue engineering. We believe this grouping serves readers who may work in more than one musculoskeletal tissue system, which is an increasingly common theme in musculoskeletal tissue engineering.

## 4. Conclusions

Throughout the last decade, the expanding ability to direct cell behavior for musculoskeletal tissue regeneration via the use of engineered ECM and ECM-based materials has greatly expanded our knowledge of musculoskeletal development, disease, and healing. This review has focused on recent and noteworthy advances that incorporate cell, ECM, and material interactions to increase the regenerative capacity of muscle, cartilage, bone, and tendon. We examined approaches focused on a single tissue, as well as studies that incorporated two or more musculoskeletal tissues, to more faithfully recreate the in vivo environment. The works we highlight suggest that single- and multi-tissue approaches can generate valuable insights into how the physical, mechanical, and biochemical attributes of the ECM, alone and in combination with innovative engineered materials of natural and synthetic origins, can be harnessed to direct cellular behavior towards maximizing the regenerative capacity of musculoskeletal tissues. Human and animal ECM both deserve continued exploration, as continuous improvements to decellularization techniques reduce the danger of systemic immune response or rejection.

We concluded with predictions of the future of engineered cell–ECM–material interactions. The implementation of advanced manufacturing techniques, such as three-dimensional printing, as well as improvements in decellularization protocols that decrease the likelihood of adverse immune reactions, have paved the way for expanded use of ECM in musculoskeletal tissue engineering. Renewed interest in harnessing the ECM both as a standalone material and in combination with other well-established biomaterials is driven by studies showing that ECM instructs cells to behave more closely to how they do in native tissue. We show the expansive promise of ECM in musculoskeletal tissue engineering, and we look forward to the advancements of the coming decade. Ultimately, we expect meaningful and exciting advances in bioprinting, improved decellularization techniques, genetic engineering of both the ECM and cells, and the further determination of how ECM mechanics at multiple length scales can be exploited to precisely direct cell behavior towards regenerative phenotypes.

**Author Contributions:** Conceptualization, S.K.T.; resources, S.K.T., writing—original draft preparation, C.L.J. and B.T.P.; writing—review and editing, C.L.J., B.T.P. and S.K.T.; supervision, S.K.T., funding acquisition, S.K.T. All authors have read and agreed to the published version of the manuscript.

**Funding:** We acknowledge support from the Institutional Development Awards (IDeA) from the National Institute of General Medical Sciences of the National Institutes of Health under Grants #P20GM103408 and P20GM109095. We also acknowledge support from The Biomolecular Research Center at Boise State, BSU-Biomolecular Research Center, RRID:SCR_019174, with funding from the National Science Foundation, Grants #0619793 and #0923535; the M. J. Murdock Charitable Trust; Lori and Duane Stueckle, and the Idaho State Board of Education.

**Institutional Review Board Statement:** Not applicable.

**Informed Consent Statement:** Not applicable.

**Data Availability Statement:** No new data were created or analyzed in this study. Data sharing is not applicable to this article.

**Acknowledgments:** The authors would like to acknowledge Paul Davis and Benjamin Bailey for their insightful discussions on micro- and nanomechanical properties and the utility of AFM.

**Conflicts of Interest:** The authors have no conflict of interest to disclose.

## References


1. AAOS. *United States Bone and Joint Decade: The Burden of Musculoskeletal Diseases in the United States*; American Academy of Orthopaedic Surgeons: Rosemont, IL, USA, 2008.
2. Assunção, M.; Dehghan-Baniani, D.; Yiu, C.H.K.; Später, T.; Beyer, S.; Blocki, A. Cell-Derived Extracellular Matrix for Tissue Engineering and Regenerative Medicine. *Front. Bioeng. Biotechnol.* **2020**, *8*, 602009. Available online: https://www.frontiersin.org/articles/10.3389/fbioe.2020.602009 (accessed on 15 December 2022). [CrossRef] [PubMed]
3. Benders, K.E.M.; van Weeren, P.R.; Badylak, S.F.; Saris, D.B.F.; Dhert, W.J.A.; Malda, J. Extracellular matrix scaffolds for cartilage and bone regeneration. *Trends Biotechnol.* **2013**, *31*, 169–176. [CrossRef] [PubMed]
4. Karageorgiou, V.; Kaplan, D.L. Porosity of 3D biomaterial scaffolds and osteogenesis. *Biomaterials* **2005**, *26*, 5474–5491. [CrossRef] [PubMed]
5. Bueno, E.M.; Ruberti, J.W. Optimizing Collagen Transport through Track-Etched Nanopores. *J. Memb. Sci.* **2008**, *321*, 250–263. [CrossRef] [PubMed]
6. You, J.-O.; Rafat, M.; Almeda, D.; Maldonado, N.; Guo, P.; Nabzdyk, C.S.; Chun, M.; LoGerfo, F.W.; Hutchinson, J.W.; Pradhan-Nabzdyk, L.K.; et al. pH-responsive scaffolds generate a pro-healing response. *Biomaterials* **2015**, *57*, 22–32. [CrossRef]
7. Cicuéndez, M.; Doadrio, J.C.; Hernández, A.; Portolés, M.T.; Izquierdo-Barba, I.; Vallet-Regí, M. Multifunctional pH sensitive 3D scaffolds for treatment and prevention of bone infection. *Acta Biomater.* **2018**, *65*, 450–461. [CrossRef]
8. Basurto, I.M.; Passipieri, J.A.; Gardner, G.M.; Smith, K.K.; Amacher, A.R.; Hansrisuk, A.I.; Christ, G.J.; Caliari, S.R. Photoreactive Hydrogel Stiffness Influences Volumetric Muscle Loss Repair. *Tissue Eng. Part A* **2022**, *28*, 312–329. [CrossRef]
9. Guvendiren, M.; Burdick, J.A. Stiffening hydrogels to probe short- and long-term cellular responses to dynamic mechanics. *Nat. Commun.* **2012**, *3*, 792. [CrossRef]
10. Stoppel, W.L.; Gao, A.E.; Greaney, A.M.; Partlow, B.P.; Bretherton, R.C.; Kaplan, D.L.; Black, L.D. Elastic, silk-cardiac extracellular matrix hydrogels exhibit time-dependent stiffening that modulates cardiac fibroblast response. *J. Biomed. Mater. Res. Part A* **2016**, *104*, 3058–3072. [CrossRef]
11. Hasturk, O.; Jordan, K.E.; Choi, J.; Kaplan, L. Enzymatically crosslinked silk and silk-gelatin hydrogels with tunable T gelation kinetics, mechanical properties and bioactivity for cell culture and encapsulation. *Biomaterials* **2020**, *232*, 119720. [CrossRef]
12. Cheng, W.; Ding, Z.; Zheng, X.; Lu, Q.; Kong, X.; Zhou, X.; Lu, G.; Kaplan, D.L. Injectable hydrogel systems with multiple biophysical and biochemical cues for bone regeneration. *Biomater. Sci.* **2020**, *8*, 2537. [CrossRef]
13. Dixon, T.A.; Cohen, E.; Cairns, D.M.; Rodriguez, M.; Mathews, J.; Jose, R.R.; Kaplan, D.L. Bioinspired Three-Dimensional Human Neuromuscular Junction Development in Suspended Hydrogel Arrays. *Tissue Eng. Part C Methods* **2018**, *24*, 346–359. [CrossRef]
14. Guo, X.; Badu-Mensah, A.; Thomas, M.C.; McAleer, C.W.; Hickman, J.J. Characterization of Functional Human Skeletal Myotubes and Neuromuscular Junction Derived—From the Same Induced Pluripotent Stem Cell Source. *Bioengineering* **2020**, *7*, 133. [CrossRef]
15. Ahn, H.; Kim, K.J.; Park, S.Y.; Huh, J.E.; Kim, H.J.; Yu, W.R. 3D braid scaffolds for regeneration of articular cartilage. *J. Mech. Behav. Biomed. Mater.* **2014**, *34*, 37–46. [CrossRef]
16. Hao, D.; Lopez, J.M.; Chen, J.; Iavorovschi, A.M.; Lelivelt, N.M.; Wang, A. Engineering Extracellular Microenvironment for Tissue Regeneration. *Bioengineering* **2022**, *9*, 202. [CrossRef]
17. Brown, M.; Li, J.; Moraes, C.; Tabrizian, M.; Li-Jessen, N.Y.K. Decellularized extracellular matrix: New promising and challenging biomaterials for regenerative medicine. *Biomaterials* **2022**, *289*, 121786. [CrossRef]
18. Kaukonen, R.; Jacquemet, G.; Hamidi, H.; Ivaska, J. Cell-derived matrices for studying cell proliferation and directional migration in a complex 3D microenvironment. *Nat. Protoc.* **2017**, *12*, 2376–2390. [CrossRef]
19. Schneider, K.H.; Aigner, P.; Holnthoner, W.; Monforte, X.; Nürnberger, S.; Rünzler, D.; Redl, H.; Teuschl, A.H. Decellularized human placenta chorion matrix as a favorable source of small-diameter vascular grafts. *Acta Biomater.* **2016**, *29*, 125–134. [CrossRef]
20. Crapo, P.M.; Gilbert, T.W.; Badylak, S.F. An overview of tissue and whole organ decellularization processes. *Biomaterials* **2011**, *32*, 3233–3243. [CrossRef]
21. Gilbert, T.W.; Stewart-Akers, A.M.; Simmons-Byrd, A.; Badylak, S.F. Degradation and remodeling of small intestinal submucosa in canine Achilles tendon repair. *J. Bone Jt. Surg. Am.* **2007**, *89*, 621–630. [CrossRef]
22. Woods, T.; Gratzer, P.F. Effectiveness of three extraction techniques in the development of a decellularized bone-anterior cruciate ligament-bone graft. *Biomaterials* **2005**, *26*, 7339–7349. [CrossRef] [PubMed]
23. Sicari, B.M.; Rubin, J.P.; Dearth, C.L.; Wolf, M.T.; Ambrosio, F.; Boninger, M.; Turner, N.J.; Weber, D.J.; Simpson, T.W.; Wyse, A.; et al. An Acellular Biologic Scaffold Promotes Skeletal Muscle Formation in Mice and Humans with Volumetric Muscle Loss. *Sci. Transl. Med.* **2014**, *6*, 234ra58. [CrossRef] [PubMed]

24. Adams, J.E.; Zobitz, M.E.; Reach, J.S.; An, K.N.; Steinmann, S.P. Rotator Cuff Repair Using an Acellular Dermal Matrix Graft: An In Vivo Study in a Canine Model. *Arthrosc. J. Arthrosc. Relat. Surg.* **2006**, *22*, 700–709. [CrossRef] [PubMed]
25. Bandzerewicz, A.; Gadomska-Gajadhur, A. Into the Tissues: Extracellular Matrix and Its Artificial Substitutes: Cell Signalling Mechanisms. *Cells* **2022**, *11*, 914. [CrossRef]
26. Xing, H.; Lee, H.; Luo, L.; Kyriakides, T.R. Extracellular matrix-derived biomaterials in engineering cell function. *Biotechnol. Adv.* **2020**, *42*, 107421. [CrossRef]
27. Kim, K.M.; Jang, H.C.; Lim, S. Differences among skeletal muscle mass indices derived from height-, weight-, and body mass index-adjusted models in assessing sarcopenia. *Korean J. Intern. Med.* **2016**, *31*, 643. [CrossRef]
28. Csapo, R.; Gumpenberger, M.; Wessner, B. Skeletal Muscle Extracellular Matrix—What Do We Know About Its Composition, Regulation, and Physiological Roles? A Narrative Review. *Front. Physiol.* **2020**, *11*, 253. Available online: https://www.frontiersin.org/articles/10.3389/fphys.2020.00253 (accessed on 21 January 2023). [CrossRef]
29. Gillies, A.R.; Lieber, R.L. Structure and function of the skeletal muscle extracellular matrix. *Muscle Nerve* **2011**, *44*, 318–331. [CrossRef]
30. Fuoco, C.; Petrilli, L.L.; Cannata, S.; Gargioli, C. Matrix scaffolding for stem cell guidance toward skeletal muscle tissue engineering. *J. Orthop. Surg.* **2016**, *11*, 86. [CrossRef]
31. Grogan, B.F.; Hsu, J.R. Volumetric muscle loss. *J. Am. Acad. Orthop. Surg.* **2011**, *19* (Suppl. S1), S35–S37. [CrossRef]
32. Stantzou, A.; Relizani, K.; Morales-Gonzalez, S.; Gallen, C.; Grassin, A.; Ferry, A.; Schuelke, M.; Amthor, H. Extracellular matrix remodelling is associated with muscle force increase in overloaded mouse plantaris muscle. *Neuropathol. Appl. Neurobiol.* **2021**, *47*, 218–235. [CrossRef]
33. Zhu, M.; Li, W.; Dong, X.; Yuan, X.; Midgley, A.C.; Chang, H.; Wang, Y.; Wang, H.; Wang, K.; Ma, P.X.; et al. In vivo engineered extracellular matrix scaffolds with instructive niches for oriented tissue regeneration. *Nat. Commun.* **2019**, *10*, 4620. [CrossRef]
34. Magarotto, F.; Sgrò, A.; Dorigo Hochuli, A.H.; Andreetta, M.; Grassi, M.; Saggioro, M.; Nogara, L.; Tolomeo, A.M.; Francescato, R.; Collino, F.; et al. Muscle functional recovery is driven by extracellular vesicles combined with muscle extracellular matrix in a volumetric muscle loss murine model. *Biomaterials* **2021**, *269*, 120653. [CrossRef]
35. Turner, N.J.; Badylak, J.S.; Weber, D.J.; Badylak, S.F. Biologic Scaffold Remodeling in a Dog Model of Complex Musculoskeletal Injury. *J. Surg. Res.* **2012**, *176*, 490–502. [CrossRef]
36. Almici, E.; Chiappini, V.; López-Márquez, A.; Badosa, C.; Blázquez, B.; Caballero, D.; Montero, J.; Natera-de Benito, D.; Nascimento, A.; Roldán, M.; et al. Personalized in vitro Extracellular Matrix Models of Collagen VI-Related Muscular Dystrophies. *Front. Bioeng. Biotechnol.* **2022**, *10*, 851825. [CrossRef]
37. Hu, M.; Ling, Z.; Ren, X. Extracellular matrix dynamics: Tracking in biological systems and their implications. *J. Biol. Eng.* **2022**, *16*, 13. [CrossRef]
38. Petrosino, J.M.; Longenecker, J.Z.; Angell, C.D.; Hinger, S.A.; Martens, C.R.; Accornero, F. CCN2 participates in overload-induced skeletal muscle hypertrophy. *Matrix Biol.* **2022**, *106*, 1–11. [CrossRef]
39. Carvalho, C.M.F.; Leonel, L.C.P.C.; Cañada, R.R.; Barreto, R.S.N.; Maria, D.A.; Del Sol, M.; Miglino, M.A.; Lobo, S.E. Comparison between placental and skeletal muscle ECM: In vivo implantation. *Connect. Tissue Res.* **2021**, *62*, 629–642. [CrossRef]
40. Chen, H.; Qian, Z.; Zhang, S.; Tang, J.; Fang, L.; Jiang, F.; Ge, D.; Chang, J.; Cao, J.; Yang, L.; et al. Silencing COX-2 blocks PDK1/TRAF4-induced AKT activation to inhibit fibrogenesis during skeletal muscle atrophy. *Redox Biol.* **2021**, *38*, 101774. [CrossRef]
41. Wang, Y.; Liu, S.; Yan, Y.; Li, S.; Tong, H. SPARCL1 promotes C2C12 cell differentiation via BMP7-mediated BMP/TGF-β cell signaling pathway. *Cell Death Dis.* **2019**, *10*, 852. [CrossRef]
42. Gutiérrez, J.; Gonzalez, D.; Escalona-Rivano, R.; Takahashi, C.; Brandan, E. Reduced RECK levels accelerate skeletal muscle differentiation, improve muscle regeneration, and decrease fibrosis. *FASEB J.* **2021**, *35*, e21503. [CrossRef] [PubMed]
43. Wood, A.J.; Lin, C.H.; Li, M.; Nishtala, K.; Alaei, S.; Rossello, F.; Sonntag, C.; Hersey, L.; Miles, L.B.; Krisp, C.; et al. FKRP-dependent glycosylation of fibronectin regulates muscle pathology in muscular dystrophy. *Nat. Commun.* **2021**, *12*, 2951. [CrossRef] [PubMed]
44. Sahani, R.; Wallace, C.H.; Jones, B.K.; Blemker, S.S. Diaphragm Muscle Fibrosis Involves Changes in Collagen Organization with Mechanical Implications in Duchenne Muscular Dystrophy. *J. Appl. Physiol.* **2022**, *132*, 653–672. [CrossRef] [PubMed]
45. Rebolledo, D.L.; González, D.; Faundez-Contreras, J.; Contreras, O.; Vio, C.P.; Murphy-Ullrich, J.E.; Lipson, K.E.; Brandan, E. Denervation-Induced Skeletal Muscle Fibrosis Is Mediated by CTGF/CCN2 Independently of TGF-β. *Matrix Biol.* **2019**, *82*, 20–37. [CrossRef]
46. Piasecki, A.; Leiva, O.; Ravid, K. Lysyl Oxidase Inhibition in Primary Myelofibrosis: A Renewed Strategy. *Arch. Stem Cell Ther.* **2020**, *1*, 23–27. [CrossRef]
47. CDC. Arthritis. Centers for Disease Control and Prevention. 2021. Available online: https://www.cdc.gov/chronicdisease/resources/publications/factsheets/arthritis.htm (accessed on 2 January 2023).
48. Donahue, R.P.; Link, J.M.; Meli, V.S.; Hu, J.C.; Liu, W.F.; Athanasiou, K.A. Stiffness- and Bioactive Factor-Mediated Protection of Self-Assembled Cartilage against Macrophage Challenge in a Novel Co-Culture System. *Cartilage* **2022**, *13*, 19476035221081466. [CrossRef]

49.  Friedemann, M.; Kalbitzer, L.; Franz, S.; Moeller, S.; Schnabelrauch, M.; Simon, J.C.; Pompe, T.; Franke, K. Instructing Human Macrophage Polarization by Stiffness and Glycosaminoglycan Functionalization in 3D Collagen Networks. *Adv. Healthc. Mater.* **2017**, *6*, 1600967. [CrossRef]
50.  Robinson, W.H.; Lepus, C.M.; Wang, Q.; Raghu, H.; Mao, R.; Lindstrom, T.M.; Sokolove, J. Low-grade inflammation as a key mediator of the pathogenesis of osteoarthritis. *Nat. Rev. Rheumatol.* **2016**, *12*, 580–592. [CrossRef]
51.  Previtera, M.L.; Sengupta, A. Substrate Stiffness Regulates Proinflammatory Mediator Production through TLR4 Activity in Macrophages. *PLoS ONE* **2015**, *10*, e0145813. [CrossRef]
52.  Blakney, A.K.; Swartzlander, M.D.; Bryant, S.J. The effects of substrate stiffness on the in vitro activation of macrophages and in vivo host response to poly(ethylene glycol)-based hydrogels. *J. Biomed. Mater. Res. A* **2012**, *100*, 1375–1386. [CrossRef]
53.  Maldonado, M.; Nam, J. The Role of Changes in Extracellular Matrix of Cartilage in the Presence of Inflammation on the Pathology of Osteoarthritis. *BioMed Res. Int.* **2013**, *2013*, 284873. [CrossRef]
54.  Guilak, F.; Nims, R.J.; Dicks, A.; Wu, C.L.; Meulenbelt, I. Osteoarthritis as a disease of the cartilage pericellular matrix. *Matrix Biol.* **2018**, *71–72*, 40–50. [CrossRef]
55.  Verzijl, N.; DeGroot, J.; Zaken, C.B.; Braun-Benjamin, O.; Maroudas, A.; Bank, R.A.; Mizrahi, J.; Schalkwijk, C.G.; Thorpe, S.R.; Baynes, J.W.; et al. Crosslinking by advanced glycation end products increases the stiffness of the collagen network in human articular cartilage: A possible mechanism through which age is a risk factor for osteoarthritis. *Arthritis Rheum.* **2002**, *46*, 114–123. [CrossRef]
56.  Nasiri, N.; Hosseini, S.; Alini, M.; Khademhosseini, A.; Baghaban Eslaminejad, M. Targeted cell delivery for articular cartilage regeneration and osteoarthritis treatment. *Drug Discov. Today* **2019**, *24*, 2212–2224. [CrossRef]
57.  Han, L.; Grodzinsky, A.J.; Ortiz, C. Nanomechanics of the Cartilage Extracellular Matrix. *Annu. Rev. Mater. Res.* **2011**, *41*, 133–168. [CrossRef]
58.  Wang, M.; Peng, Z. Investigation of the nano-mechanical properties and surface topographies of wear particles and human knee cartilages. *Wear* **2015**, *324–325*, 74–79. [CrossRef]
59.  Okamoto, T.; Takagi, Y.; Kawamoto, E.; Park, E.J.; Usuda, H.; Wada, K.; Shimaoka, M. Reduced Substrate Stiffness Promotes M2-like Macrophage Activation and Enhances Peroxisome Proliferator-Activated Receptor γ Expression. *Exp. Cell Res.* **2018**, *367*, 264–273. [CrossRef]
60.  Ricard-Blum, S. The Collagen Family. *Cold Spring Harb. Perspect. Biol.* **2011**, *3*, a004978. [CrossRef]
61.  Akkiraju, H.; Nohe, A. Role of Chondrocytes in Cartilage Formation, Progression of Osteoarthritis and Cartilage Regeneration. *J. Dev. Biol.* **2015**, *3*, 177–192. [CrossRef]
62.  Peng, Z.; Sun, H.; Bunpetch, V.; Koh, Y.; Wen, Y.; Wu, D.; Ouyang, H. The Regulation of Cartilage Extracellular Matrix Homeostasis in Joint Cartilage Degeneration and Regeneration. *Biomaterials* **2021**, *268*, 120555. [CrossRef]
63.  Zhao, Z.; Li, Y.; Wang, M.; Zhao, S.; Zhao, Z.; Fang, J. Mechanotransduction pathways in the regulation of cartilage chondrocyte homoeostasis. *J. Cell. Mol. Med.* **2020**, *24*, 5408–5419. [CrossRef] [PubMed]
64.  Chen, J.L.; Duan, L.; Zhu, W.; Xiong, J.; Wang, D. Extracellular matrix production in vitro in cartilage tissue engineering. *J. Transl. Med.* **2014**, *12*, 88. [CrossRef] [PubMed]
65.  Plainfossé, M.; Hatton, P.V.; Crawford, A.; Jin, Z.M.; Fisher, J. Influence of the extracellular matrix on the frictional properties of tissue-engineered cartilage. *Biochem. Soc. Trans.* **2007**, *35*, 677–679. [CrossRef] [PubMed]
66.  Peng, G.; McNary, S.M.; Athanasiou, K.A.; Reddi, A.H. Superficial Zone Extracellular Matrix Extracts Enhance Boundary Lubrication of Self-Assembled Articular Cartilage. *Cartilage* **2016**, *7*, 256. [CrossRef] [PubMed]
67.  Mahmood, H.; Eckold, D.; Stead, I.; Shepherd, D.E.T.; Espino, D.M.; Dearn, K.D. A method for the assessment of the coefficient of friction of articular cartilage and a replacement biomaterial. *J. Mech. Behav. Biomed. Mater.* **2020**, *103*, 103580. [CrossRef]
68.  Nie, X.; Chuah, Y.J.; Zhu, W.; He, P.; Peck, Y.; Wang, D.A. Decellularized tissue engineered hyaline cartilage graft for articular cartilage repair. *Biomaterials* **2020**, *235*, 119821. [CrossRef]
69.  Yang, Q.; Peng, J.; Guo, Q.; Huang, J.; Zhang, L.; Yao, J.; Yang, F.; Wang, S.; Xu, W.; Wang, A.; et al. A cartilage ECM-derived 3-D porous acellular matrix scaffold for in vivo cartilage tissue engineering with PKH26-labeled chondrogenic bone marrow-derived mesenchymal stem cells. *Biomaterials* **2008**, *29*, 2378–2387. [CrossRef]
70.  Hanai, H.; Jacob, G.; Nakagawa, S.; Tuan, R.S.; Nakamura, N.; Shimomura, K. Potential of Soluble Decellularized Extracellular Matrix for Musculoskeletal Tissue Engineering—Comparison of Various Mesenchymal Tissues. *Front. Cell Dev. Biol.* **2020**, *8*, 581972. Available online: https://www.frontiersin.org/articles/10.3389/fcell.2020.581972 (accessed on 21 January 2023). [CrossRef]
71.  Yang, Y.; Lin, H.; Shen, H.; Wang, B.; Lei, G.; Tuan, R.S. Mesenchymal stem cell-derived extracellular matrix enhances chondrogenic phenotype of and cartilage formation by encapsulated chondrocytes in vitro and in vivo. *Acta Biomater.* **2018**, *69*, 71–82. [CrossRef]
72.  Chen, M.; Li, Y.; Liu, S.; Feng, Z.; Wang, H.; Yang, D.; Guo, W.; Yuan, Z.; Gao, S.; Zhang, Y.; et al. Hierarchical macro-microporous WPU-ECM scaffolds combined with Microfracture Promote in Situ Articular Cartilage Regeneration in Rabbits. *Bioact. Mater.* **2021**, *6*, 1932–1944. [CrossRef]
73.  Rim, Y.A.; Ju, J.H. The Role of Fibrosis in Osteoarthritis Progression. *Life* **2020**, *11*, 3. [CrossRef]
74.  Islam, M.S.; Afrin, S.; Singh, B.; Jayes, F.L.; Brennan, J.T.; Borahay, M.A.; Leppert, P.C.; Segars, J.H. Extracellular matrix and Hippo signaling as therapeutic targets of antifibrotic compounds for uterine fibroids. *Clin. Transl. Med.* **2021**, *11*, e475. [CrossRef]

75. Wang, C.; Qu, L. The anti-fibrotic agent nintedanib protects chondrocytes against tumor necrosis factor-α (TNF-α)-induced extracellular matrix degradation. *Bioengineered* **2022**, *13*, 5318–5329. [CrossRef]
76. Liu, Y.; Shah, K.M.; Luo, J. Strategies for Articular Cartilage Repair and Regeneration. *Front. Bioeng. Biotechnol.* **2021**, *9*, 770655. [CrossRef]
77. Sutherland, A.J.; Beck, E.C.; Dennis, S.C.; Converse, G.L.; Hopkins, R.A.; Berkland, C.J.; Detamore, M.S. Decellularized Cartilage May Be a Chondroinductive Material for Osteochondral Tissue Engineering. *PLoS ONE* **2015**, *10*, e0121966. [CrossRef]
78. Mohan, N.; Gupta, V.; Sridharan, B.; Sutherland, A.; Detamore, M.S. The potential of encapsulating "raw materials" in 3D osteochondral gradient scaffolds: Chondroitin Sulfate and Bioactive Glass as Raw Materials. *Biotechnol. Bioeng.* **2014**, *111*, 829–841. [CrossRef]
79. Gao, G.; Chen, S.; Pei, Y.A.; Pei, M. Impact of perlecan, a core component of basement membrane, on regeneration of cartilaginous tissues. *Acta Biomater.* **2021**, *135*, 13–26. [CrossRef]
80. Han, Y.; Lian, M.; Zhang, C.; Jia, B.; Wu, Q.; Sun, B.; Qiao, Z.; Sun, B.; Dai, K. Study on bioactive PEGDA/ECM hybrid bi-layered hydrogel scaffolds fabricated by electro-writing for cartilage regeneration. *Appl. Mater. Today* **2022**, *28*, 101547. [CrossRef]
81. Beketov, E.E.; Isaeva, E.V.; Yakovleva, N.D.; Demyashkin, G.A.; Arguchinskaya, N.V.; Kisel, A.A.; Lagoda, T.S.; Malakhov, E.P.; Kharlov, V.I.; Osidak, E.O.; et al. Bioprinting of Cartilage with Bioink Based on High-Concentration Collagen and Chondrocytes. *Int. J. Mol. Sci.* **2021**, *22*, 11351. [CrossRef]
82. Govindharaj, M.; Hashimi, N.A.; Soman, S.S.; Kanwar, S.; Vijayavenkataraman, S. 3D Bioprinting of human Mesenchymal Stem Cells in a novel tunic decellularized ECM bioink for Cartilage Tissue Engineering. *Materialia* **2022**, *23*, 101457. [CrossRef]
83. Screen, H.R.C.; Birk, D.E.; Kadler, K.E.; Ramirez, F.; Young, M.F. Tendon Functional Extracellular Matrix. *J. Orthop. Res. Off. Publ. Orthop. Res. Soc.* **2015**, *33*, 793–799. [CrossRef] [PubMed]
84. Xu, Y.; Murrell, G.A.C. The Basic Science of Tendinopathy. *Clin. Orthop.* **2008**, *466*, 1528–1538. [CrossRef] [PubMed]
85. Thorpe, C.T.; Birch, H.L.; Clegg, P.D.; Screen, H.R.C. The role of the non-collagenous matrix in tendon function. *Int. J. Exp. Pathol.* **2013**, *94*, 248–259. [CrossRef] [PubMed]
86. Screen, H.R.C.; Chhaya, V.H.; Greenwald, S.E.; Bader, D.L.; Lee, D.A.; Shelton, J.C. The influence of swelling and matrix degradation on the microstructural integrity of tendon. *Acta Biomater.* **2006**, *2*, 505–513. [CrossRef] [PubMed]
87. Li, W.; Midgley, A.C.; Bai, Y.; Zhu, M.; Chang, H.; Zhu, W.; Wang, L.; Wang, Y.; Wang, H.; Kong, D. Subcutaneously engineered autologous extracellular matrix scaffolds with aligned microchannels for enhanced tendon regeneration. *Biomaterials* **2019**, *224*, 119488. [CrossRef]
88. Magnusson, S.P.; Langberg, H.; Kjaer, M. The pathogenesis of tendinopathy: Balancing the response to loading. *Nat. Rev. Rheumatol.* **2010**, *6*, 262–268. [CrossRef]
89. Theodossiou, S.K.; Schiele, N.R. Models of tendon development and injury. *BMC Biomed. Eng.* **2019**, *1*, 1–24. [CrossRef]
90. Theodossiou, S.K.; Pancheri, N.M.; Bozeman, A.L.; Brumley, M.R.; Raveling, A.R.; Schiele, N.R. Spinal Cord Transection Disrupts Neonatal Locomotion and Tendon Mechanical Properties. In Proceedings of the Biomedical Engineering Society (BMES) Annual Meeting (2019), Philadelphia, PA, USA, 11–14 October 2019.
91. Chatterjee, M.; Muljadi, P.M.; Andarawis-Puri, N. The role of the tendon ECM in mechanotransduction: Disruption and repair following overuse. *Connect. Tissue Res.* **2022**, *63*, 28–42. [CrossRef]
92. Subramanian, A.; Kanzaki, L.F.; Galloway, J.L.; Schilling, T.F. Mechanical force regulates tendon extracellular matrix organization and tenocyte morphogenesis through TGFbeta signaling. *eLife* **2018**, *7*, e38069. [CrossRef]
93. Birk, D.E.; Trelstad, R.L. Extracellular compartments in tendon morphogenesis: Collagen fibril, bundle, and macroaggregate formation. *J. Cell Biol.* **1986**, *103*, 231–240. [CrossRef]
94. Scott, J.E. Elasticity in extracellular matrix "shape modules" of tendon, cartilage, etc. A sliding proteoglycan-filament model. *J. Physiol.* **2003**, *553 Pt 2*, 335–343. [CrossRef]
95. Samiric, T.; Ilic, M.Z.; Handley, C.J. Characterisation of proteoglycans and their catabolic products in tendon and explant cultures of tendon. *Matrix Biol.* **2004**, *23*, 127–140. [CrossRef]
96. Fessel, G.; Snedeker, J.G. Evidence against proteoglycan mediated collagen fibril load transmission and dynamic viscoelasticity in tendon. *Matrix Biol.* **2009**, *28*, 503–510. [CrossRef]
97. Birch, H.L.; Thorpe, C.T.; Rumian, A.P. Specialisation of extracellular matrix for function in tendons and ligaments. *Muscles Ligaments Tendons J.* **2013**, *3*, 12–22. [CrossRef]
98. Theodossiou, S.K.; Tokle, J.; Schiele, N.R. TGFβ2-induced tenogenesis impacts cadherin and connexin cell-cell junction proteins in mesenchymal stem cells. *Biochem. Biophys. Res. Commun.* **2019**, *508*, 889–893. [CrossRef]
99. Rao, Y.; Zhu, C.; Suen, H.C.; Huang, S.; Liao, J.; Ker, D.F.E.; Tuan, R.S.; Wang, D. Tenogenic induction of human adipose-derived stem cells by soluble tendon extracellular matrix: Composition and transcriptomic analyses. *Stem Cell Res. Ther.* **2022**, *13*, 380. [CrossRef]
100. Kjaer, M.; Magnusson, P.; Krogsgaard, M.; Boysen Møller, J.; Olesen, J.; Heinemeier, K.; Hansen, M.; Haraldsson, B.; Koskinen, S.; Esmarck, B.; et al. Extracellular matrix adaptation of tendon and skeletal muscle to exercise. *J. Anat.* **2006**, *208*, 445–450. [CrossRef]
101. Kiyotake, E.A.; Cheng, M.E.; Thomas, E.E.; Detamore, M.S. The Rheology and Printability of Cartilage Matrix-Only Biomaterials. *Biomolecules* **2022**, *12*, 846. [CrossRef]
102. Hauptstein, J.; Forster, L.; Nadernezhad, A.; Groll, J.; Teßmar, J.; Blunk, T. Tethered TGF-B1 in a Hyaluronic Acid-Based Bioink for Bioprinting Cartilaginous Tissues. *Int. J. Mol. Sci.* **2022**, *23*, 924. [CrossRef]

103. Behan, K.; Dufour, A.; Garcia, O.; Kelly, D. Methacrylated Cartilage ECM-Based Hydrogels as Injectables and Bioinks for Cartilage Tissue Engineering. *Biomolecules* **2022**, *12*, 216. [CrossRef]
104. Lien, S.-M.; Ko, L.-Y.; Huang, T.-J. Effect of Pore Size on ECM Secretion and Cell Growth in Gelatin Scaffold for Articular Cartilage Tissue Engineering. *Acta Biomater.* **2009**, *5*, 670–679. [CrossRef] [PubMed]
105. Tao, M.; Liang, F.; He, J.; Ye, W.; Javed, R.; Wang, W.; Yu, T.; Fan, J.; Tian, X.; Wang, X.; et al. Decellularized tendon matrix membranes prevent post-surgical tendon adhesion and promote functional repair. *Acta Biomater.* **2021**, *134*, 160–176. [CrossRef] [PubMed]
106. Liu, C.; Bai, J.; Yu, K.; Liu, G.; Tian, S.; Tian, D. Biological Amnion Prevents Flexor Tendon Adhesion in Zone II: A Controlled, Multicentre Clinical Trial. *BioMed Res. Int.* **2019**, *2019*, 2354325. [CrossRef] [PubMed]
107. Lee, Y.J.; Ryoo, H.J.; Shim, H.S. Prevention of postoperative adhesions after flexor tendon repair with acellular dermal matrix in Zones III, IV, and V of the hand: A randomized controlled (CONSORT-compliant) trial. *Medicine* **2022**, *101*, e28630. [CrossRef]
108. Shim, H.S.; Park, K.S.; Kim, S.W. Preventing postoperative adhesions after hand tendon repair using acellular dermal matrix. *J. Wound Care* **2021**, *30*, 890–895. [CrossRef]
109. Eisner, L.E.; Rosario, R.; Andarawis-Puri, N.; Arruda, E.M. The Role of the Non-Collagenous Extracellular Matrix in Tendon and Ligament Mechanical Behavior: A Review. *J. Biomech. Eng.* **2022**, *144*, 050801. [CrossRef]
110. Bi, Y.; Ehirchiou, D.; Kilts, T.M.; Inkson, C.A.; Embree, M.C.; Sonoyama, W.; Li, L.; Leet, A.I.; Seo, B.M.; Zhang, L.; et al. Identification of tendon stem/progenitor cells and the role of the extracellular matrix in their niche. *Nat. Med.* **2007**, *13*, 1219–1227. [CrossRef]
111. Zhang, C.; Zhu, J.; Zhou, Y.; Thampatty, B.P.; Wang, J.H.C. Tendon Stem/Progenitor Cells and Their Interactions with Extracellular Matrix and Mechanical Loading. *Stem Cells Int.* **2019**, *2019*, 3674647. [CrossRef]
112. Lu, J.; Yang, X.; He, C.; Chen, Y.; Li, C.; Li, S.; Chen, Y.; Wu, Y.; Xiang, Z.; Kang, J.; et al. Rejuvenation of tendon stem/progenitor cells for functional tendon regeneration through platelet-derived exosomes loaded with recombinant Yap1. *Acta Biomater.* **2023**, *161*, 80–99. [CrossRef]
113. Regard, J.B.; Zhong, Z.; Williams, B.O.; Yang, Y. Wnt Signaling in Bone Development and Disease: Making Stronger Bone with Wnts. *Cold Spring Harb. Perspect. Biol.* **2012**, *4*, a007997. [CrossRef]
114. Schaffler, M.B.; Kennedy, O.D. Osteocyte Signaling in Bone. *Curr. Osteoporos Rep.* **2012**, *10*, 118–125. [CrossRef]
115. Salari, N.; Ghasemi, H.; Mohammadi, L.; Behzadi, M.h.; Rabieenia, E.; Shohaimi, S.; Mohammadi, M. The global prevalence of osteoporosis in the world: A comprehensive systematic review and meta-analysis. *J. Orthop. Surg.* **2021**, *16*, 609. [CrossRef]
116. Conti, V.; Russomanno, G.; Corbi, G.; Toro, G.; Simeon, V.; Filippelli, W.; Ferrara, N.; Grimaldi, M.; D'Argenio, V.; Maffulli, N.; et al. A Polymorphism at the Translation Start Site of the Vitamin D Receptor Gene Is Associated with the Response to Anti-Osteoporotic Therapy in Postmenopausal Women from Southern Italy. *Int. J. Mol. Sci.* **2015**, *16*, 5452–5466. [CrossRef]
117. Yan, S.; Zhang, Y.; Lu, D.; Dong, F.; Lian, Y. ECM-receptor interaction as a prognostic indicator for clinical outcome of primary osteoporosis. *Int. J. Clin. Exp. Med.* **2016**, *9*, 9–20.
118. Prakoso, A.T.; Basri, H.; Adanta, D.; Yani, I.; Ammarullah, M.I.; Akbar, I.; Ghazali, F.A.; Syahrom, A.; Kamarul, T. The Effect of Tortuosity on Permeability of Porous Scaffold. *Biomedicines* **2023**, *11*, 427. [CrossRef]
119. Freeman, F.E.; Browe, D.C.; Nulty, J.; Von Euw, S.; Grayson, W.L.; Kelly, D.J. Biofabrication of multiscale bone extracellular matrix scaffolds for bone tissue engineering. *Eur. Cell Mater.* **2019**, *38*, 168–187. [CrossRef]
120. Li, W.; Xu, R.; Huang, J.; Bao, X.; Zhao, B. Treatment of rabbit growth plate injuries with oriented ECM scaffold and autologous BMSCs. *Sci Rep.* **2017**, *7*, 44140. [CrossRef]
121. Research Progress Related to Growth Plate Injuries. *Growth Plate Injuries*; National Institute of Arthritis and Musculoskeletal and Skin Diseases: Bethesda, MD, USA, 2017. Available online: https://www.niams.nih.gov/health-topics/growth-plate-injuries (accessed on 20 January 2023).
122. Sun, T.; Meng, C.; Ding, Q.; Yu, K.; Zhang, X.; Zhang, W.; Tian, W.; Zhang, Q.; Guo, X.; Wu, B.; et al. In situ bone regeneration with sequential delivery of aptamer and BMP2 from an ECM-based scaffold fabricated by cryogenic free-form extrusion. *Bioact. Mater.* **2021**, *6*, 4163–4175. [CrossRef]
123. Putra, R.U.; Basri, H.; Prakoso, A.T.; Chandra, H.; Ammarullah, M.I.; Akbar, I.; Syahrom, A.; Kamarul, T. Level of Activity Changes Increases the Fatigue Life of the Porous Magnesium Scaffold, as Observed in Dynamic Immersion Tests, over Time. *Sustainability* **2023**, *15*, 823. [CrossRef]
124. Kim, D.-S.; Lee, J.-K.; Jung, J.-W.; Baek, S.-W.; Kim, J.H.; Heo, Y.; Kim, T.-H.; Han, D.K. Promotion of Bone Regeneration Using Bioinspired PLGA/MH/ECM Scaffold Combined with Bioactive PDRN. *Materials* **2021**, *14*, 4149. [CrossRef]
125. Ventura, R.D.; Padalhin, A.R.; Lee, B.T. Functionalization of extracellular matrix (ECM) on multichannel biphasic calcium phosphate (BCP) granules for improved bone regeneration. *Mater. Des.* **2020**, *192*, 108653. [CrossRef]
126. Park, Y.; Cheong, E.; Kwak, J.G.; Carpenter, R.; Shim, J.H.; Lee, J. Trabecular bone organoid model for studying the regulation of localized bone remodeling. *Sci. Adv.* **2021**, *7*, eabd6495. [CrossRef] [PubMed]
127. Neupane, S.; Berardinelli, S.J.; Cameron, D.C.; Grady, R.C.; Komatsu, D.E.; Percival, C.J.; Takeuchi, M.; Ito, A.; Liu, T.-W.; Nairn, A.V.; et al. O-fucosylation of thrombospondin type 1 repeats is essential for ECM remodeling and signaling during bone development. *Matrix Biol.* **2022**, *107*, 77–96. [CrossRef] [PubMed]
128. Li, Z.; Yue, M.; Liu, X.; Liu, Y.; Lv, L.; Zhang, P.; Zhou, Y. The PCK2-glycolysis axis assists three-dimensional-stiffness maintaining stem cell osteogenesis. *Bioact. Mater.* **2022**, *18*, 492–506. [CrossRef]

129. Huber, A.K.; Patel, N.; Pagani, C.A.; Marini, S.; Padmanabhan, K.R.; Matera, D.L.; Said, M.; Hwang, C.; Hsu, G.C.-Y.; Poli, A.A.; et al. Immobilization after injury alters extracellular matrix and stem cell fate. *J. Clin. Investig.* **2020**, *130*, 5444–5460. [CrossRef]
130. Barruet, E.; Garcia, S.M.; Wu, J.; Morales, B.M.; Tamaki, S.; Moody, T.; Pomerantz, J.H.; Hsiao, E.C. Modeling the ACVR1R206H mutation in human skeletal muscle stem cells. *eLife* **2021**, *10*, e66107. [CrossRef]
131. Wang, B.; Ke, W.; Wang, K.; Li, G.; Ma, L.; Lu, S.; Xiang, Q.; Liao, Z.; Luo, R.; Song, Y.; et al. Mechanosensitive Ion Channel Piezo1 Activated by Matrix Stiffness Regulates Oxidative Stress-Induced Senescence and Apoptosis in Human Intervertebral Disc Degeneration. *Oxidative Med. Cell. Longev.* **2021**, *2021*, 8884922. [CrossRef]
132. Tan, G.; Chen, R.; Tu, X.; Guo, L.; Guo, L.; Xu, J.; Zhang, C.; Zou, T.; Sun, S.; Jiang, Q. Research on the osteogenesis and biosafety of ECM–Loaded 3D–Printed Gel/SA/58sBG scaffolds. *Front. Bioeng. Biotechnol.* **2022**, *10*, 973886. [CrossRef]
133. Padhi, A.; Nain, A.S. ECM in Differentiation: A Review of Matrix Structure, Composition and Mechanical Properties. *Ann. Biomed. Eng.* **2020**, *48*, 1071–1089. [CrossRef]
134. Humphrey, J.D.; Dufresne, E.R.; Schwartz, M.A. Mechanotransduction and extracellular matrix homeostasis. *Nat. Rev. Mol. Cell Biol.* **2014**, *15*, 802–812. [CrossRef]
135. Sun, Z.; Guo, S.S.; Fässler, R. Integrin-mediated mechanotransduction. *J. Cell Biol.* **2016**, *215*, 445–456. [CrossRef]
136. Young, J.L.; Holle, A.W.; Spatz, J.P. Nanoscale and mechanical properties of the physiological cell–ECM microenvironment. *Exp. Cell Res.* **2016**, *343*, 3–6. [CrossRef]
137. Murphy, C.M.; O'Brien, F.J. Understanding the effect of mean pore size on cell activity in collagen-glycosaminoglycan scaffolds. *Cell Adhes. Migr.* **2010**, *4*, 377–381. [CrossRef]
138. Feltz, K.P.; Kalaf, E.A.G.; Chen, C.; Martin, R.S.; Sell, S.A. A review of electrospinning manipulation techniques to direct fiber deposition and maximize pore size. *Electrospinning* **2017**, *1*, 46–61. [CrossRef]
139. Variola, F. Atomic force microscopy in biomaterials surface science. *Phys. Chem. Chem. Phys.* **2015**, *17*, 2950–2959. [CrossRef]
140. Norman, M.D.A.; Ferreira, S.A.; Jowett, G.M.; Bozec, L.; Gentleman, E. Measuring the elastic modulus of soft culture surfaces and three-dimensional hydrogels using atomic force microscopy. *Nat. Protoc.* **2021**, *16*, 2418–2449. [CrossRef]
141. Jowett, G.M.; Norman, M.D.A.; Yu, T.T.L.; Rosell Arévalo, P.; Hoogland, D.; Lust, S.T.; Read, E.; Hamrud, E.; Walters, N.J.; Niazi, U.; et al. ILC1 drive intestinal epithelial and matrix remodelling. *Nat. Mater.* **2021**, *20*, 250–259. [CrossRef]
142. Grad, S.; Loparic, M.; Peter, R.; Stolz, M.; Aebi, U.; Alini, M. Sliding motion modulates stiffness and friction coefficient at the surface of tissue engineered cartilage. *Osteoarthr. Cartil.* **2012**, *20*, 288–295. [CrossRef]
143. Davis-Hall, D.; Thomas, E.E.; Peña, B.; Magin, C.M. 3D-bioprinted, phototunable hydrogel models for studying adventitial fibroblast activation in pulmonary arterial hypertension. *Biofabrication* **2022**, *15*, 015017. Available online: http://iopscience.iop.org/article/10.1088/1758-5090/aca8cf (accessed on 16 December 2022).
144. Park, J.H.; Gillispie, G.J.; Copus, J.S.; Zhang, W.; Atala, A.; Yoo, J.J.; Yelick, P.C.; Lee, S.J. The Effect of BMP-Mimetic Peptide Tethering Bioinks on the Differentiation of Dental Pulp Stem Cells (DPSCs) in 3D Bioprinted Dental Constructs. *Biofabrication* **2020**, *12*, 035029. [CrossRef]
145. Valentin, J.E.; Stewart-Akers, A.M.; Gilbert, T.W.; Badylak, S.F. Macrophage Participation in the Degradation and Remodeling of Extracellular Matrix Scaffolds. *Tissue Eng. Part A* **2009**, *15*, 1687–1694. [CrossRef]
146. Freeberg, M.A.T.; Perelas, A.; Rebman, J.K.; Phipps, R.P.; Thatcher, T.H.; Sime, P.J. Mechanical Feed-Forward Loops Contribute to Idiopathic Pulmonary Fibrosis. *Am. J. Pathol.* **2021**, *191*, 18–25. [CrossRef] [PubMed]
147. Nichols, A.E.C.; Best, K.T.; Loiselle, A.E. The Cellular Basis of Fibrotic Tendon Healing: Challenges and Opportunities. *Transl. Res.* **2019**, *209*, 156–168. [CrossRef] [PubMed]
148. Keane, T.J.; Horejs, C.-M.; Stevens, M.M. Scarring vs. Functional Repair: Matrix-Based Strategies to Regulate Tissue Healing. *Adv. Drug Deliv. Rev.* **2018**, *129*, 407–419. [CrossRef] [PubMed]
149. Yates, C.C.; Bodnar, R.; Wells, A. Matrix Control of Scarring. *Cell Mol. Life Sci.* **2011**, *68*, 1871–1881. [CrossRef]
150. Loomis, T.; Hu, L.-Y.; Wohlgemuth, R.P.; Chellakudam, R.R.; Muralidharan, P.D.; Smith, L.R. Matrix Stiffness and Architecture Drive Fibro-Adipogenic Progenitors' Activation into Myofibroblasts. *Sci. Rep.* **2022**, *12*, 13582. [CrossRef]

 *bioengineering*

*Article*

# Human-Origin iPSC-Based Recellularization of Decellularized Whole Rat Livers

Aylin Acun [1,2,3], Ruben Oganesyan [1,2], Maria Jaramillo [1,2], Martin L. Yarmush [1,2,4] and Basak E. Uygun [1,2,*]

[1] Center for Engineering in Medicine and Surgery, Massachusetts General Hospital, Harvard Medical School, Shriners Hospitals for Children, Boston, MA 02114, USA; aacun@widener.edu (A.A.); oganesyan.rv@gmail.com (R.O.); maja983@gmail.com (M.J.); ireis@sbi.org (M.L.Y.)
[2] Department of Surgery, Massachusetts General Hospital, Boston, MA 02114, USA
[3] Department of Biomedical Engineering, Widener University, Chester, PA 19013, USA
[4] Department of Biomedical Engineering, Rutgers University, Piscataway, NJ 08854, USA
* Correspondence: basakuygun@mgh.harvard.edu; Tel.: +1-617-726-3474; Fax: +1-617-573-9471

**Abstract:** End-stage liver diseases lead to mortality of millions of patients, as the only treatment available is liver transplantation and donor scarcity means that patients have to wait long periods before receiving a new liver. In order to minimize donor organ scarcity, a promising bioengineering approach is to decellularize livers that do not qualify for transplantation. Through decellularization, these organs can be used as scaffolds for developing new functional organs. In this process, the original cells of the organ are removed and ideally should be replaced by patient-specific cells to eliminate the risk of immune rejection. Induced pluripotent stem cells (iPSCs) are ideal candidates for developing patient-specific organs, yet the maturity and functionality of iPSC-derived cells do not match those of primary cells. In this study, we introduced iPSCs into decellularized rat liver scaffolds prior to the start of differentiation into hepatic lineages to maximize the exposure of iPSCs to native liver matrices. Through exposure to the unique composition and native 3D organization of the liver microenvironment, as well as the more efficient perfusion culture throughout the differentiation process, iPSC differentiation into hepatocyte-like cells was enhanced. The resulting cells showed significantly higher expression of mature hepatocyte markers, including important CYP450 enzymes, along with lower expression of fetal markers, such as AFP. Importantly, the gene expression profile throughout the different stages of differentiation was more similar to native development. Our study shows that the native 3D liver microenvironment has a pivotal role to play in the development of human-origin hepatocyte-like cells with more mature characteristics.

**Keywords:** decellularization; liver bioengineering; iPSCs; recellularization

**Citation:** Acun, A.; Oganesyan, R.; Jaramillo, M.; Yarmush, M.L.; Uygun, B.E. Human-Origin iPSC-Based Recellularization of Decellularized Whole Rat Livers. *Bioengineering* **2022**, *9*, 219. https://doi.org/10.3390/bioengineering9050219

Academic Editors: Ngan F. Huang and Brandon J. Tefft

Received: 29 April 2022
Accepted: 17 May 2022
Published: 19 May 2022



## 1. Introduction

Late-stage liver diseases, such as cirrhosis, acute hepatitis, and liver cancer, were reported to lead to approximately 4% of all deaths globally in 2010 [1]. In such severe cases, the current gold-standard treatment is orthotopic liver transplantation. Although liver transplantation has high success rates and has been immensely improved since its first application in 1963 [2], many patients suffer from long waiting times as there is a large gap between the number of organs needed and the number of those that are available for transplantation. It was reported that in 2019 only 8896 patients received a new liver out of a total of 17,000 in need of a liver transplant, showing that 48% of the patients in the waiting list could not receive a new liver [3]. Alternative clinical approaches are being explored to reduce donor scarcity, such as split liver transplantation and living donor transplantation; however, these approaches alone are not enough to minimize the gap.

In order to increase the supply of livers available for transplantation, an approach that yields a high number of functional liver substitutes is needed. In the past decade, organ engineering through decellularization has emerged as a promising approach. This approach makes use of the extracellular matrix (ECM) scaffolds of native organs by removing all cellular materials from the organs [4,5]. The resulting scaffolds maintain the original overall shape and ultrastructure of the native organ. In addition, the makeup of the scaffold is not heavily altered, leading to the preservation of organ-specific ECM–cell signaling. An important consideration in translating this approach to clinical settings is to successfully recellularize these scaffolds to reobtain the native functions. Primary hepatocytes have been strongly preferred in such attempts; however, lack of patient-specificity, limited sources, and proliferation potential challenge their large-scale clinical translatability. As an alternative, induced pluripotent stem cells (iPSCs) provide a patient-specific, easily accessible, and expendable cell source [6]. The combination of decellularized natural scaffolds and patient-specific iPSCs would enable the development of readily available, tailor-made liver substitutes for patients.

The differentiation of iPSCs into hepatocyte-like cells has been demonstrated by several groups [7–13]. An important challenge in the clinical translation of iPSC-derived cells, however, is their immature phenotype, as indicated by significantly lower expression of mature hepatic markers, such as P450 enzymes, as well as higher expression of fetal markers, such as alpha fetoprotein (AFP), compared to primary hepatocytes. In attempts to induce maturation of these cells, our group [14] and others [15–17] have used decellularized liver matrices as substrates and showed an improvement in mature hepatic functions in response to interactions with the liver matrix. Importantly, our group has shown that if iPSCs interact with a liver matrix in a 2D culture setting, starting from the early stages of differentiation, the resulting hepatocyte-like cells have a more mature phenotype [14]. Recellularization of decellularized whole rat livers with iPSC-derived hepatocytes was shown in a study by Park et al., where fully differentiated cells were seeded using perfusion [17]. However, the effects of performing a full differentiation within a whole decellularized liver have not been shown.

In this study, we recellularized decellularized rat livers with undifferentiated iPSCs and performed the differentiation of hepatocytes using a perfusion bioreactor system (Supplementary Figure S1). We show that the 3D native liver matrix and perfusion culture conditions improved the differentiation efficiency, as evidenced by the drastic increase in the expression of markers specific to each stage of differentiation. We also showed that differentiation within the liver matrix leads to a more developmentally similar expression pattern throughout the differentiation compared to the gold-standard geltrex substrates and static culture conditions. In addition, we added a 7-day long maturation step to the differentiation process which yielded higher albumin and urea secretion, along with a significant decrease in AFP expression, showing the improved maturity of iPSC-origin livers.

## 2. Materials and Methods

### 2.1. Rat Liver Procurement and Decellularization

Livers of 3-month-old female F. Lewis rats ($N$ = 18) were procured with cannulas attached to the portal vein in accordance with the Institutional Animal Care and Use Committee (IACUC) at Massachusetts General Hospital. Rat livers were decellularized using an adaptation of a previously reported protocol [18]. Briefly, the rat livers were attached via the portal vein cannula to a single-pass perfusion system composed of a peristaltic pump and a bubble trap and perfused with deionized (DI) water for 16 h. Following DI water perfusion, the livers were perfused with 0.1% sodium dodecyl sulfate (SDS) (Sigma-Aldrich, Burlington, MA, USA) for 24 h and with 0.2% and 0.5% SDS for 1 h each. Finally, the livers were washed with DI water and Triton X-100 (Sigma-Aldrich) for 1.5 h each and with PBS for 3 h. Throughout the decellularization process a constant flow rate of 1.6 mL/min was used. The decellularized livers were maintained in PBS at 4 °C until use.

The day before recellularization, the decellularized rat livers were sterilized using PBS supplemented with 0.1% ($v/v$) peracetic acid (Pfaltz and Bauer, Waterbury, CT, USA) and 4% ($v/v$) ethanol. Sterilization was performed by perfusing the liver with 50 mL of peracetic acid solution and incubation in the same solution for 3 h. Following this treatment, the livers were washed by perfusion with 50 mL of PBS followed by perfusion with 50 mL of PBS supplemented with 2% penicillin/streptomycin (Invitrogen, Waltham, MA, USA) and 2.5 µg/mL amphotericin B (Sigma-Aldrich), then incubated in this solution overnight until cell seeding.

### 2.2. Cell Culture

Human skin fibroblast-derived iPSC line hIPS-K3 cells were kindly provided by Dr. Stephen Duncan (Medical College of Wisconsin, Milwaukee, WI, USA). The cells were maintained in Geltrex™ (LDEV-Free Reduced Growth Factor Basement Membrane Matrix, Gibco, Waltham, MA, USA) coated culture flasks in mTeSR plus culture medium (Stemcell Technologies, Vancouver, BC, Canada) with daily media changes. Once the cells reached 80% confluency, they were collected using ReLeSR (Stemcell Technologies). Throughout the culture the pluripotency of the cells was examined by daily observation of the colony phenotype.

### 2.3. Recellularization of Decellularized Rat Livers

After sterilization, the decellularized rat livers were connected to a bioreactor system (Harvard Apparatus) (Supplementary Figure S2) through the portal vein cannula and perfused with mTeSR plus media supplemented with 10 µM ROCK inhibitor (Stemcell Technologies) for 30 min. Then, $15 \times 10^6$ iPSCs in 4 mL mTeSR plus media were injected directly into the bubble trap and allowed to reach the liver through perfusion at 8 mL/min. At this point the flow was stopped and the liver was incubated statically for 15 min at RT. Then, the bioreactor system was maintained in an incubator (5% $CO_2$, 37 °C) with perfusion at 0.1 mL/min flow rate for 1.5 h. The cell seeding steps were repeated 3 more times, reaching a total cell number of $60 \times 10^6$ per liver. Once the last 1.5 h perfusion at 0.1 mL/min flow rate was completed, the flow rate was increased to 2 mL/min. The next day, the culture media was replaced with fresh mTeSR plus without ROCK inhibitor. The livers were maintained with mTeSR plus culture medium for a total of 4 days to induce proliferation of iPSCs prior to the initiation of differentiation, with media changes every other day.

### 2.4. Perfusion Differentiation of iPSCs in Decellularized Rat Livers

The differentiation of iPSCs to a hepatic lineage was achieved by adapting a previously established protocol [7] (Figure 1A). Briefly, differentiation was initiated by introducing RPMI (1640, Invitrogen) media supplemented with 50 ng/mL Activin A (ActA) (Peprotech, Cranbury, NJ, USA) and B27 without insulin (B27(-)) (Invitrogen) for 5 days. During the first 2 days, the media was also supplemented with 10 ng/mL bone morphogenic protein 4 (BMP4) (Peprotech) and 20 ng/mL fibroblast growth factor 2 (FGF-2) (Peprotech). The first 5 days of differentiation aimed to drive definitive endoderm (DE). Next, the media in the bioreactor was replaced with RPMI containing B27 with insulin supplemented with 20 ng/mL BMP4 (Peprotech) and 10 ng/mL FGF-2 (Peprotech) for 5 days for hepatic specification (HS). Over the following 5 days, the media was replaced with RPMI containing B27 with insulin supplemented with 20 ng/mL hepatocyte growth factor (HGF) (Peprotech) in RPMI/B27 to achieve hepatoblast expansion (HE). Finally, for 5 days the liver was perfused with Hepatocyte Basal Media (HBM) (Lonza, Rockville, MD, USA) supplemented with SingleQuots (without EGF) supplemented with 20 ng/mL Oncostatin-M (Onc) (R&D Systems, Minneapolis, MN) to achieve immature hepatocyte derivation (IHC). In order to induce a more mature phenotype of iPSC-derived cells, we added another step, namely, hepatocyte maturation (HM), by which the livers were perfused with hepatocyte growth media (C + H) (DMEM supplemented with 10% fetal bovine serum, 0.5 U/mL insulin, 7 ng/mL glucagon, 20 ng/mL epidermal growth factor, 7.5 µg/mL hydrocortisone, 200 U/mL penicillin/streptomycin, and 50 µg/mL gentamycin) for 7 days. Throughout the differentiation the bioreactor was oxygenated and maintained at 37 °C supplemented with 5% $CO_2$ with daily media changes.

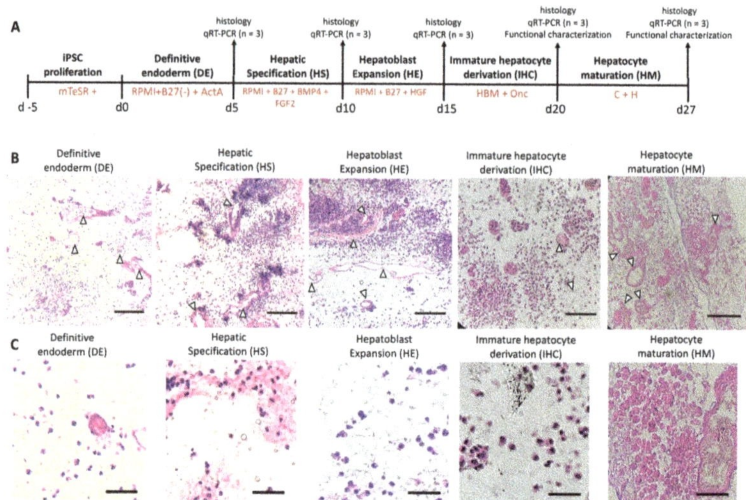

**Figure 1.** The perfusion differentiation of iPSCs to hepatocyte-like cells in rat livers. (**A**) The differentiation protocol and timeline showing the different analyses performed at each stage of differentiation. (**B**) Histological analysis of decellularized rat livers recellularized with iPSCs throughout the differentiation through H&E staining. White triangles show that vessels in the rat liver remained open following perfusion seeding and differentiation. (Scale bars = 200 μm.) (**C**) High magnification histological images showing cell morphology at different stages of differentiation. (Scale bars = 50 μm).

## 2.5. Quantitative Real Time PCR (qRT-PCR)

In order to determine the mRNA expression levels of specific markers at each stage of differentiation, livers were sacrificed at the end of each stage and approximately 2/3 of the livers were used for RNA extraction ($n = 3$ per stage). The liver tissues were flash frozen in liquid nitrogen and ground using a mortar and pestle. The resulting tissue was then used for RNA isolation using a PureLink RNA isolation kit (Thermo Fisher Scientific, Waltham, MA, USA), following the manufacturer's instructions. The resulting RNA was used for cDNA synthesis using an iScript cDNA synthesis kit (Bio-Rad, Hercules, CA, USA), following the manufacturer's instructions. The cDNA was then used in qRT-PCR analysis using a ViiA7 Real time PCR system (Thermo Fisher Scientific) and a power SYBR Green PCR master mix kit (Thermo Fisher Scientific), according to the manufacturer's instructions. The list of primers used is provided in Supplementary Table S1. All expression levels were normalized to GAPDH expression. Results for pluripotency and endoderm marker expression were represented relative to undifferentiated iPSCs cultured on geltrex-coated well plates. Results for early and mature hepatic marker expression were represented relative to cells differentiated on geltrex-coated well plates unless stated otherwise.

## 2.6. Albumin and Urea Quantification

For albumin and urea quantification, media were collected at the end of the IHC stage and daily during the HM stage. For all experiments, the same volume of media was collected (10 mL). The level of albumin secreted in the livers was determined using a Human Albumin ELISA kit (Abcam, Cambridge, UK), following the manufacturer's instructions. The urea nitrogen direct kit (Stanbio, Boerne, TX, USA) was used, following the manufacturer's instructions, to determine the amount of urea secreted. Both albumin and urea contents were represented as micrograms secreted per liver.

### 2.7. Histological Analysis and Immunohistochemistry

For histological analysis, the recellularized rat liver tissues were collected at the end of each stage of differentiation and fixed with 10% formalin for 24 to 48 h at room temperature (RT) and then maintained in 70% ethanol at 4 °C. The tissues were then dehydrated and embedded in paraffin. The tissues were microsectioned to 5 μm thick slices and stained with hematoxylin (Leica, Wetzlar, Germany) and eosin (Leica) (H&E) to visualize the ECM and cell nuclei. The stained sections were imaged using a Nikon Eclipse E800 (Tokyo, Japan).

Immunohistochemistry analysis was performed at the Histopathology Research Core at Massachusetts General Hospital. Briefly, the recellularized rat liver tissues were embedded in Tissue-Tek O.C.T. compound (Sakura Finetek, Torrance, CA, USA) and frozen at −80 °C. The tissues were cryo-sectioned and labeled with antibodies against HNF-4a (HNF4A Monoclonal Antibody (F.674.9), Thermo Fisher Scientific), AFP (AFP monoclonal antibody (35436), Thermo Fisher Scientific), and Ki67 (Ki-67 Monoclonal Antibody (SolA15), eBioscience). The sections were then labeled with species-appropriate secondary antibodies as well as DAPI and imaged using an Olympus Nanozoomer slide scanner at 488 nm.

The quantitative analysis of HNF-4a, AFP, and Ki67 expression was performed by measuring the fluorescence intensity of the respective immunohistochemistry images. For the purpose of the analysis, the backgrounds of the extracellular matrices in the images were removed, using the same noise settings for each target, by means of ImageJ software (Image J 1.51). The results are represented as fluorescence intensity, arbitrary units (A.U.).

### 2.8. Statistical Analysis

For all statistical analyses Microsoft Excel Office 365 (Version 16.39, Redmond, WA, USA) and GraphPad Prism (Version 8.3.1, San Diego, CA, USA) were used. The Student's *t*-test with Welch's correction and one-way-ANOVA analysis were used, and statistical difference was defined as $p < 0.05$. All results are represented as averages $\pm$ standard deviation of 3 different liver recellularization experiments using the same cell line but different cultures within passages 21–27.

## 3. Results

### 3.1. iPSCs Attach to and Differentiate within Decellularized Rat Livers

In order to develop a functional and patient-specific liver, the use of human-origin iPSC-derived cells is a promising approach. We used a closed bioreactor system to recellularize the decellularized rat livers (Supplementary Figure S2A,B). The iPSCs were introduced into the parenchymas of the livers in four steps through the portal veins. The cells were maintained for five days prior to differentiation to allow for cell proliferation. At the end of the first stage of differentiation, we observed that the cells penetrated the parenchymas of the livers, with minimal numbers of cells remaining in the vasculature. With the start of differentiation, we sacrificed the livers at the end of each stage of differentiation and assessed cell localization and changes in cell number qualitatively (Figure 1B). We observed that until the HE stage there was an increase in cell number, as shown by histological analysis, which stabilized after this stage. We also observed macroscopically that the livers appeared opaquer at the end of the differentiation period compared to immediately after decellularization, showing that the cells populated the parenchymas uniformly throughout the livers (Supplementary Figure S3A,B).

### 3.2. iPSC Differentiation Efficiency Is Induced by the Native Liver Microenvironment

Following the attachment and proliferation of iPSCs within decellularized rat livers, we started the differentiation process, following a previously established protocol [7]. We determined the expression of pluripotency, endoderm, developmental hepatoblast, and mature hepatocyte markers at the corresponding stages throughout the differentiation. We compared the results to cells differentiated on conventional geltrex-coated culture well plates and normalized all results to undifferentiated iPSCs (Figure 2A,B; Supplementary Figure S4A–D) or to cells differentiated on geltrex (Figure 2C–F). We observed that with the induction of the endoderm lineage, the expression of pluripotency marker OCT-4 significantly decreased in both geltrex and decellularized rat liver groups (Figure 2A). The decrease in expression of NANOG was significant in decellularized rat livers compared to undifferentiated iPSCs; however, the decrease in the geltrex group was not significant ($p = 0.24$). In addition, the expression of the endoderm-specific markers SOX17 and FOXA2 significantly increased in both groups. Importantly, using decellularized rat livers as substrates significantly improved the endoderm induction efficiency compared to differentiation on geltrex, as shown by significantly higher expressions of SOX17 ($p = 0.015$), GATA4 ($p < 0.0001$), and FOXA2 ($p < 0.001$) (Figure 2B), revealing the effects of native liver matrices at early stages of differentiation.

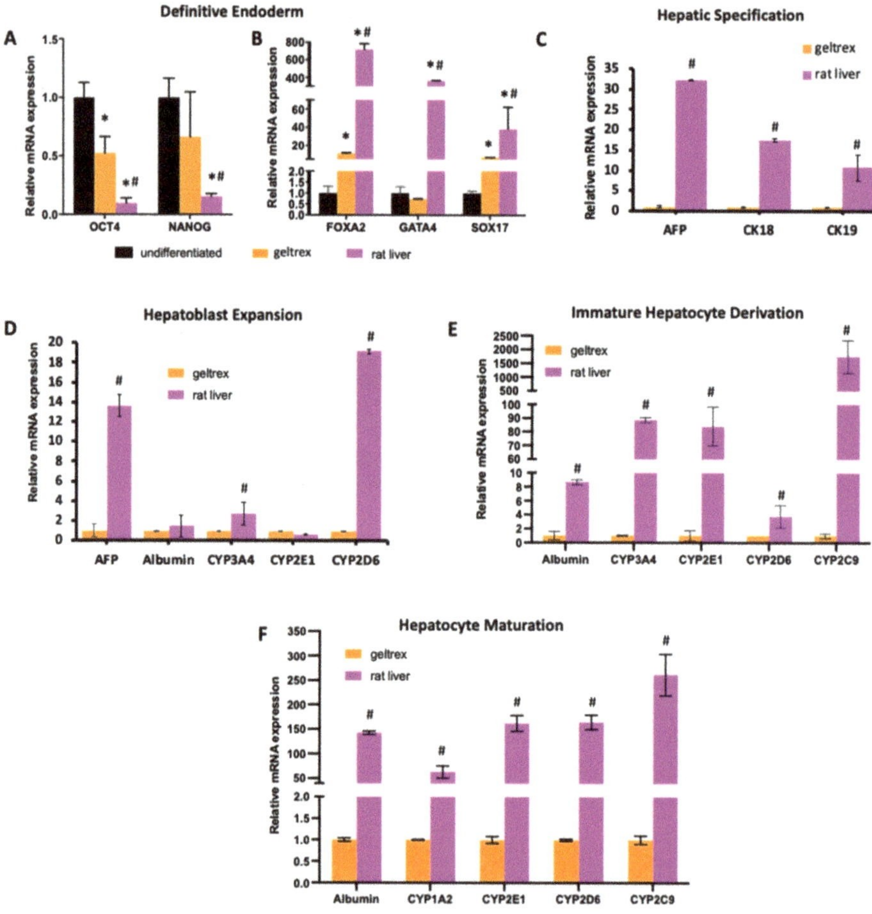

**Figure 2.** mRNA expression levels of key markers specific to each differentiation stage. qRT-PCR analysis showing mRNA expression of (**A**) pluripotency and (**B**) endoderm markers at the DE stage. (**C**) mRNA expression of early hepatic markers at the HS stage. mRNA expression of mature hepatic markers at (**D**) the HE stage, (**E**) the IHC stage, and (**F**) the HM stage, relative to undifferentiated iPSCs. (* indicates statistically significant differences compared to undifferentiated iPSCs ($p < 0.05$); # indicates statistically significant differences in the decellularized rat liver group compared to the geltrex group ($p < 0.05$).) For rat liver data, samples from three scaffolds at DE, HS, HE, IHC, and IHC+ stages were analyzed. For geltrex data, at least three different differentiations at each stage were used in the analysis.

At the hepatic specification stage, we observed that the markers expressed during early development of the liver, namely, AFP, CK18, and CK19, were induced in both the geltrex and decellularized rat liver groups. Expression of these markers was significantly higher in the decellularized rat liver group compared to the geltrex group (AFP: 32.2 $\pm$ 0.03 fold, CK18: 17.4 $\pm$ 0.3 fold, and CK19: 10.7 $\pm$ 3.2 fold increase) (Figure 2C). Undifferentiated iPSCs showed no expression of these hepatic markers (Supplementary Figure S4A).

At the end of the hepatoblast expansion stage of differentiation, we determined the expression of both developmental and mature hepatocyte markers. The expression of AFP was 13.6 $\pm$ 1.1-fold higher in the decellularized rat livers compared to the geltrex group ($p < 0.001$) (Figure 2D). In addition, expression levels of CYP3A4 and CYP2D6 were significantly higher in decellularized rat livers compared to the geltrex group ($p = 0.026$ for CYP3A4, and $p < 0.0001$ for CYP2D6), whereas the expressions of albumin and CYP2E1 were similar.

We compared the expression of only mature hepatocyte markers at the last two stages of differentiation (Figure 2E,F). The most significant difference in expression between the decellularized rat livers and the geltrex group was observed in these stages. For each marker, the expression levels were significantly higher in the decellularized rat livers. At the IHC stage, the expression of CYP2C9 was over 1745 $\pm$ 603-fold higher in decellularized rat livers compared to the geltrex group (Figure 2E). In addition, the expression levels of CYP3A4 and CYP2E1 showed an over 80-fold difference between geltrex and decellularized rat livers (CYP3A4:88.7 $\pm$ 1.8 and CYP2E1: 84 $\pm$ 14.2). With the completion of the hepatocyte maturation step, the drastic difference in the expression of mature hepatocyte markers was maintained (Figure 2F). The expression levels of albumin, CYP1A2, CYP2E1, CYP2D6, and CYP2C9 in decellularized rat livers were, respectively, 142.6 $\pm$ 3.3-, 63.3 $\pm$ 12.4-, 162.3 $\pm$ 15.9-, 164.7 $\pm$ 14.6-, and 262.1 $\pm$ 42.1-fold higher, showing the importance of the contribution of the native liver environment to the late stages of differentiation. We did not observe any significant expression of the mature hepatic markers in the undifferentiated cells, as expected (Supplementary Figure S4B–D).

### 3.3. Differentiation within Decellularized Rat Livers Better Mimics Native Development

In order to assess the role of native liver micro- and macroenvironments on iPSC differentiation, we determined the changes in expression trends of some key markers throughout the differentiation process. We observed that the expression of the pluripotency markers OCT4 and NANOG decreased significantly at the start of differentiation within rat livers; however, there was a more gradual decrease when iPSCs were differentiated on geltrex (Figure 3A,B). Within decellularized rat livers, the lowest levels of OCT4 and NANOG expression were reached at the DE stage and maintained at similar levels throughout. In the geltrex group, the lowest expression for the markers was achieved at the HE stage, with OCT4 expression gradually decreasing at each stage and NANOG expression remaining constant during DE and showing a significant drop in the HS stage, followed by a milder decrease in the HE stage. When we traced the expression of the endoderm markers SOX17, GATA4, and FOXA2 in iPSCs differentiated in decellularized rat livers, we observed that the highest expression level for each marker was reached in the DE stage, as expected (Figure 3C–E). In the geltrex samples, however, the peak for each marker was observed in the HS stage. The peak values for GATA4 and FOXA2 in decellularized rat liver and geltrex samples did not show any significant differences (Figure 3D,E). The peak expression levels for SOX17, however, were significantly higher ($p = 0.015$) in decellularized rat livers (38.3 $\pm$ 9.1) compared to geltrex samples (15.5 $\pm$ 3.1) (Figure 3C).

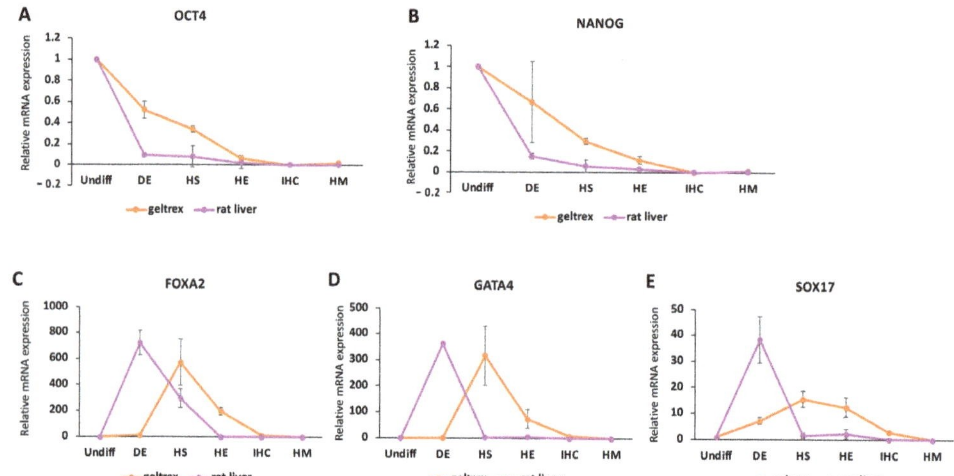

**Figure 3.** Changes in the expression patterns of pluripotency and endoderm markers in iPSCs throughout differentiation in native liver ECM. qRT−PCR analysis results showing the mRNA expression levels of (**A**) OCT4, (**B**) NANOG, (**C**) FOXA2, (**D**) GATA4, and (**E**) SOX17 relative to the respective expression levels in undifferentiated iPSCs throughout the differentiation in the geltrex and decellularized rat liver (rat liver) groups. For rat liver data, samples from three scaffolds at the DE, HS, HE, IHC, and IHC+ stages were analyzed. For geltrex data, at least three different differentiations at each stage were used in the analysis.

We determined expression patterns during early hepatic specification through tracking the expression levels of CK18, CK19, and AFP. Both CK18 and CK19 reached their highest expression levels at the HS stage in decellularized rat livers (Figure 4A,B). In geltrex samples, the highest expressions of CK18 and CK19 were reached at the HE and HS stages, respectively. For both markers the peak expression was significantly higher ($p < 0.001$) in decellularized rat livers (CK18: 114.7 ± 2.1, CK19: 144.1 ± 42.4) compared to geltrex (CK18: 10.7 ± 3.9, CK19: 13.4 ± 0.7). The expression of AFP, however, showed a different trend, with the highest values being recorded in the IHC stage in the geltrex group and in the HS stage in decellularized rat livers (Figure 4C). In addition to the delayed increase in expression, the peak values reached in the geltrex group (27,394 ± 3254) were significantly higher ($p < 0.001$) compared to the decellularized rat livers (435.3 ± 0.4).

Finally, we determined the expression patterns of the mature hepatic markers albumin, CYP2D6, CYP2E1, and CYP3A4 (Figure 4D–G). All of these markers followed a similar trend, showing increased expression at the later stages of differentiation. For all markers, peak expression levels were recorded at the end of the HM stage regardless of the cell attachment substrate. In the decellularized rat livers, however, the highest expression levels reached were significantly higher compared to the highest levels reached in the geltrex samples (albumin: $p < 0.001$, CYP2D6: $p = 0.004$, CYP2E1: $p < 0.001$, CYP3A4: $p < 0.001$) Interestingly, the expression levels of CYP2D6, CYP2E1, and CYP3A4 started to increase in the HS stage in decellularized rat livers, while this response was delayed until the IHC stage in the geltrex group.

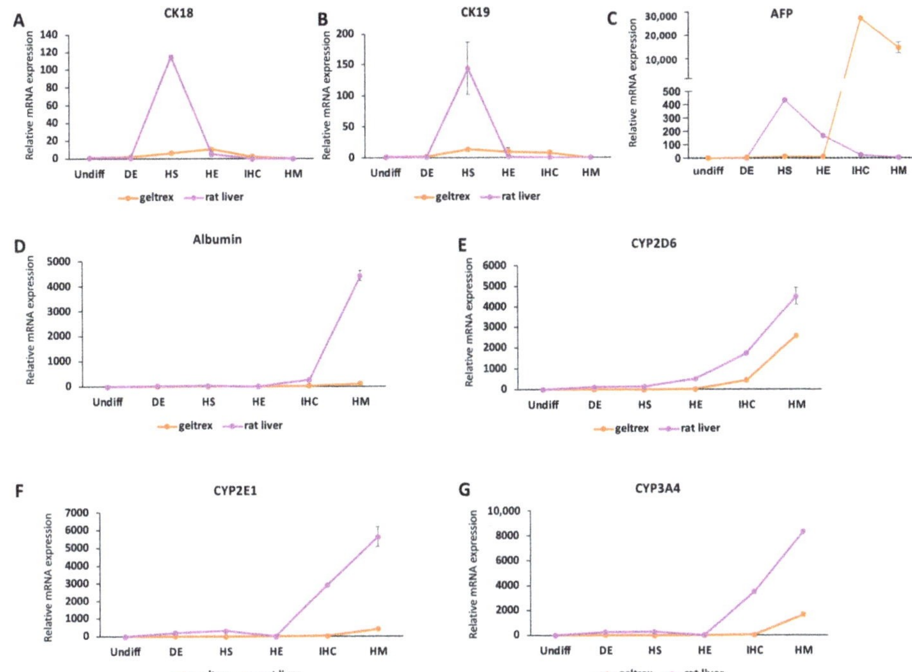

**Figure 4.** The changes in expression patterns of early and mature hepatic markers in iPSCs throughout differentiation in native liver ECM. qRT−PCR analysis results showing the mRNA expression levels of (**A**) CK18, (**B**) CK19, (**C**) AFP, (**D**) albumin, (**E**) CYP2D6, (**F**) CYP2E1, and (**G**) CYP3A4 relative to the respective expression levels in undifferentiated iPSCs throughout the differentiation in the geltrex and decellularized rat liver (rat liver) groups. For rat liver data, samples from three scaffolds at the DE, HS, HE, IHC, and IHC+ stages were analyzed. For geltrex data, at least three different differentiations at each stage were used in analysis.

### 3.4. A Functional Human iPSC-Derived Hepatocyte-Based Liver Is Developed

We determined the functional maturation of the livers at the protein level by assessing the albumin and urea levels secreted by the iPSC-derived hepatocytes during the HM stage (Figure 5A,B). We collected media samples from the bioreactors at days 1, 3, 5, and 7 of culture during the HM stage. Our results showed a significant increase in both urea and albumin secretion over 7 days compared to day 1 ($p = 0.002$ for urea, $p < 0.001$ for albumin). The urea secretion was $3.8 \pm 0.2$ µg/mL on day 1 of the HM stage, whereas it reached $6.0 \pm 0.1$ µg/mL on day 7 (Figure 5A). The albumin secretion on day 1 of the HM stage was $31 \pm 3$ µg per liver and increased to $252 \pm 31$ µg per liver at the end of the HM stage. The increase in urea secretion followed a different trend compared to albumin secretion such that the increase in urea secretion peaked on day 3 of culture followed by a stabilization over the following days. The albumin secretion, however, showed a gradual increase in the early days of the HM stage, while increasing significantly from day 5 of culture (Figure 5B).

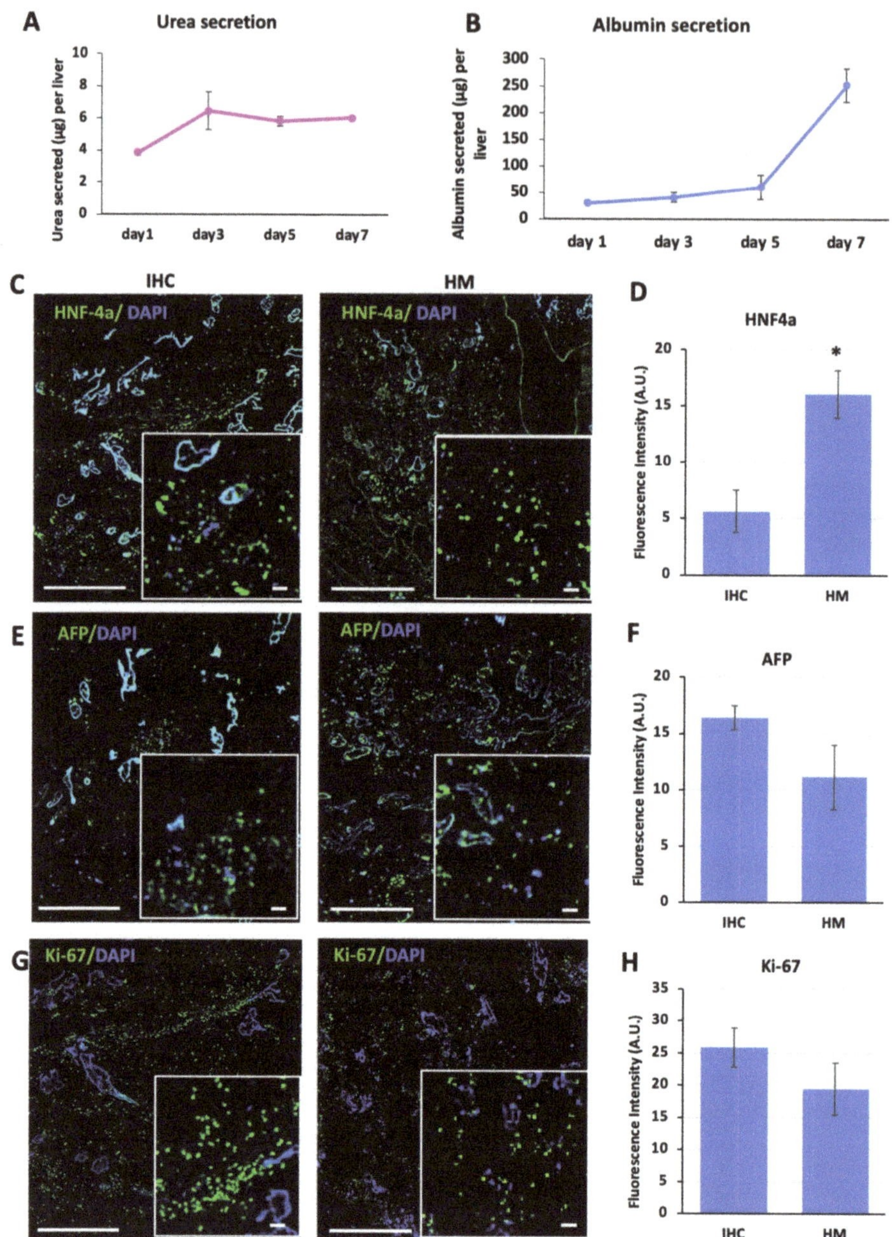

**Figure 5.** The functional characterization of recellularized rat livers. The amounts of (**A**) urea and (**B**) albumin secreted by the recellularized livers determined on days 1, 3, 5, and 7 were determined for the HM stage. Images of the immunohistochemistry analysis for (**C**) HNF4a, (**E**) AFP, and (**G**) Ki-67 in recellularized rat livers in the IHC and HM stages. Fluorescence intensity quantification of the immunohistochemistry images showing changes in the expression levels of (**D**) HNF4a, (**F**) AFP, and (**H**) Ki-67 in the IHC and HM stages. Three scaffolds at the IHC stage and three scaffolds in the IHC+ stages were analyzed for quantification. (* indicates statistically significant differences ($p < 0.05$); scale bars = 200 μm; scale bars for insets = 50 μm.)

In addition, we determined the protein expression of HNF4a, AFP, and Ki-67 in the livers at the end of the IHC and HM stages through immunohistochemistry (Figure 5C–G). We observed that the iPSC-derived hepatocyte-like cells were positive for all markers at both the IHC and HM stages. The expression of HNF4a showed a significant increase ($p = 0.034$), indicating that the maturation stage induced more mature cells (Figure 5C,D). In addition, there was a decrease in AFP protein levels at the HM stage compared to the IHC stage in decellularized rat livers, although this decrease was not significant ($p = 0.133$) (Figure 5E,F). This decrease was in line with the decrease observed in AFP mRNA expression levels (Figure 4C). The mRNA expression of AFP was $24.2 \pm 16.7$ at the IHC stage and $5.1 \pm 0.5$ at the HM stage, showing a decrease with maturation, although the difference was not significant ($p = 0.124$). We also determined the expression of the proliferation marker Ki-67. In response to the maturation stage, we observed a slight decrease in Ki-67 levels; however, this decrease was not significant ($p = 0.215$) (Figure 5G,H).

## 4. Discussion

The use of decellularized scaffolds in clinic is a promising alternative to replacing damaged or diseased tissues and organs. The successful decellularization of rodent to human livers has been shown by our group and others [18–21]. The next challenge in carrying this approach to clinic is the recellularization of these native scaffolds with patient-specific cells to develop functional tissues/organs. iPSCs, since their discovery, have been the main focus for personalized treatments. Although promising, the current differentiation protocols yield hepatocyte-like cells with fetal-like phenotypes, leading to inferior functionality compared to primary cells [22]. To tackle this problem our group and others have used native liver scaffolds as differentiation substrates and demonstrated improved phenotypes. Wang and colleagues [16] cultured iPSC-hepatocytes on 500 μm thick decellularized rat liver ECM discs and used a poly-L-lactic acid (PLLA)–collagen mix as a control. They observed that the liver ECM induced higher mRNA expression of P450 enzymes and higher albumin secretion compared to the controls. After 14 days of culture on the liver matrix, expression levels of the fetal markers AFP and CYP3A7 were significantly lower compared to controls, showing improved maturity in response to ECM–cell interactions. In another study, Park et al. [17] investigated the effect of porcine liver matrices on porcine iPSC-derived hepatocyte-like cells. For this, they supplemented the differentiation medium with solubilized porcine liver matrix. The highest albumin expression was achieved when the supplementation was performed at the last stage of differentiation. In addition, they seeded the median lobes of decellularized rat livers with porcine iPSC-derived cells via perfusion. They showed that there was albumin and urea secretion by day 3 of culture, yet the apoptosis rate significantly increased by day 7 of culture in the recellularized scaffolds. In a previous study, our group differentiated human iPSCs on human decellularized liver matrix and determined the stage of differentiation at which plating the cells on human liver matrix would yield the best differentiation results [14]. The results showed that exposing the cells to liver matrix from the earliest stage of differentiation induced higher expression of the markers specific to each stage at the respective stages. Guided by these findings, we investigated how differentiation would be affected by the 3D native liver environment under perfusion culture, showin—for the first time, to our knowledge—full differentiation within the whole decellularized rat livers.

Rat livers were selected as our platform as their smaller size and wider availability rendered them suitable for our proof-of-concept study. It is important to minimize the presence of residual detergents in decellularized scaffolds due to their cytotoxicity. Different methods of residual SDS removal from decellularized scaffolds have been shown [23] and in this study we have performed an extensive wash to remove residual detergent presence. We have adapted a well-studied method for decellularizing rat livers in this study. Since the successful outcome of the adapted method has been shown numerous times in our previous studies [24–26], we do not provide here a detailed characterization of the decellularization process. The iPSCs reached the parenchymas of the livers and attached to the liver matrices in agreement with reports showing iPSC attachment to 2D gels made of liver matrix [14,17]. Histological analysis showed that the livers were more heavily populated at the HS and HE stages, suggesting growth in cell number at earlier stages of differentiation, as well as further confirming that there was no cytotoxicity due to residual detergents.

ECM–cell interactions have been shown to have an important role in cell differentiation and function in various tissues [27,28]. The specific effect of decellularized liver matrices on the change of expression patterns in iPSCs throughout their differentiation towards hepatic lineages, as well as their effect on differentiated cell functions [14,16,17], have been shown in 2D settings. However, the structural organization and other advantages of a 3D microenvironment are overlooked in such 2D settings and the direct effect of the 3D native liver environment has not been shown. In one study, when differentiated using a sandwich culture method with human decellularized matrix, iPSC expression patterns showed higher expression of markers specific to each stage compared to cells differentiated on Matrigel [14]. At the definitive endoderm stage, the endoderm markers GATA-4, FOXA2, and SOX17 were expressed at significantly higher levels compared to the Matrigel groups. However, in this culture setting the difference was 10-fold or less. In our study, the differentiation within whole decellularized rat livers yielded an over 65-fold increase in FOXA2 and an over 500-fold increase in GATA4 expression compared to the geltrex samples. These drastic differences were observed at the later stages of differentiation in the expression levels of mature hepatocyte markers. Overall, the induction of expression of markers at this level was greater than the increase reported for 2D liver matrix substrates, indicating the importance of the 3D microenvironment in addition to the components of the ECM. This drastic increase was likely induced by a combination of factors, including the increased cell–ECM interactions provided in the whole liver setting, the ECM remaining intact without going through a digestion step that was used to develop the 2D substrates, and the perfusion culture delivering fresh and oxygenated media continuously. Our results are consistent with those of Sassi et al. [29], who showed that the perfusion culture of primary human hepatocytes in a bioreactor system induced significantly higher cell viability and functioning compared to static cultures within the same scaffolds.

In addition to comparing the expression of markers at the respective differentiation stages at which they are expected to peak, we also observed the expression patterns throughout the differentiation so as to understand the overall expression changes in the 3D environment. Interestingly, we observed an immediate decrease in pluripotency markers in decellularized rat livers, while a more gradual decrease took place in the geltrex group. This could potentially point to a lower differentiation efficiency in the geltrex group compared to the decellularized rat livers. This is in line with the previous observations of Jaramillo et al., who reported that decreases in the expression of OCT4 and NANOG were more gradual when cells were differentiated on Matrigel surfaces compared to 2D human liver matrix gels [14]. We also observed that the endoderm markers SOX17, GATA4, and FOXA2 reached peak expression levels at the HS stage in the geltrex group as opposed to the DE stage in decellularized rat livers. A similar expression trend was noted in the work of Jaramillo et al., although it was reported for FOXA2 expression only. In the next stage of differentiation, we observed large differences in the expression levels of the developmental hepatic markers CK18 and CK19 between the decellularized rat liver and geltrex groups. The delayed increase in the expression of endoderm markers is in line with a lower efficiency in the induction of hepatic markers at the HS stage.

A commonly used hepatic differentiation protocol consists of four stages that constitute a 20 day-long protocol [7]. Just as is observed in hepatocyte differentiation, other cell lineages also face the problem of resulting iPSC-derived cells exhibiting fetal expression patterns and associated inferior functionality [30,31]. The 3D environment and the native liver microenvironment are two factors that are shown to be effective in improving the mature phenotypes of cells [16,32,33]. Another factor was shown to be an extended culture with appropriate culture media [34,35]. In this study we added a final stage that we named hepatocyte maturation, in which the cells were cultured for an extra 7 days with a medium used for primary hepatocytes. We hypothesized that this additional step, along with the 3D native microenvironment and physiologically relevant flow conditions, would contribute to further maturation of the iPSC-derived hepatocyte-like cells. Although we did not observe any changes in the organizational structure of hepatocytes within the parenchymas of the livers histologically, at mRNA or protein levels the cells showed higher expression of mature markers, including important CYP450 enzymes, albumin, and HNF4a [36], after the maturation step compared to the IHC stage. The albumin secretion achieved at the end of the HM stage was superior to other reports, such as Park et al.'s, in which it was reported that about 4 µg albumin per liver was secreted, while above 250 µg was secreted through our protocol. In addition, Sassi et al. reported approximately 40 µg albumin secretion in the lateral left lobe of rat livers populated with primary human hepatocytes after 11 days or longer of culture under perfusion [29]. It should be noted that in the study by Park et al. only the median lobes of livers were populated and that the cells interacted with decellularized liver ECMs only after full differentiation. Similarly, in the study by Sassi et al., only the left lateral lobes of rat livers were used and characterized for albumin secretion under perfusion. Even though the albumin secretion analysis in our study showed superior results, the urea secretion recorded by Sassi et al. in the left lateral lobes of primary hepatocyte-populated rat livers was higher. Overall, although inferior compared to those for native liver functions and primary hepatocyte urea secretion levels, our results still suggest an improvement compared to other reports of iPSC-hepatocyte-populated livers, with potential benefits owing to cell–ECM interactions provided throughout the differentiation process. Feldhoff et al. reported albumin secretion levels for a male Sprague Dawley rat of 540 µg per g of liver per hour [37], which is much higher than the levels reached in our setting. Through improving the number of cells seeded in decellularized livers, both albumin and urea synthesis rates can be improved.

An important observation we made was that expression of the fetal marker AFP was significantly lower following the maturation stage. This is in line with the natural development of the liver, as AFP expression has been reported to be higher in the fetal liver compared to the adult liver [36]. This decrease suggests a more mature-like expression profile. The observation of a similar trend in the geltrex group further suggests that the HM stage added here induced maturation of iPSC-hepatocytes and that it is important to include it in regular practices. Another indicator of maturity in hepatocytes is quiescence. In the healthy adult liver, the proliferation capacity of hepatocytes is extremely low as they are quiescent [38]. The proliferation of hepatocytes is only triggered by injury in the mature state [39]. Although not significant, we observed a slight decrease in Ki67-positive cells after the maturation stage. Taken together, the increase in the expression of mature hepatocyte markers, the increases in albumin and urea secretion, and the decreases in AFP and Ki67 expression suggest an improved maturity evoked by the 3D microenvironment and the maturation stage that we incorporated into our protocol.

## 5. Conclusions

Overall, we have reported, for the first time, the perfusion differentiation of iPSCs within whole decellularized rat livers. In this report, we have shown the importance of the native 3D microenvironment in developing human-origin hepatocyte-like cells with more mature characteristics and the possibility of further improving hepatic functioning through prolonged perfusion culture. We believe that although the native liver microenvironment used here is of rat origin, the results are clinically relevant, as we have shown in our previous studies that rat and human decellularized liver matrices have similar contents [21]. We emphasize that further optimization and improvements are required, as well as the addition of non-parenchymal liver cells, to produce tissues of clinically applicable sizes and functionalities and suggest that the native microenvironment and structural organization are crucial to achieve this.

**Supplementary Materials:** The following supporting information can be downloaded at: https://www.mdpi.com/article/10.3390/bioengineering9050219/s1, Figure S1: Schematic representation of the study; Figure S2: (A) The bioreactor system used for recellularization of decellularized rat livers. (B) The inside of the bioreactor chamber with a rat liver being perfused with culture media prior to cell seeding; Figure S3: (A) The rat liver at the end of decellularization. (B) The picture of a recellularized rat liver at the end of IHC stage; Figure S4: The mRNA expression levels of key markers specific to early and mature hepatic markers as normalized to expression in undifferentiated iPSCs; Table S1: Primers for qRT-PCR.

**Author Contributions:** Conceptualization, A.A. and B.E.U.; formal analysis, A.A. and B.E.U.; funding acquisition, A.A., M.L.Y. and B.E.U.; investigation, A.A., R.O. and M.J.; methodology, A.A. and B.E.U.; supervision, M.L.Y. and B.E.U.; validation, A.A. and B.E.U.; visualization, A.A.; writing—original draft, A.A.; writing—review and editing, A.A., M.L.Y. and B.E.U. All authors have read and agreed to the published version of the manuscript.

**Funding:** This study was supported by the National Institutes of Health (grant number: R01DK084053, M.L.Y. and B.E.U.) and Shriners Hospitals for Children (grant number: 84702, A.A.).

**Institutional Review Board Statement:** The study was conducted according to the guidelines of the Institutional Animal Care and Use Committee (IACUC) at Massachusetts General Hospital.

**Informed Consent Statement:** Not applicable.

**Data Availability Statement:** The data presented in this study are available on request from the corresponding author.

**Acknowledgments:** We would like to acknowledge Yibin Chen at Massachusetts General Hospital for procurement of the rat livers. We would also like to acknowledge the Histopathology Research Core at Massachusetts General hospital for performing the immunohistology staining of the recellularized liver sections.

**Conflicts of Interest:** The authors declare no conflict of interest.

# References

1. Byass, P. The global burden of liver disease: A challenge for methods and for public health. *BMC Med.* **2014**, *12*, 159. [CrossRef] [PubMed]
2. Starzl, T.E.; Marchioro, T.L.; Kaulla, K.N.V.; Hermann, G.; Brittain, R.S.; Waddell, W.R. Homotransplantation of the liver in humans. *Surg. Gynecol. Obstet.* **1963**, *117*, 659–676. [PubMed]
3. United Network for Organ Sharing (UNOS). Transplant Trends. Available online: https://unos.org/data/transplant-trends/ (accessed on 4 May 2020).
4. Badylak, S.F.; Taylor, D.; Uygun, K. Whole-organ tissue engineering: Decellularization and recellularization of three-dimensional matrix scaffolds. *Annu. Rev. Biomed. Eng.* **2011**, *13*, 27–53. [CrossRef] [PubMed]
5. Sabetkish, S.; Kajbafzadeh, A.-M.; Sabetkish, N.; Khorramirouz, R.; Akbarzadeh, A.; Seyedian, S.L.; Pasalar, P.; Orangian, S.; Beigi, R.S.H.; Aryan, Z.; et al. Whole-organ tissue engineering: Decellularization and recellularization of three-dimensional matrix liver scaffolds. *J. Biomed. Mater. Res. A* **2015**, *103*, 1498–1508. [CrossRef]
6. Takahashi, K.; Tanabe, K.; Ohnuki, M.; Narita, M.; Ichisaka, T.; Tomoda, K.; Yamanaka, S. Induction of Pluripotent Stem Cells from Adult Human Fibroblasts by Defined Factors. *Cell* **2007**, *131*, 861–872. [CrossRef]
7. Si-Tayeb, K.; Noto, F.K.; Nagaoka, M.; Li, J.; Battle, M.A.; Duris, C.; North, P.E.; Dalton, S.; Duncan, S.A. Highly efficient generation of human hepatocyte–like cells from induced pluripotent stem cells. *Hepatology* **2010**, *51*, 297–305. [CrossRef]
8. Sullivan, G.J.; Hay, D.C.; Park, I.-H.; Fletcher, J.; Hannoun, Z.; Payne, C.M.; Dalgetty, D.; Black, J.R.; Ross, J.A.; Samuel, K.; et al. Generation of functional human hepatic endoderm from human induced pluripotent stem cells. *Hepatology* **2010**, *51*, 329–335. [CrossRef]
9. Liu, H.; Ye, Z.; Kim, Y.; Sharkis, S.; Jang, Y.-Y. Generation of endoderm-derived human induced pluripotent stem cells from primary hepatocytes. *Hepatology* **2010**, *51*, 1810–1819. [CrossRef]
10. Song, Z.; Cai, J.; Liu, Y.; Zhao, D.; Yong, J.; Duo, S.; Song, X.; Guo, Y.; Zhao, Y.; Qin, H.; et al. Efficient generation of hepatocyte-like cells from human induced pluripotent stem cells. *Cell Res.* **2009**, *19*, 1233–1242. [CrossRef]
11. Espejel, S.; Roll, G.R.; McLaughlin, K.J.; Lee, A.Y.; Zhang, J.Y.; Laird, D.J.; Okita, K.; Yamanaka, S.; Willenbring, H. Induced pluripotent stem cell-derived hepatocytes have the functional and proliferative capabilities needed for liver regeneration in mice. *J. Clin. Investig.* **2010**, *120*, 3120–3126. [CrossRef]
12. Toba, Y.; Kiso, A.; Nakamae, S.; Sakurai, F.; Takayama, K.; Mizuguchi, H. FGF signal is not required for hepatoblast differentiation of human iPS cells. *Sci. Rep.* **2019**, *9*, 3713. [CrossRef]

13. Gao, X.; Li, R.; Cahan, P.; Zhao, Y.; Yourick, J.J.; Sprando, R.L. Hepatocyte-like cells derived from human induced pluripotent stem cells using small molecules: Implications of a transcriptomic study. *Stem Cell Res. Ther.* **2020**, *11*, 393. [CrossRef]

14. Jaramillo, M.; Yeh, H.; Yarmush, M.L.; Uygun, B.E. Decellularized human liver extracellular matrix (hDLM)-mediated hepatic differentiation of human induced pluripotent stem cells (hIPSCs). *J. Tissue Eng. Regen. Med.* **2018**, *12*, e1962–e1973. [CrossRef]

15. Kanagavel, V.; Ren, S.; Irudayam, J.I.; Xiong, W.; Talavera, D.; Klein, A.; French, S.; Arumugaswami, V. Organ engineering using decellularized liver scaffold recellularized with iPSC-derived hepatocytes (398.10). *FASEB J.* **2014**, *28*, 398.10. [CrossRef]

16. Wang, B.; Jakus, A.E.; Baptista, P.M.; Soker, S.; Soto-Gutierrez, A.; Abecassis, M.M.; Shah, R.N.; Wertheim, J.A. Functional Maturation of Induced Pluripotent Stem Cell Hepatocytes in Extracellular Matrix—A Comparative Analysis of Bioartificial Liver Microenvironments. *Stem Cells Transl. Med.* **2016**, *5*, 1257–1267. [CrossRef]

17. Park, K.-M.; Hussein, K.H.; Hong, S.-H.; Ahn, C.; Yang, S.-R.; Park, S.-M.; Kweon, O.-K.; Kim, B.-M.; Woo, H.-M. Decellularized Liver Extracellular Matrix as Promising Tools for Transplantable Bioengineered Liver Promotes Hepatic Lineage Commitments of Induced Pluripotent Stem Cells. *Tissue Eng. Part A* **2016**, *22*, 449–460. [CrossRef]

18. Uygun, B.E.; Soto-Gutierrez, A.; Yagi, H.; Izamis, M.-L.; Guzzardi, M.A.; Shulman, C.; Milwid, J.; Kobayashi, N.; Tilles, A.; Berthiaume, F.; et al. Organ reengineering through development of a transplantable recellularized liver graft using decellularized liver matrix. *Nat. Med.* **2010**, *16*, 814–820. [CrossRef]

19. Bühler, N.E.M.; Schulze-Osthoff, K.; Königsrainer, A.; Schenk, M. Controlled processing of a full-sized porcine liver to a decellularized matrix in 24 h. *J. Biosci. Bioeng.* **2015**, *119*, 609–613. [CrossRef]

20. Mazza, G.; Rombouts, K.; Hall, A.R.; Urbani, L.; Luong, T.V.; Al-Akkad, W.; Longato, L.; Brown, D.; Maghsoudlou, P.; Dhillon, A.P.; et al. Decellularized human liver as a natural 3D-scaffold for liver bioengineering and transplantation. *Sci. Rep.* **2015**, *5*, 13079. [CrossRef]

21. Acun, A.; Oganesyan, R.; Uygun, K.; Yeh, H.; Yarmush, M.L.; Uygun, B.E. Liver donor age affects hepatocyte function through age-dependent changes in decellularized liver matrix. *Biomaterials* **2021**, *270*, 120689. [CrossRef]

22. Nghiem-Rao, T.H.; Pfeifer, C.; Asuncion, M.; Nord, J.; Schill, D.; Pulakanti, K.; Patel, S.B.; Cirillo, L.A.; Rao, S. Human induced pluripotent stem cell derived hepatocytes provide insights on parenteral nutrition associated cholestasis in the immature liver. *Sci. Rep.* **2021**, *11*, 12386. [CrossRef]

23. Alizadeh, M.; Rezakhani, L.; Soleimannejad, M.; Sharifi, E.; Anjomshoa, M.; Alizadeh, A. Evaluation of vacuum washing in the removal of SDS from decellularized bovine pericardium: Method and device description. *Heliyon* **2019**, *5*, e02253. [CrossRef]

24. Chen, Y.; Devalliere, J.; Bulutoglu, B.; Yarmush, M.L.; Uygun, B.E. Repopulation of intrahepatic bile ducts in engineered rat liver grafts. *Technology* **2019**, *7*, 46–55. [CrossRef]

25. Chen, Y.; Geerts, S.; Jaramillo, M.; Uygun, B.E. Preparation of Decellularized Liver Scaffolds and Recellularized Liver Grafts. *Methods Mol. Biol.* **2018**, *1577*, 255–270. [CrossRef]

26. Uygun, B.E.; Price, G.; Saeidi, N.; Izamis, M.-L.; Berendsen, T.; Yarmush, M.; Uygun, K. Decellularization and Recellularization of Whole Livers. *JoVE* **2011**, *48*, e2394. [CrossRef]

27. Agmon, G.; Christman, K.L. Controlling stem cell behavior with decellularized extracellular matrix scaffolds. *Curr. Opin. Solid State Mater. Sci.* **2016**, *20*, 193–201. [CrossRef]

28. Hynes, R.O. The extracellular matrix: Not just pretty fibrils. *Science* **2009**, *326*, 1216–1219. [CrossRef] [PubMed]

29. Sassi, L.; Ajayi, O.; Campinoti, S.; Natarajan, D.; McQuitty, C.; Siena, R.; Mantero, S.; De Coppi, P.; Pellegata, A.; Chokshi, S.; et al. A Perfusion Bioreactor for Longitudinal Monitoring of Bioengineered Liver Constructs. *Nanomaterials* **2021**, *11*, 275. [CrossRef] [PubMed]

30. Xia, N.; Zhang, P.; Fang, F.; Wang, Z.; Rothstein, M.; Angulo, B.; Chiang, R.; Taylor, J.; Pera, R.A.R. Transcriptional comparison of human induced and primary midbrain dopaminergic neurons. *Sci. Rep.* **2016**, *6*, 20270. [CrossRef] [PubMed]

31. Doss, M.X.; Sachinidis, A. Current Challenges of iPSC-Based Disease Modeling and Therapeutic Implications. *Cells* **2019**, *8*, 403. [CrossRef]

32. Iii, R.L.G.; Hannan, N.; Bort, R.; Hanley, N.; Drake, R.A.L.; Cameron, G.W.W.; Wynn, T.; Vallier, L. Maturation of Induced Pluripotent Stem Cell Derived Hepatocytes by 3D-Culture. *PLoS ONE* **2014**, *9*, e86372. [CrossRef]

33. Asai, A.; Aihara, E.; Watson, C.; Mourya, R.; Mizuochi, T.; Shivakumar, P.; Phelan, K.; Mayhew, C.; Helmrath, M.; Takebe, T.; et al. Paracrine signals regulate human liver organoid maturation from induced pluripotent stem cells. *Development* **2017**, *144*, 1056–1064. [CrossRef]

34. Piccini, I.; Rao, J.; Seebohm, G.; Greber, B. Human pluripotent stem cell-derived cardiomyocytes: Genome-wide expression profiling of long-term in vitro maturation in comparison to human heart tissue. *Genom. Data* **2015**, *4*, 69–72. [CrossRef]

35. Feyen, D.A.; McKeithan, W.L.; Bruyneel, A.A.; Spiering, S.; Hörmann, L.; Ulmer, B.; Zhang, H.; Briganti, F.; Schweizer, M.; Hegyi, B.; et al. Metabolic Maturation Media Improve Physiological Function of Human iPSC-Derived Cardiomyocytes. *Cell Rep.* **2020**, *32*, 107925. [CrossRef]

36. Zabulica, M.; Srinivasan, R.C.; Vosough, M.; Hammarstedt, C.; Wu, T.; Gramignoli, R.; Ellis, E.; Kannisto, K.; De L'Hortet, A.C.; Takeishi, K.; et al. Guide to the Assessment of Mature Liver Gene Expression in Stem Cell-Derived Hepatocytes. *Stem Cells Dev.* **2019**, *28*, 907–919. [CrossRef]

37. Feldhoff, R.C.; Taylor, J.M.; Jefferson, L.S. Synthesis and secretion of rat albumin in vivo, in perfused liver, and in isolated hepatocytes. Effects of hypophysectomy and growth hormone treatment. *J. Biol. Chem.* **1977**, *252*, 3611–3616. [CrossRef]

38. Fabris, G.; Dumortier, O.; Pisani, D.F.; Gautier, N.; van Obberghen, E. Amino acid-induced regulation of hepatocyte growth: Possible role of Drosha. *Cell Death Dis.* **2019**, *10*, 566. [CrossRef]
39. Mangnall, D.; Bird, N.C.; Majeed, A.W. The molecular physiology of liver regeneration following partial hepatectomy. *Liver Int.* **2003**, *23*, 124–138. [CrossRef]

 *bioengineering*

*Article*

# Three-Dimensional Bioprinting with Alginate by Freeform Reversible Embedding of Suspended Hydrogels with Tunable Physical Properties and Cell Proliferation

Yuanjia Zhu [1,2], Charles J. Stark [1], Sarah Madira [1], Sidarth Ethiraj [1], Akshay Venkatesh [1], Shreya Anilkumar [1], Jinsuh Jung [1], Seunghyun Lee [1], Catherine A. Wu [1], Sabrina K. Walsh [1], Gabriel A. Stankovich [1] and Yi-Ping Joseph Woo [1,2,*]

[1]  Department of Cardiothoracic Surgery, Stanford University, Stanford, CA 94305, USA
[2]  Department of Bioengineering, Stanford University, Stanford, CA 94305, USA
*   Correspondence: joswoo@stanford.edu

**Abstract:** Extrusion-based three-dimensional (3D) bioprinting is an emerging technology that allows for rapid bio-fabrication of scaffolds with live cells. Alginate is a soft biomaterial that has been studied extensively as a bio-ink to support cell growth in 3D constructs. However, native alginate is a bio-inert material that requires modifications to allow for cell adhesion and cell growth. Cells grown in modified alginates with the RGD (arginine-glycine-aspartate) motif, a naturally existing tripeptide sequence that is crucial to cell adhesion and proliferation, demonstrate enhanced cell adhesion, spreading, and differentiation. Recently, the bioprinting technique using freeform reversible embedding of suspended hydrogels (FRESH) has revolutionized 3D bioprinting, enabling the use of soft bio-inks that would otherwise collapse in air. However, the printability of RGD-modified alginates using the FRESH technique has not been evaluated. The associated physical properties and bioactivity of 3D bio-printed alginates after RGD modification remains unclear. In this study, we characterized the physical properties, printability, and cellular proliferation of native and RGD-modified alginate after extrusion-based 3D bioprinting in FRESH. We demonstrated tunable physical properties of native and RGD-modified alginates after FRESH 3D bioprinting. Sodium alginate with RGD modification, especially at a high concentration, was associated with greatly improved cell viability and integrin clustering, which further enhanced cell proliferation.

**Keywords:** 3D bioprinting; FRESH; hydrogel; RGD; alginate; tunable; physical property

**Citation:** Zhu, Y.; Stark, C.J.; Madira, S.; Ethiraj, S.; Venkatesh, A.; Anilkumar, S.; Jung, J.; Lee, S.; Wu, C.A.; Walsh, S.K.; et al. Three-Dimensional Bioprinting with Alginate by Freeform Reversible Embedding of Suspended Hydrogels with Tunable Physical Properties and Cell Proliferation. *Bioengineering* **2022**, *9*, 807. https://doi.org/10.3390/bioengineering9120807

Academic Editor: Gary Chinga Carrasco

Received: 21 October 2022
Accepted: 12 December 2022
Published: 15 December 2022



## 1. Introduction

Three-dimensional (3D) bioprinting is an emerging technology used to rapidly fabricate engineered tissues in vitro using biomaterials combined with live cells and growth factors [1–4]. Extrusion-based bioprinting works by depositing cell-laden bio-inks onto printing platforms in a layer-by-layer fashion to generate 3D constructs [5]. Biocompatibility, printability, and mechanical properties of the bio-ink are important factors to consider when selecting a suitable bio-ink for extrusion-based 3D bioprinting. Alginate is a soft biomaterial that has been frequently used as a bio-ink to support cell growth in 3D as a matrix scaffold due to its biocompatibility and tunable physical properties [3,6–10]. Although 3D bioprinting has made significant advances in creating engineered functional tissues, limitations exist when printing living cells with soft biomaterials such as alginate [11,12]. Additionally, unmodified alginate is a bioinert material that requires modifications to allow for cell adhesion [3,13].

RGD (arginine-glycine-aspartate) is a tripeptide sequence that naturally exists in extracellular matrix proteins such as fibronectin and laminin [14,15]. The RGD sequence is crucial to cell adhesion and proliferation as it binds cell-surface integrin receptors to form focal adhesions [16]. Native alginate can be chemically modified to include RGD sequences to promote cell adhesion [16–19]. A previous study demonstrated that RGD-coupled alginate enhanced cell adhesion, spreading, and differentiation of myoblast cells in hydrogels [17].

Freeform reversible embedding of suspended hydrogels (FRESH) was recently developed to allow for bio-ink extrusion within a thermo-reversible support bath [11,20]. This support bath is composed of gelatin microparticles in a slurry consistency, can provide mechanical support during printing and cross-linking, and is removable at 37 °C. FRESH printing of soft biomaterials such as alginate mitigates the effects of gravity by embedding bio-inks in the aqueous phase of the bath to allow for freeform printing of unsupported structures [20]. Although FRESH printing of native alginates has been reported, the printability and physical properties of RGD-modified alginates printed using the FRESH method have not been studied. The physical properties of RGD-modified alginates are different from those of native alginates [16,19], as with any type of modification, but the impact of this change is unclear in extrusion-based 3D printing using FRESH. Lastly, bioactivity such as cell viability, proliferation, and morphology using the RGD-modified alginate after extrusion-based 3D printing in FRESH remains unclear. Thus, we sought to characterize the physical properties and cellular proliferation of native and RGD-modified alginate after extrusion-based 3D printing in FRESH.

## 2. Materials and Methods

### 2.1. Cell Culture of Adult Human Dermal Fibroblasts

Adult human dermal fibroblasts (HDFa, catalog no. C0135C, ThermoFisher Scientific, Waltham, MA, USA) were cryopreserved as primary cultures. When needed, they were thawed and immediately plated in fibroblast growth media which consisted of human fibroblast expansion basal medium (catalog no. M106500, Gibco, Waltham, MA, USA), 2% $(v/v)$ low serum growth supplement (catalog no. S00310, ThermoFisher Scientific, Waltham, MA, USA), and 1% $(v/v)$ penicillin-streptomycin (catalog no. 15140148, ThermoFisher Scientific, Waltham, MA, USA). Cells were cultured in a cell culture incubator (catalog no. 50116047, Thermo Scientific, Waltham, MA, USA) and set to 5% $CO_2$ at 37 °C. The media was changed every other day, and the cells were passaged when confluent.

### 2.2. Alginate Bio-Ink Preparation

Native sodium alginate (catalog no. ALG, Allevi Inc., Philadelphia, PA, USA, MW = 295,000 g/mol, M:G ratio = 1.29) and RGD-modified sodium alginate (catalog no. 4270425, NovaMatrix, Sandvika, Bærum, Norway, MW = 188,000 g/mol, M:G ratio = 1.22) were first sterilized using ethylene oxide. The RGD-modified sodium alginate had a RGD peptide substitution of 0.255%, and the RGD:alginate ratio was 0.013 µM/mg prior to printing. The sterilized alginates were dissolved in 50 mM HEPES (catalog no. 15630080, Gibco, Waltham, MA, USA) under stirring at 60 °C to create 2% $(w/v)$ native sodium alginate (2% SA), 5% $(w/v)$ native sodium alginate (5% SA), 2% $(w/v)$ RGD-modified sodium alginate (2% RGD-SA), and 5% $(w/v)$ RGD-modified sodium alginate solutions (5% RGD-SA). When alginates were fully dissolved, the solution was allowed to cool to 37 °C while stirring. The alginate solutions were stored at 4 °C for up to 48 h if not used immediately. When ready for bioprinting, 2 mL of the alginate solutions were loaded into a 5 mL sterile syringe (catalog no. PSYR5, Allevi Inc., Philadelphia, PA, USA) using sterile technique. Using another 5 mL sterile syringe and a syringe coupler (catalog no. SYRCOUP, Allevi Inc., Philadelphia, PA, USA), 0.01% $(w/v)$ Alcian Blue 8GX (catalog no. J60122.14, Thermo Scientific, Waltham, MA, USA) was finally mixed in the alginate solutions. Alcian Blue allowed for visualization of the printed constructs prior to cross-linking to avoid loss of constructs during post-print processing. The final alginate solutions were kept in one of the 5 mL sterile syringes, and the other syringe was discarded. Air was removed from the printing syringe by centrifuging the printing syringe at $300\times g$ for 1 min, inverting the syringe without introducing air into the alginate solution, and extruding the air from the printing syringe. The syringe was covered with a syringe cap (catalog no. SYRCAP, Allevi Inc., Philadelphia, PA, USA) and stored at 4 °C until ready for printing.

To print cell-laden alginates, HDFa cells of desired concentration were suspended in 10 μL of fibroblast growth media described above. The alginate solutions were warmed to 37 °C prior to cell mixing. This concentrated HDFa mixture was added into a 5 mL sterile syringe with 245 μL of air-free alginate solution using a P20 pipette. Another 5 mL syringe with 245 μL of air-free alginate solution was also prepared and intended for mixing only. Using the syringe coupler, HDFa cells were homogenously suspended in the alginate solutions by moving the plunger back and forth between two sterile, air-free 5 mL syringes. This allowed for mixing without reintroducing air into the solution. After thorough mixing, the cell-laden alginate solutions were kept in one syringe to be used immediately for bioprinting.

To cross-link alginates, $CaCl_2$ solution was used. 100 mM sterile $CaCl_2$ cross-linking solution was made by dissolving $CaCl_2$ (catalog no. CACL, Allevi Inc., Philadelphia, PA, USA) in 50 mM HEPES. This cross-linking solution was then stored in 4 °C for up to 48 h if not used immediately.

### 2.3. FRESH Support Bath Preparation

FRESH support bath is a slurry of spherical gelatin microparticles with an average diameter of 25 μm and reduced polydispersity generated via coacervation. FRESH functions as a thermo-reversible support bath for bioprinting of bio-inks with gelation mechanisms orthogonal to gelatin [11]. The FRESH support bath was prepared for bioprinting according to the recommended protocol with some modifications [11]. Briefly, 20 mg of sterile $CaCl_2$ was dissolved in 20 mL of 50 mM HEPES to generate a 0.1% $(w/v)$ $CaCl_2$ solution. This $CaCl_2$ solution was then transferred to a 50 mL microcentrifuge conical tube containing 1g of sterile LifeSupport (catalog no. LIFES, Allevi, Philadelphia, PA, USA). Specifically, the LifeSupport powder is composed of dehydrated gelatin microparticles of defined size and shape. The LifeSupport powder was first rehydrated by dissolving the powder in the $CaCl_2$ solution by vigorous mixing using a spatula for 1 min to ensure all powder was fully resuspended. The LifeSupport was then kept at 4 °C undisturbed to fully rehydrate. Then the LifeSupport solution was centrifuged at $2000 \times g$ for 7 min until LifeSupport was compacted, and the resulting supernatant was discarded. The resulting FRESH support bath was transported to well plates for 3D printing. All support baths were used within 12 h.

### 2.4. Extrusion-Based Bioprinting

Allevi 3 Bioprinter (Allevi, Inc., Philadelphia, PA, USA) was used to bioprint alginates in FRESH. The pre-loaded alginate syringe, prepared as described above, was attached to one of the extruders with a 30G, 0.25 in plastic tipped needle attached. After automatic calibration according to the standard protocol, pressure was applied to test extrusion until alginate was dispensed. The extrusion pressures for 2% SA, 5% SA, 2% RGD-SA, and 5% RGD-SA were 7.5 psi, 15 psi, 7.5 psi, and 15 psi, respectively. For construct fidelity testing, 3 mm × 3 mm × 3 mm open single filament boxes were printed. For cell viability and proliferation assessment, 2 mm cylinders of 4 mm in diameters were printed with 100,000 cells/mL in concentration. For cell morphology evaluation, 1 mm cylinders with a 6 mm diameter were printed with 1 million cells/mL in concentration. Cell-laden constructs had an infill grid pattern with an infill distance of 0.16 mm and a layer height of 0.15 mm. All constructs were completed at 22 °C with a print speed of 6 mm/s in 12-well plates (catalog no. 351143, Corning Inc., Corning, NY, USA).

After printing, all constructs were cross-linked with 1 mL of cross-linking $CaCl_2$ solution for 30 min at 37 °C. Then, the cross-linking solution was removed, and constructs were washed twice with 1 mL of 50 mM HEPES at 22 °C. Constructs were again cross-linked again with 1 mL of cross-linking solution for another 30 min at 37 °C and washed twice with 50 mM HEPES at 22 °C. Constructs were cross-linked a final time with 1 mL of cross-linking solution for 7 min at 37 °C. Constructs were washed one last time and then cultured in 2 mL of fibroblast growth media in a 5% $CO_2$ 37 °C cell culture incubator.

### 2.5. Cell Culture of 3D-Printed Construct

3D-printed constructs were cultured in the fibroblast growth media in a cell culture incubator (catalog no. 50116047, Thermo Scientific, Waltham, MA, USA) set to 5% $CO_2$ at 37 °C for up to 28 days. Three days after bioprinting, the growth media was carefully removed using a micropipette without disturbing the constructs. Thereafter, 2 mL of new growth media was added. The media was changed every 3 days until the final timepoint was reached.

### 2.6. Mechanical Tensile Testing

A two-piece mold was 3D-printed in UMA 90 using a 3D Carbon Printer (Carbon, Redwood City, CA, USA). Alginate solutions prepared as described above were cast in the mold to yield alginate samples with ASTM D412-C specified dimensions (Figure 1A,B). Cross-linking $CaCl_2$ solution was then sprayed onto the alginate once every 10 min for 30 min. Next, the alginate-containing molds were placed in cross-linking $CaCl_2$ solution for 12 h at room temperature to complete cross-linking. Alginate specimens were removed from their molds immediately before analysis with a tensile tester. The alginate specimens were affixed to an Instron 5565 Microtester (Norwood, MA, USA) equipped with a 100 N load cell (Figure 1C). The Instron then tensioned the alginate specimens three times at 5 mm/min until 3 mm of displacement was achieved. Following the third measurement up to 3 mm of displacement, each alginate specimen was tensioned at 5 mm/min until failure. For each alginate specimen (2% SA, 5% SA, 2% RGD-SA, and 5% RGD-SA), the Young's modulus was calculated from the slope of the linear portion of the stress-strain curve.

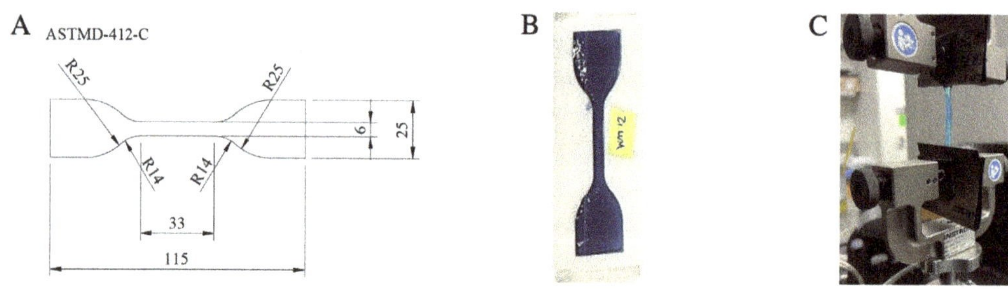


**Figure 1.** Alginate specimens molded for mechanical tensile testing and swelling ratio assessment. (**A**) ASTMD-412-C specified dimensions. All units are in millimeters. (**B**) 3D-printed molds were used to cast alginate specimens with ASTM D412-C specified dimensions. (**C**) The alginate specimens were secured to an Instron tester for Young's modulus measurement.

### 2.7. Swelling Properties

Alginate strips after cross-linking as described above were carefully blotted with filter paper to remove any excess liquid from the surface. Baseline weight was obtained. The alginate strips were placed in fibroblast growth media in a 37 °C incubator. At 1, 2, 3, 4, 6, 8, 20, and 24 h after incubation, the alginate strips were removed from the growth media, blotted, and weighed again. The swelling ratio was then determined using the following equation:

$$Swelling\ Ratio\ (\%) = \frac{W_t - W_0}{W_0} \times 100$$

where $W_t$ is the weight at each time point, and $W_0$ is the baseline weight of each sample. For each alginate, 12 independent samples were used.

### 2.8. Construct Fidelity

To determine the printed constructs' fidelity, 3 mm × 3 mm × 3 mm open single filament boxes without infill were 3D bio-printed in FRESH and cross-linked. Immediately upon cross-linking completion, we performed fidelity assessment while the constructs were submerged in the cross-linking $CaCl_2$ solution. The largest height of the box was measured. The wall thickness was also measured to assess the diameter of a single filament of alginates. For each alginate, 9 independent samples were used.

### 2.9. Cell Viability and Proliferation Assays

To assess cell viability after bioprinting with each alginate bio-ink, a fragment of each printed construct was taken 1, 5, 7, 14, and 28 days after printing. The construct fragments were stained with live/dead viability/cytotoxicity kit for mammalian cells (catalog no. L3224, Invitrogen, Carlsbad, CA, USA) according to the standard protocol. After staining for 30 min, the construct fragments were placed on a concave glass microscope slide (catalog no. 7104, Chang Bioscience, Fremont, CA, USA) and imaged with a Leica DMi8 microscope (Leica Microsystems, Wetzlar, Hesse, Germany). Images were further analyzed using ImageJ (National Institutes of Health, Bethesda, MD, USA). For each alginate at each time point, 6 samples were measured.

To assess cellular proliferation, proliferation assays were performed 1, 5, 7, 14, and 28 days after printing. At each time point, alginate constructs were moved to a new 12-well plate and submerged in 2 mL of fibroblast growth media. Next, 200 μL of Cell Counting Kit-8 (CCK8, catalog no. CK0411, Dojindo Molecular Technologies, Rockville, MD, USA) was added to each well. The constructs were incubated at 37 °C for 3 h. Using BioTek Synergy 2 (Agilent Technologies, Santa Clara, CA, USA) and BioTek Gen 5 Software (Agilent Technologies, Santa Clara, CA, USA), the optical density at 450 nm ($OD_{450}$) of each well was obtained.

### 2.10. Cell Morphology

To assess cellular morphology, constructs were taken 1 and 7 days after printing for staining. The printed constructs were first fixed at room temperature using 4% paraformaldehyde (catalog no. J19943.K2, Thermo Fisher, Waltham, MA, USA) and 1% ($w/v$) $CaCl_2$ in 50 mM HEPES for 30 min. The constructs were then washed with 50 mM HEPES. Next, the constructs were permeabilized by treating them with 0.1% ($v/v$) Triton X-100 (catalog no. X-100, Sigma-Aldrich, St. Louis, MO, USA) and 1% ($w/v$) $CalCl_2$ in 50 mM HEPES for 30 min at room temperature followed by another wash with 50 mM HEPES. The constructs were then blocked in 1% ($v/v$) BSA (catalog no. A1595, Sigma-Aldrich, St. Louis, MO, USA) and 1% ($w/v$) $CaCl_2$ in 50 mM HEPES for 1 h at room temperature. After that, the constructs were washed in 50 mM HEPES before incubating overnight at 4 °C with (1:200) anti-rabbit actin antibody (catalog no. PIMA532479, Thermo Fisher, Waltham, MA, USA), (1:200) anti-mouse integrin beta 1/CD29 antibody (catalog no. MAB1778, Thermo Fisher, Waltham, MA, USA), and 1% ($w/v$) $CaCl_2$ in 50 mM HEPES. After washing the constructs in 50 mM HEPES, the constructs were then incubated for 2 h at room temperature in (1:200) Alexa Fluor 594 Goat Anti-Rabbit IgG H&L (catalog no. ab150080, Abcam, Waltham, MA, USA), (1:200) Alexa Fluor 488 Goat Anti-Mouse IgG H&L (catalog no. ab150113, Abcam, Waltham, MA, USA), and 1% ($w/v$) $CaCl_2$ in 50 mM HEPES. Following another wash in 50 mM HEPES, DAPI (catalog no. NC9524612, Fisher Scientific, Waltham, MA, USA) was added to stain the constructs. The stained constructs were then imaged using a Zeiss LSM 780 inverted confocal microscope and Zeiss Xen software at 20×/0.5 (ZEISS Microscopy, Jena, Germany).

### 2.11. Statistical Analysis

Continuous variables were reported as mean ± standard deviation unless specified otherwise. Sample variance was assessed using F-test. To compare each metric (at each time point if applicable) among the different alginate bio-inks, analysis of variance (ANOVA) with Tukey post-hoc correction was performed. A $p$ value of less than 0.05 was considered statistically significant.

## 3. Results

### 3.1. Mechanical Properties of Alginates

Native alginates and RGD-modified alginates at different concentrations demonstrated a broad range of mechanical properties ($p < 0.0001$, Figure 2A). In general, alginates at higher concentrations were stiffer, with the 2% vs. 5% SA having a Young's modulus of $492.3 \pm 13.0$ kPa vs. $1816.8 \pm 61.0$ kPa ($p < 0.0001$), respectively; and 2% vs. 5% RGD-SA with Young's modulus measuring $531.8 \pm 26.8$ kPa vs. $1767.0 \pm 57.5$ kPa ($p < 0.0001$), respectively. RGD modification did not significantly alter alginates' stiffness. The force vs. displacement plots are shown in Figure S1.

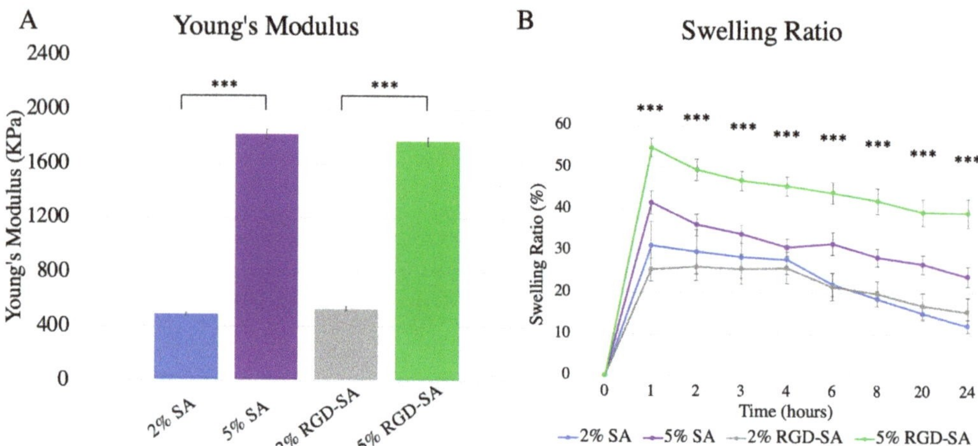


**Figure 2.** Physical properties of alginates. (**A**) Mechanical tensile testing performed for four different alginates demonstrated significantly different Young's Modulus ($p < 0.0001$) with 5% ($w/v$) sodium alginates, with or without RGD modification, demonstrating statistically significantly higher Young's Modulus compared to 2% ($w/v$) alginates with or without RGD modification. (**B**) Swelling ratio increased within the first hour for all alginates, but steadily decreased over the course of 24 h. At each time point, swelling ratios of different alginates were significantly different from each other with 5% RGD-SA demonstrating the highest swelling ratio amongst all alginates. *** denotes $p < 0.001$. SA: sodium alginate; RGD-SA: RGD-modified sodium alginate. Error bars represent standard error.

### 3.2. Swelling Ratio of Alginates

Swelling ratios differed significantly between the alginates at each time point from after 1 h ($p < 0.0001$), 2 h ($p = 0.0002$), 3 h ($p = 0.0004$), 4 h ($p < 0.0001$), 6 h ($p < 0.0001$), 8 h ($p < 0.0001$), 20 h ($p < 0.0001$), and 24 h ($p < 0.0001$) of incubation (Figure 2B). Significant swelling appeared within the first hour of incubation. Specifically, 5% RGD-SA compared to 2% SA as well as compared to 2% RGD-SA demonstrated a significantly higher swelling ratio at all time points ($p < 0.01$). Furthermore, 5% RGD-SA maintained a significantly higher swelling ratio compared to 5% SA starting at 2 h after incubation ($p < 0.05$). There was no statistically significant difference in swelling ratio between 2% SA and 2% RGD-SA at any time point.

### 3.3. Alginate Bioprinting Fidelity

3D bioprinting using alginates in FRESH resulted in generally excellent fidelity in shape, and constructs maintained a squared box geometry in 3D (Figure 3A–D). The largest height of the box for 2% SA, 5% SA, 2% RGD-SA, and 5% RGD-SA were $3.7 \pm 0.5$ mm,

$3.2 \pm 0.1$ mm, $3.3 \pm 0.2$ mm, and $3.1 \pm 0.1$ mm, respectively ($p = 0.0001$, Figure 3E). Specifically, 2% SA constructs were larger compared to constructs printed with 2% RGD-SA ($p = 0.02$), 5% SA ($p = 0.002$), and 5% RGD-SA ($p = 0.001$). Compared to the desired construct size, the final 3D constructs using 2% SA, 5% SA, 2% RGD-SA, and 5% RGD-SA were $23.7 \pm 16.7\%$, $6.3 \pm 3.3\%$, $10.9 \pm 6.5\%$, and $2.6 \pm 2.2\%$ larger in maximal height ($p = 0.0001$). In terms of filament sizes, constructs printed with 2% SA, 5% SA, 2% RGD-SA, and 5% RGD-SA had wall thickness measured $1.8 \pm 0.8$ mm, $1.0 \pm 0.1$ mm, $1.3 \pm 0.4$ mm, and $0.8 \pm 0.1$ mm, respectively ($p = 0.0004$, Figure 3E).

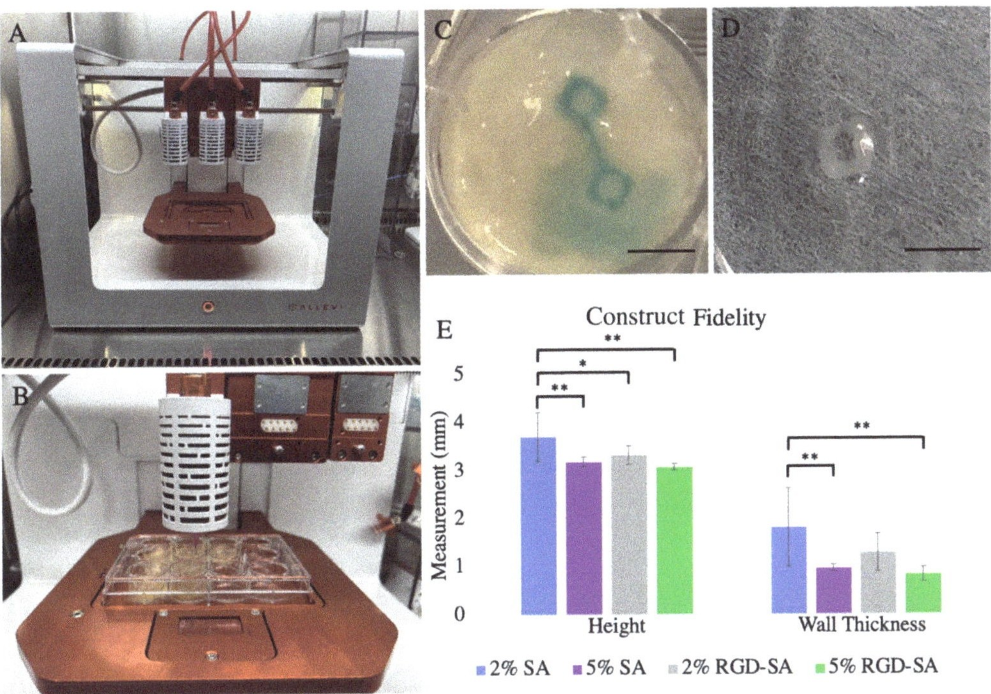


**Figure 3.** 3D bioprinting using the Allevi 3 bioprinter and alginates as bio-ink in freeform reversible embedding of suspended hydrogels (FRESH) for construct fidelity testing (**A**) Photograph of the Allevi 3 3D bioprinter with all 3 extruders attached. (**B**) Demonstration of 3D bioprinting of alginates in FRESH. (**C**) Open, single filament alginate boxes of 3 mm $\times$ 3 mm $\times$ 3 mm in size were printed in FRESH prior to cross-linking. Scale bar = 5 mm. (**D**) An exemplary open, single filament alginate box after cross-linking. Scale bar = 5 mm. (**E**) Construct fidelity analysis compared box maximal height and wall thickness printed using four different alginates. Mixture of 2% ($w/v$) sodium alginate (SA) demonstrated the largest box height and wall thickness, whereas 5% ($w/v$) SA with or without RGD modification demonstrated smaller box height and wall thickness. * denotes $p < 0.05$; ** denotes $p < 0.01$. SA: sodium alginate; RGD-SA: RGD-modified sodium alginate. Error bars represent standard deviation.

### 3.4. HDFa Cell Viability in 3D Bio-Printed Alginates

3D bioprinting using HDFa cell-laden alginates as bio-ink was successful (Figures 4A,B and S2). HDFa cell viability at day 1 after bioprinting remained high with an average viability of $88.3 \pm 1.0\%$, $85.6 \pm 3.5\%$, $84.1 \pm 3.5\%$, and $84.0 \pm 1.4\%$ using 2% SA, 5% SA, 2% RGD-SA, and 5% RGD-SA, respectively ($p = 0.04$). Throughout the 28-day incubation period (Figure 4C), cell viability in 2% SA remained the lowest compared to 2% RGD-SA and 5% RGD-SA at day 7 ($69.9 \pm 2.8\%$ vs. $81.3 \pm 3.4\%$ and $82.8 \pm 4.4\%$, $p = 0.01$ and 0.004), day 14 ($66.6 \pm 3.3\%$ vs. $85.9 \pm 11.3\%$ and $81.1 \pm 6.5\%$, $p = 0.01$ and 0.03), and day 28 (56.2 ± 6.3% vs. 84.3 ± 3.9% and 80.1 ± 6.3%, $p$ = 0.001 and 0.001). At day 28, cells printed in 5% SA also demonstrated significantly lowered viability compared to 2% RGD-SA (69.4 ± 6.1% vs. 84.3 ± 3.9%, $p$ = 0.02), though still higher than 2% SA ($p$ = 0.01). Additionally, cell viability in SA regardless of the concentration continued to decrease over time. There was no difference in cell viability printed in 2% or 5% RGD-SA throughout the 28-day incubation period and the proportion of viable cells in RGD-SA remained stable over time.

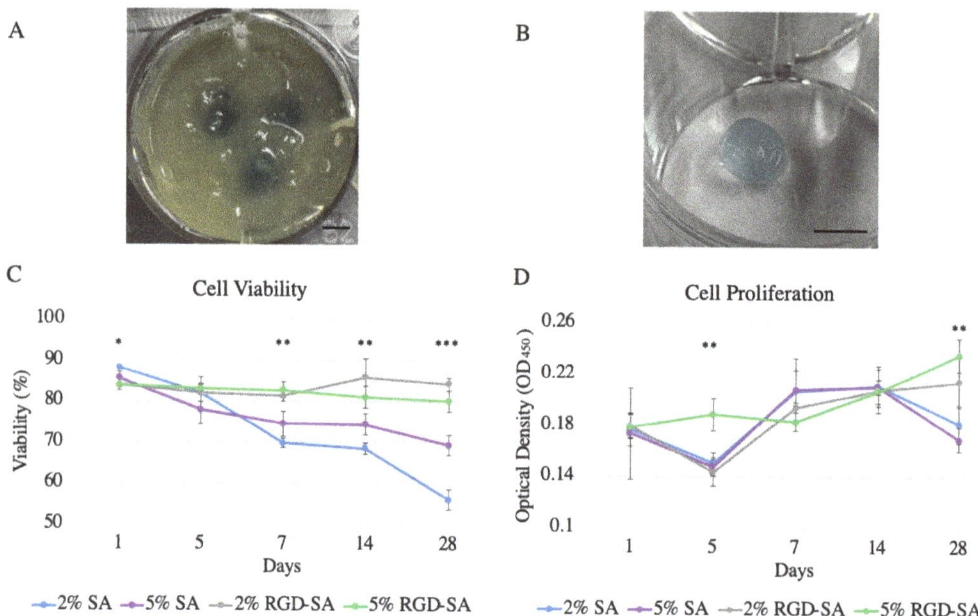

**Figure 4.** Adult human dermal fibroblast cell viability and proliferation after 3D bioprinting in different alginate bio-inks over 28 days. (**A**) 6 mm cylinders were 3D printed with adult human dermal fibroblast-laden alginates in freeform reversible embedding of suspended hydrogels (FRESH). Scale bar = 5 mm. (**B**) After FRESH removal and cross-linking, the 3D printed cylinder was successfully harvested for further cell viability and proliferation testing. Scale bar = 5 mm. (**C**) Cell viability in 2% ($w/v$) sodium alginate (SA) demonstrated the lowest cell viability compared to RGD-modified SA (RGD-SA) after 7 days. Cell viability in 2% SA continued to decrease over time. Cells printed in RGD-SA remained viable over time. (**D**) Cells after 3D bioprinting in different alginates demonstrated varied proliferation. 5% RGD-SA demonstrated the highest cell proliferation compared to 2% RGD-SA at day 5 and compared to SA at day 28. Cells were able to successfully proliferate in RGD-SA with 5% RGD-SA supported better cell proliferation compared to 2% RGD-SA. * denotes $p < 0.05$; ** denotes $p < 0.01$; *** denotes $p < 0.001$. Error bars represent standard error.

HDFa cell proliferation measured after bioprinting in alginates varied over time (Figure 4D). Specifically, 5% RGD-SA consistently supported the highest cell proliferation compared to 2% RGD-SA at day 5 (0.19 ± 0.03 vs. 0.14 ± 0.01, $p$ = 0.001) compared to 2% SA (0.24 ± 0.05 vs. 0.18 ± 0.01, $p$ = 0.04) and 5% SA (0.24 ± 0.05 vs. 0.17 ± 0.01, $p$ = 0.004) at day 28, respectively. Furthermore, both RGD-SAs supported an increase in cell growth over 28 days, but 5% RGD-SA demonstrated better cell proliferation over time compared to 2% RGD-SA.

### 3.5. HDFa Cell Morphology in 3D Bio-Printed Alginates

HDFa cells were well-visualized in the alginates after 3D bioprinting (Figure 5). There were no significant visual differences in cell morphology on day 1. However, slightly enhanced but faint integrin signals were observed throughout cell cytoplasm in cells printed in 2% and 5% RGD-SAs compared to those printed in native alginates (Figure S3). After 7 days of culture, cell morphology remained similar in different alginates. However, actin signals were more pronounced compared to cells 1 day after bioprinting (Figure S4). The integrin signal was only observed in the cells printed in 5% RGD-SA. Area of integrin aggregation and clustering around the cell surface was also observed (Figure S3). However, the integrin signal disappeared in cells printed in 2% RGD-SA, similar to cells printed in native alginates after 7 days of culture.

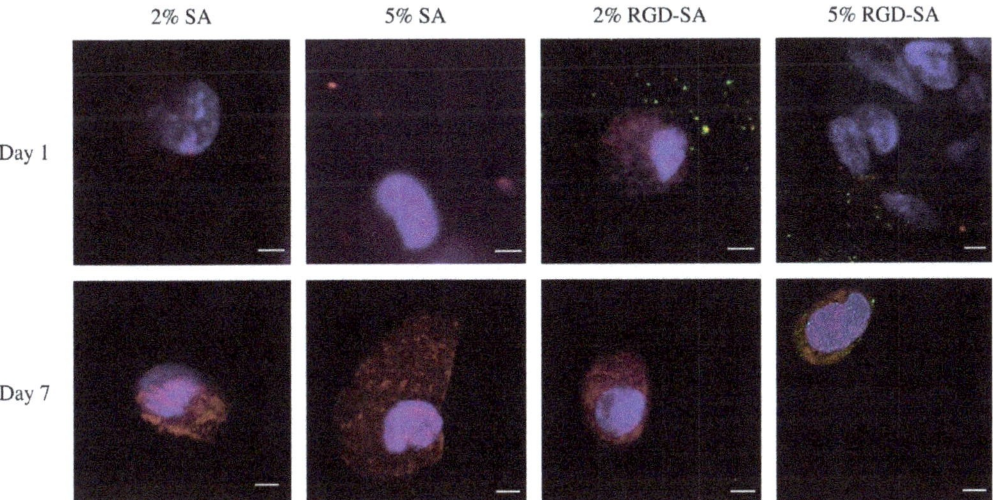


**Figure 5.** Cell morphology 1 and 7 days after bioprinting in alginates. Cells appeared similar in appearance 1 day after bioprinting, but only RGD-modified alginates (RGd-SAs) showed integrin signals. However, 7 days after bioprinting, only cells printed in 5% RGD-SA demonstrated integrin signals with clustering. Cell morphologies were otherwise similar in different alginates 7 days after bioprinting. SA: sodium alginate. Scale bar = 5 μm.

## 4. Discussion

FRESH 3D bioprinting is a significant advancement for the bio-fabrication of soft materials such as alginates. Although FRESH bioprinting native alginates and physical properties of RGD-modified alginates have been reported independently, the physical properties of 3D bio-printed RGD-modified alginates using the FRESH technique have not been reported previously. This represents a knowledge gap that requires further investigation. Active research is being conducted to modify alginates and printing methodologies to improve scaffold physical properties, printability, and bioactivity [5]. In this study, we characterized the physical properties, printability, and cellular proliferation of native and RGD-modified alginate after extrusion-based 3D bioprinting in FRESH. Given that RGD-modified alginates are attractive biomaterials for cell growth and differentiation [16], and that FRESH 3D bioprinting is a popular printing technique to generate scaffolds in complex geometry with high precision [11], we believe this study may have an important impact in the field of 3D bioprinting to facilitate further biomedical research in biomaterials and tissue engineering.

In terms of physical properties, bio-inks with a higher concentration of alginates were found to be associated with a higher Young's Modulus. This was in accordance with previous literature published on native alginates [16,17,21]. Previous literature also reported a significant effect on mechanical properties from adjusting the molecular weight of alginate [22–24]. Interestingly, the mechanical property can be further tuned by modifying alginates with RGD. We also found that a higher concentration of alginates was associated with a higher swelling ratio for 5% RGD-SA, which demonstrated the highest swelling ratio amongst the four alginates investigated in this study. Water absorption is an important feature for hydrogels and tissue engineering scaffolds as it reflects the scaffold's ability to absorb body fluids and transport water and nutrients [25]. All alginates after FRESH 3D bioprinting had relatively high swelling ratios. The varying swelling ratios observed in this study may be due to the different degrees of crosslinking because of different alginate concentrations and the addition of RGD motifs [25].

In terms of bio-ink printability and construct fidelity, 5% RGD-SA demonstrated the best fidelity with highest precision amongst the different alginates used in this study. Printability of a bio-ink is impacted by the mechanical stability of the first layer, rheological properties, as well as cross-linking mechanisms [5]. For lower concentration of alginate, the viscosity was lower, and the bio-ink was more spread out after extrusion. This further impacts the rheology properties of alginates, as reflected by the higher extrusion pressure required to print alginates at higher concentrations. The width of the printed filament was the finest using 5% RGD-SA, suggesting the best construct fidelity amongst the alginates tested in this study. In fact, a previous study has shown slightly increased viscosity of RGD-SA compared to native SA of the same concentration [26]. This slight change of viscosity, though not significant enough to affect extrusion pressure required for bioprinting, allowed for finer filament deposition.

Throughout the printing process and the post-print processing, many factors can impact cell viability. These include but are not limited to shear stress during extrusion-based bioprinting, fabrication time, gelling condition, cross-linking methodology, and duration [10]. In this study, we controlled for variables to isolate the impact of bio-ink alone on cell viability. We showed that cell viability in all alginates remained high until 7 days post-bioprinting when RGD-modified alginates demonstrated superior viability. We hypothesize that the presence of RGD provides a favorable 3D micro-environment to allow for cell adhesion and further facilitates enhanced cell growth and proliferation over an extended period. Our study demonstrated significantly augmented cell proliferation associated with 5% RGD-modified alginates compared to native alginates at the 28-day time point. Prior studies showed that RGD-modified alginates were able to support cell proliferation to maintain 80% of viability for at least 2 days [27]. Cell migration and organization have also been observed [26]. The findings from this study again suggest that the presence of RGD motif in alginates not only helps maintain cell viability but also supports cell growth and proliferation. Although the cell-matrix interaction gets less important as time passes as cell-cell contacts and interactions develop, the initial 3D micro-environment when the cells were printed can be a critical factor to determine cell fate [28]. In our study, we found that cell reorganization with actin polymerization was prominent 7 days after bioprinting across different alginates. Since prior literature observed cytoskeleton remodeling after cells being exposed to shear stress [29,30], we hypothesize that the extrusion-based 3D bioprinting process can activate cytoskeleton. Interestingly, integrin was initially present in both 2% and 5% RGD-modified alginates 1 day after bioprinting, but after 7 days of culture, integrin was only observed on the cell surface in those printed in 5% RGD-SA. These cells also demonstrated increased integrin density and aggregation after 7 days of culture. This might explain why cell proliferation was significantly improved when cells were printed in 5% RGD-SA as integrin binding to the extracellular matrix can contribute to the cell's decision to proliferate, migrate, or die [31]. The RGD peptide density may also impact cell proliferation and viability, given that we showed cells printed in 2% RGD-SA lost integrin after 7 days culture, whereas cells printed in 5% RGD-SA presented enhanced integrin with focal aggregates after 7 days of culture. We hypothesize that with lower RGD peptide density, integrin clusters were unable to form, thereby affecting integrin-extracellular matrix bonds, adhesive force distribution, cell signaling systems, migration, and proliferation [32–34].

## 5. Conclusions

In conclusion, we described tunable physical properties of native and RGD-modified alginates after FRESH 3D bioprinting. Sodium alginate with RGD modification, especially at a high concentration, was associated with greatly improved cell viability and integrin clustering, which further enhanced cell proliferation. A few limitations should be mentioned. While construct fidelity was assessed immediately upon cross-linking completion, this should be more comprehensively evaluated over time and in physiologically relevant conditions. Future studies should also analyze and quantify RGD substitution in alginates after printing and in cell culture to confirm the presence or absence of RGD motif release from the cell-alginate matrix during cell culture and incubation. Further analysis should also be conducted to evaluate the mechanism of cytoskeleton activation upon 3D bioprinting.

**Supplementary Materials:** The following supporting information can be downloaded at: https://www.mdpi.com/article/10.3390/bioengineering9120807/s1, Figure S1: Exemplary live dead images of adult human dermal fibroblasts in different alginate bio-inks after 3D printing over 28 days. Scale bar = 100 m. SA: sodium alginate; RGD-SA: RGD-modified sodium alginate; Figure S2: Integrin signals of cells printed in alginates 1 and 7 days after bioprinting. Integrin was present in both RGD-modified alginates 1 day after bioprinting, but after 7 days of culturing, only 5% RGD-SA demonstrated enhanced integrin signals with clustering. SA: sodium alginate. Scale bar = 5 $\mu$m; Figure S3: Actin signals of cells printed in alginates 1 and 7 days after bioprinting. Actin signals were well visualized in different alginates at both 1 and 7 days after bioprinting. More pronounced actin formalization was observed after 7 days compared to 1 day after bioprinting. SA: sodium alginate. RGD-SA: RGD-modified sodium alginate. Scale bar = 5 $\mu$m; Figure S4: DAPI signals of cells printed in alginates 1 and 7 days after bioprinting. No changes in DAPI signals were observed in different alginates at 1 and 7 days after bioprinting. SA: sodium alginate. RGD-SA: RGD-modified sodium alginate. Scale bar = 5 $\mu$m.

**Author Contributions:** Y.Z.: conception and design of the work, data acquisition, data analysis, interpretation of data, and manuscript drafting; C.J.S.: design of the work, data acquisition, and manuscript drafting and revision; S.M.: design of the work, data acquisition, data analysis, and manuscript revision; S.E.: data acquisition, data analysis, and manuscript drafting and revision; A.V.: data acquisition and manuscript drafting and revision; S.A.: data acquisition and manuscript revision; J.J.: data acquisition and manuscript revision; S.L.: data acquisition and manuscript revision; C.A.W.: data acquisition and manuscript revision; S.K.W.: data acquisition and manuscript revision; G.A.S.: data acquisition and manuscript revision; Y.-P.J.W.: conception and design of the work, interpretation of data, and manuscript revision. All authors have approved the submitted version and have agreed to be personally accountable for the author's own contribution. All authors have read and agreed to the published version of the manuscript.

**Funding:** This work was supported by the National Institutes of Health (NIH F32 HL158151, Y.Z.) and the Thoracic Surgery Foundation Resident Research Fellowship (Y.Z.).

**Institutional Review Board Statement:** Not applicable.

**Informed Consent Statement:** Not applicable.

**Data Availability Statement:** Data will be made available upon reasonable request to the corresponding author.

**Acknowledgments:** The manuscript contents are solely the responsibility of the authors and do not necessarily represent the official views of the funders. The authors would also like to thank Kevin Taweel for the generous donation for this research.

**Conflicts of Interest:** The authors declare that they have no competing interest.

## References

1. Zopf, D.A.; Hollister, S.J.; Nelson, M.E.; Ohye, R.G.; Green, G.E. Bioresorbable airway splint created with a three-dimensional printer. *N. Engl. J. Med.* **2013**, *368*, 2043–2045. [CrossRef] [PubMed]

2. Nakamura, M.; Iwanaga, S.; Henmi, C.; Arai, K.; Nishiyama, Y. Biomatrices and biomaterials for future developments of bioprinting and biofabrication. *Biofabrication* **2010**, *2*, 014110. [CrossRef] [PubMed]
3. Jia, J.; Richards, D.J.; Pollard, S.; Tan, Y.; Rodriguez, J.; Visconti, R.P.; Trusk, T.C.; Yost, M.J.; Yao, H.; Markwald, R.R.; et al. Engineering alginate as bioink for bioprinting. *Acta Biomater.* **2014**, *10*, 4323–4331. [CrossRef]
4. Murphy, S.V.; Atala, A. 3D bioprinting of tissues and organs. *Nat. Biotechnol.* **2014**, *32*, 773–785. [CrossRef]
5. You, F.; Eames, B.F.; Chen, X. Application of Extrusion-Based Hydrogel Bioprinting for Cartilage Tissue Engineering. *Int. J. Mol. Sci.* **2017**, *18*, 1597. [CrossRef] [PubMed]
6. Stevens, M.M.; Marini, R.P.; Schaefer, D.; Aronson, J.; Langer, R.; Shastri, V.P. In vivo engineering of organs: The bone bioreactor. *Proc. Natl. Acad. Sci. USA* **2005**, *102*, 11450–11455. [CrossRef] [PubMed]
7. Fedorovich, N.E.; Alblas, J.; De Wijn, J.R.; Hennink, W.E.; Verbout, A.B.J.; Dhert, W.J.A. Hydrogels as extracellular matrices for skeletal tissue engineering: State-of-the-art and novel application in organ printing. *Tissue Eng.* **2007**, *13*, 1905–1925. [CrossRef] [PubMed]
8. Cohen, D.L.; Malone, E.; Lipson, H.; Bonassar, L.J. Direct freeform fabrication of seeded hydrogels in arbitrary geometries. *Tissue Eng.* **2006**, *12*, 1325–1335. [CrossRef]
9. Ahn, S.; Lee, H.; Bonassar, L.J.; Kim, G. Cells (MC3T3-E1)-laden alginate scaffolds fabricated by a modified solid-freeform fabrication process supplemented with an aerosol spraying. *Biomacromolecules* **2012**, *13*, 2997–3003. [CrossRef]
10. Freeman, F.E.; Kelly, D.J. Tuning Alginate Bioink Stiffness and Composition for Controlled Growth Factor Delivery and to Spatially Direct MSC Fate within Bioprinted Tissues. *Sci. Rep.* **2017**, *71*, 17042. [CrossRef]
11. Lee, A.; Hudson, A.R.; Shiwarski, D.J.; Tashman, J.W.; Hinton, T.J.; Yerneni, S.; Bliley, J.M.; Campbell, P.G.; Feinberg, A.W. 3D bioprinting of collagen to rebuild components of the human heart. *Science* **2019**, *365*, 482–487. [CrossRef] [PubMed]
12. Schuurman, W.; Khristov, V.; Pot, M.W.; Van Weeren, P.R.; Dhert, W.J.A.; Malda, J. Bioprinting of hybrid tissue constructs with tailorable mechanical properties. *Biofabrication* **2011**, *3*, 021001. [CrossRef] [PubMed]
13. Gepp, M.M.; Fischer, B.; Schulz, A.; Dobringer, J.; Gentile, L.; Vasquez, J.A.; Neubauer, J.C.; Zimmermann, H. Bioactive surfaces from seaweed-derived alginates for the cultivation of human stem cells. *J. Appl. Phycol.* **2017**, *29*, 2451–2461. [CrossRef]
14. Sasaki, M.; Kleinman, H.K.; Huber, H.; Deutzmann, R.; Yamada, Y. Laminin, a multidomain protein. The A chain has a unique globular domain and homology with the basement membrane proteoglycan and the laminin B chains. *J. Biol. Chem.* **1988**, *263*, 16536–16544. [CrossRef]
15. Pierschbacher, M.D.; Ruoslahti, E. Cell attachment activity of fibronectin can be duplicated by small synthetic fragments of the molecule. *Nature* **1984**, *309*, 30–33. [CrossRef]
16. Neves, M.I.; Moroni, L.; Barrias, C.C. Modulating Alginate Hydrogels for Improved Biological Performance as Cellular 3D Microenvironments. *Front. Bioeng. Biotechnol.* **2020**, *8*, 665. [CrossRef]
17. Rowley, J.A.; Madlambayan, G.; Mooney, D.J. Alginate hydrogels as synthetic extracellular matrix materials. *Biomaterials* **1999**, *20*, 45–53. [CrossRef]
18. Ho, S.S.; Murphy, K.C.; Binder, B.Y.; Vissers, C.B.; Leach, J.K. Increased Survival and Function of Mesenchymal Stem Cell Spheroids Entrapped in Instructive Alginate Hydrogels. *Stem Cells Transl. Med.* **2016**, *5*, 773–781. [CrossRef]
19. Hazur, J.; Detsch, R.; Karakaya, E.; Kaschta, J.; Teßmar, J.; Schneidereit, D.; Friedrich, O.; Schubert, D.W.; Boccaccini, A.R. Improving alginate printability for biofabrication: Establishment of a universal and homogeneous pre-crosslinking technique. *Biofabrication* **2020**, *12*, 045004. [CrossRef]
20. Mirdamadi, E.; Tashman, J.W.; Shiwarski, D.J.; Palchesko, R.N.; Feinberg, A.W. FRESH 3D Bioprinting a Full-Size Model of the Human Heart. *ACS Biomater. Sci. Eng.* **2020**, *6*, 6453–6459. [CrossRef]
21. Kaklamani, G.; Cheneler, D.; Grover, L.M.; Adams, M.J.; Bowen, J. Mechanical properties of alginate hydrogels manufactured using external gelation. *J. Mech. Behav. Biomed. Mater.* **2014**, *36*, 135–142. [CrossRef] [PubMed]
22. Kuo, C.K.; Ma, P.X. Ionically crosslinked alginate hydrogels as scaffolds for tissue engineering: Part 1. Structure, gelation rate and mechanical properties. *Biomaterials* **2001**, *22*, 511–521. [CrossRef] [PubMed]
23. Boontheekul, T.; Kong, H.J.; Mooney, D.J. Controlling alginate gel degradation utilizing partial oxidation and bimodal molecular weight distribution. *Biomaterials* **2005**, *26*, 2455–2465. [CrossRef] [PubMed]
24. Kong, H.J.; Kaigler, D.; Kim, K.; Mooney, D.J. Controlling rigidity and degradation of alginate hydrogels via molecular weight distribution. *Biomacromolecules* **2004**, *5*, 1720–1727. [CrossRef]
25. Pan, T.; Song, W.; Cao, X.; Wang, Y. 3D Bioplotting of Gelatin/Alginate Scaffolds for Tissue Engineering: Influence of Crosslinking Degree and Pore Architecture on Physicochemical Properties. *J. Mater. Sci. Technol.* **2016**, *32*, 889–900. [CrossRef]
26. Duan, P.; Kandemir, N.; Wang, J.; Chen, J. Rheological Characterization of Alginate Based Hydrogels for Tissue Engineering. *MRS Adv.* **2017**, *2*, 1309–1314. [CrossRef]
27. Bidarra, S.J.; Barrias, C.C.; Fonseca, K.B.; Barbosa, M.A.; Soares, R.A.; Granja, P.L. Injectable in situ crosslinkable RGD-modified alginate matrix for endothelial cells delivery. *Biomaterials* **2011**, *32*, 7897–7904. [CrossRef]
28. Maia, F.R.; Fonseca, K.B.; Rodrigues, G.; Granja, P.L.; Barrias, C.C. Matrix-driven formation of mesenchymal stem cell-extracellular matrix microtissues on soft alginate hydrogels. *Acta Biomater.* **2014**, *10*, 3197–3208. [CrossRef]
29. Osborn, E.A.; Rabodzey, A.; Dewey, C.F., Jr.; Hartwig, J.H. Endothelial actin cytoskeleton remodeling during mechanostimulation with fluid shear stress. *Am. J. Physiol. Cell Physiol.* **2006**, *290*, C444–C452. [CrossRef]

30. Birukov, K.G.; Birukova, A.A.; Dudek, S.M.; Verin, A.D.; Crow, M.T.; Zhan, X.; DePaola, N.; Garcia, J.G. Shear stress-mediated cytoskeletal remodeling and cortactin translocation in pulmonary endothelial cells. *Am. J. Respir. Cell Mol. Biol.* **2002**, *26*, 453–464. [CrossRef]

31. Stupack, D.G.; Cheresh, D.A. Apoptotic cues from the extracellular matrix: Regulators of angiogenesis. *Oncogene* **2003**, *22*, 9022–9029. [CrossRef] [PubMed]

32. Goffin, J.M.; Pittet, P.; Csucs, G.; Lussi, J.W.; Meister, J.J.; Hinz, B. Focal adhesion size controls tension-dependent recruitment of alpha-smooth muscle actin to stress fibers. *J. Cell Biol.* **2006**, *172*, 259–268. [CrossRef] [PubMed]

33. Wiseman, P.W.; Brown, C.M.; Webb, D.J.; Hebert, B.; Johnson, N.L.; Squier, J.A.; Ellisman, M.H.; Horwitz, A.F. Spatial mapping of integrin interactions and dynamics during cell migration by image correlation microscopy. *J. Cell Sci.* **2004**, *117*, 5521–5534. [CrossRef] [PubMed]

34. Welf, E.S.; Naik, U.P.; Ogunnaike, B.A. A Spatial Model for Integrin Clustering as a Result of Feedback between Integrin Activation and Integrin Binding. *Biophys. J.* **2012**, *103*, 1379–1389. [CrossRef]

MDPI

St. Alban-Anlage 66

4052 Basel

Switzerland

www.mdpi.com

*Bioengineering* Editorial Office

E-mail: bioengineering@mdpi.com

www.mdpi.com/journal/bioengineering